2013

STEEL TITAN

STEEL TITAN

The Life of Charles M. Schwab

Robert Hessen

New York

Oxford University Press

1975

Dedicated to my wife, Bea

Acknowledgments

During the research and writing of this book I received assistance and encouragement from many people. I am happy to offer each of them my thanks.

Jean Wesner, the Chief Librarian of the Bethlehem Steel Corporation, was extremely generous with her time during each of my research trips and made many valuable suggestions.

Marshall D. Post, the Manager of Bethlehem's News Media Division, graciously provided me with unrestricted access to Schwab's papers and other related materials. Most of the photographs in this book were kindly supplied by the Bethlehem Steel Corporation.

Six people who knew Schwab agreed to let me probe their memories at length; I am grateful for their patience and candor. My deepest thanks go to Edward H. Schwab, the brother of Charles Schwab, as well as to Rana Ward (Mrs. J. B. Weiler), Alfred D. McKelvey, John C. Long, and Arch B. Johnston, Jr., and his wife.

Margaret Fuller and Agnes Galban expedited my use of the American Iron and Steel Institute's excellent collection of periodicals and books.

The librarians and archivists at Columbia University, Princeton University, Harvard University, Yale University, Penn State University, the Library of Congress, and the National Archives were all immensely cooperative, as was Joel Buckwald of the Federal Archives and Record Center in New York.

The Intercollegiate Studies Institute and the Lincoln Educational Foundation each granted me a Fellowship which facilitated my research.

The Faculty Research Fund of Columbia University's Graduate School of Business provided a timely grant which enabled me to complete the research.

Two of my friends (and former students), David H. Rogers and Robin Glackin, patiently tracked down some old steel industry periodicals and some elusive government reports and hearings.

Professors John A. Garraty and Stuart Bruchey, of Columbia University's Department of History, have earned my thanks for their detailed criticisms of an earlier draft of this work.

Although I only know Professor Joseph Frazier Wall of Grinnell College through his biography of Andrew Carnegie, I hope that my book reflects how much I have learned from his: it is a masterpiece of scholarship, exposition, and interpretation.

W. Glenn Campbell, the Director of the Hoover Institution on War, Revolution and Peace, Stanford University, deserves thanks not only for creating an atmosphere so conducive to scholarship, but also for permitting me to devote part of my time as a Research Fellow to the final revisions of the manuscript and its preparation for the press.

Martin Anderson, a Senior Fellow at the Hoover Institution, has given me encouragement and invaluable advice not only during the writing of this book, but throughout my academic career.

Parts of this biography were previously published as articles in various historical journals. I am grateful to the editors for their suggestions and criticisms. They include: *Pennsylvania History*, William G. Shade; *History Today*, Alan Hodge; *Pennsylvania Magazine of History and Biography*, Nicholas B. Wainwright; *Business History Review*, Glenn Porter; and *Labor History*, Milton Cantor.

Thanks are also due to Sheldon Meyer, Vice President and Executive Trade Editor of Oxford University Press, for assembling an organization of experts and thus making the publication of my first book so pleasant and trouble-free.

My editor, Caroline Taylor, made brilliant and tireless efforts to improve the manuscript.

No one should assume that anyone named above agrees with any or all of my interpretations and conclusions. I bear sole responsibility for any errors.

The philosophy and works of Ayn Rand have profoundly influenced my thinking. She did not have an opportunity to read the manuscript prior to its publication; however, I want to acknowledge my intellectual indebtedness to her, without in any way implying that she endorses this work.

Finally, and above all, I owe the greatest thanks to my wife, Bea, to whom I dedicate this book with love. She assisted me at every stage, including research, typed the first and second drafts of the manuscript, and compiled the index. However, she made her greatest contribution by being my most severe critic. She has an uncanny ability to detect flaws and inconsistencies, as well as to ask probing questions which forced me continually to re-think and revise what I had written. Her high standards of excellence—which are revealed in her own writings—make her my foremost friend and ally. We hope our children, Laurie and Johnny, will someday understand why Charlie Schwab kept us busy for so long.

R.H.

Stanford, Calif.
June 1975

Contents

Prologue

Charles Schwab was one of a generation of industrial giants, men who helped to transform America's potential wealth into an actuality. Like his contemporaries—Andrew Carnegie, John D. Rockefeller, James J. Hill, Gustavus Swift, and numerous others—he learned early in life that a fortune awaited any man who could excel in business, and that ability was far more important than family background, academic credentials, or seniority. And like them, he combined intelligence with daring and independent judgment. They were men with the courage and determination to persevere when navigating in uncharted waters—men who, in Joseph Schumpeter's famous phrase, were able to "act with confidence beyond the range of familiar beacons."

These men were among the pioneers of America's economic development, but, insofar as they are remembered at all today, it usually is because of their philanthropic activities, not their productive achievements. Their monuments chiefly consist of concert halls, libraries, museums, or research institutes which they, or their heirs, gave to a city or a university. Such gifts have conferred a degree of immortality upon their names, but not upon their careers. Those careers are little known, except to a comparative handful of people who are interested in understanding how America attained its industrial preeminence. School-age children are taught a great deal about the conquerors and killers in history, from Alexander and Caesar to Hitler and Stalin, but little or nothing about those to whom we are indebted for creating our industrial civilization. The names and achievements of the innovators in business are either consigned to oblivion or only resurrected so that they can be denounced as "exploiters" and "parasites."

The leading industrialists of late nineteenth- and early twentieth-century America have been attacked as "Robber Barons." This label contains concealed premises: that steel mills, oil refineries, and railroads came into existence merely because they were needed, and that the men who conceived and directed these enterprises performed no useful or productive function, but merely siphoned off the profits for themselves. This was the view of big business popularized earlier in this century by Thorstein Veblen; it seemed like a heresy then, but it is closer to an orthodoxy today. Nonetheless, such a belief is wholly unwarranted.

An industrial enterprise is not a spontaneous network of men, machines, and raw materials. A leader is necessary—in fact, indispensable—to create and sustain a sense of unity, purpose, and direction. This fact, which is seldom disputed in other areas of human activity, is often overlooked or even denied when applied to the realm of production. In *every* organization, whether it be a team of athletes or explorers, an army or an orchestra, someone must be the coach, leader, general, or conductor. In precisely the same way, someone must unite and direct the components of a business enterprise. That person is the entrepreneur, the man who defines the firm's fundamental nature, goals, and operating methods and directs its response to unforeseen problems and challenges.

The men who led America's Industrial Revolution were willing to assume vast responsibilities and incur great risks in order to create something great and enduring—an industrial empire. But often these same men, whose wealth derived mainly from their productive ability, apologized for their fortunes by claiming that they were only functioning for the benefit of others.

An early steelmaker, Peter Cooper, declared: "I do not recognize myself as owner . . . of one dollar of the wealth which has come into my hands. I am simply responsible for the management of an estate which belongs to humanity." And Schwab's mentor, Andrew Carnegie, made a similar statement in his famous essay on wealth:

> the duty of the man of wealth . . . [is] to consider all surplus revenues which come to him simply as trust funds, which he is called upon to administer . . . to produce the most beneficial results for the community—the man of wealth thus becoming the mere agent and trustee for his poorer brethren,

bringing to their service his superior wisdom, experience, and ability to administer, doing for them better than they would or could do for themselves.

Carnegie was almost pathetically pleading for freedom to function when
he wrote: "It will be a great mistake for the community to shoot the
millionaires, for they are the bees that make the most honey, and con-
tribute most to the hive even after they have gorged themselves full.
. . . Under the present conditions the millionaire who toils on is the
cheapest article which the community secures at the price it pays for
him, namely, his shelter, clothing, and food."

Schwab, by contrast, did not preach a "Gospel of Wealth," an altruis-
tic justification for his fortune. He did not claim, as Rockefeller did, that
"God gave me my money." Nor did he try to justify private property as
an institution of divine origin, or invoke any Darwinian imagery about
the "survival of the fittest."

In fact, Schwab said remarkably little about the causes of his success
or about any duties or "social responsibilities" imposed upon him by his
great wealth. In 1935 he said:

> I disagreed with Carnegie's ideas on how best to distribute his wealth. I
> spent mine! Spending creates more wealth for everybody. I say that with
> due regard for Carnegie's theories and the unquestioned hope in his mind
> that he was being a benefactor of mankind. But I believe that, let alone,
> wealth will distribute itself.

Schwab spent a fortune in order to enjoy his life to the fullest, but he
also gave away substantial sums to help others. When he did so, however, it was not out of a sense of duty or unchosen obligation, but rather
out of genuine benevolence and generosity. He did not regard himself as
a "golden goose" or a "gorged bee," but as a producer who had earned
the right to spend his money in any way he chose to. But he never artic-
ulated this belief; it was merely implicit in his actions.

Although Schwab was neither tongue-tied nor camera-shy, he never
courted controversy by proclaiming—either quietly or defiantly—that he
had earned his money, that no one was ever compelled to work for him
or to buy his products, that he had never possessed the power to tax or
expropriate anyone's wealth or to conscript or enslave anyone, and that
business activity consists of voluntary transactions for the mutual benefit
of all. And his disinterest in philosophical ideas and his fear of contro-

versy prevented him from discovering or stating the fact that, in a capitalist civilization, any benefit which one's work confers upon others is of secondary importance—that it is the *consequence* of freedom, but not its justification. Yet, even though he was not a philosophical spokesman for capitalism, he was nonetheless a brilliant example of the explosive release of energy which occurs when a man of titanic ability and ambition is able to function in a climate of economic freedom.

STEEL TITAN

1

An Unlikely Background

Late in 1932, Charles M. Schwab tried to explain why he had been so extraordinarily successful in the world of business. "As I sit and look back over those fifty years, I cannot for the life of me understand the whole thing. All I can do is wonder how it all happened. Here I am, a not over-good business man, a second-rate engineer. I can make poor mechanical drawings. I play the piano after a fashion. In fact, I am one of those proverbial Jack-of-all-trades who are usually failures. Why I am not, I can't tell you." [1]

When Schwab took his first job in the steel industry in 1879, he began as a day laborer. He had shown no previous interest in metallurgy, management, or finance, though he had taken a high school course in surveying and engineering. But within six months, at the age of eighteen, he was the acting chief engineer of the largest steel mill in America. Ten years later he was the foremost production expert in the American steel industry. Three years after that, in 1892, a bloody strike crippled the Homestead works of the Carnegie Steel Company. Only one man seemed capable of pacifying and conciliating the strikers, so Andrew Carnegie cabled an order from his retreat in Scotland: send in Mr. Schwab. Then, when Schwab was only thirty-five, he was named president of the Carnegie Steel Company. Four years later, in 1901, he served as the intermediary between Carnegie and J. P. Morgan, helping to bring about the merger which created the first billion-dollar company in America, the United States Steel Corporation. He was U.S. Steel's first president.

A few years later Schwab achieved his greatest coup: he founded the modern Bethlehem Steel Corporation. He took over a small company and built it, in a single decade, into the second largest steel-producing firm in the world. For the next twenty years his company dominated the structural steel market; he had risked his fortune and the future of Bethlehem Steel on a new and untried invention—and he succeeded brilliantly. In 1914, when World War I began, he obtained from England the largest order for munitions and submarines ever made to that date. Then America entered the war; Schwab was one of the men who made the nation the arsenal and breadbasket of the Allies: as Director-General of the Emergency Fleet Corporation, he accelerated the building of merchant ships which carried to Europe the products of American factories and farms.

When Schwab expressed bewilderment about his successful career, was he truly unaware of the special skills and talents which had made his achievements possible, or was he merely reticent about them? Must the explanation of his success ultimately rest upon some vague or indefinable concept such as luck, or fate, or accident, or destiny? Or was there in fact a solid reason for his achievements?

Schwab, as a child and an adolescent, gave clear evidence that he was likely to be successful in later life, but the evidence points to success in show business, not in the steel industry. Young Schwab was attracted to footlights, not to blast furnaces and forges—yet the world he was born into was unlikely to produce either a showman or a steelman.

In 1830, his grandfather, an emigrant from Baden-Baden, Germany, settled in a sparsely populated region in southwestern Pennsylvania. Grandfather Schwab, who was also named Charles, was the only son of an impoverished Catholic farmer. He viewed life as a high adventure and actively sought new experiences: in his late teens, he "walked all over France" simply to see how the people of that country lived. When he was twenty-one, he left his family and native land for good; he took passage on a ship which would carry him to America. At first he settled among his compatriots and co-religionists, seeking the security of their company until he had remedied his two major handicaps: he lacked a trade and he could not speak or write English.

Grandfather Schwab became a weaver. Four years after he first settled in Bedford, Pennsylvania, he headed north into Cambria County, in search of a place where he could use his trade to support himself. He settled in Loretto, and within a few months he got married to Ellen Myers.[2] Ellen—she was also called Elinor—was eight years younger than her husband. She was born on the Atlantic Ocean in 1817, during her parents' voyage to America. Her father, John B. Myers, had been a schoolteacher in his native Germany. There he had married a distant cousin, Catherine Myers. In America, John became a farmer; he raised crops while Catherine raised children, eight of whom survived infancy.[3]

In 1834, when Charles and Ellen Schwab were newlyweds, Loretto was an isolated and inaccessible village on the crest of the Allegheny Mountains, in a region which barely supported subsistence farming. The inhabitants of Loretto had come from many countries, but they shared a common faith: the village was virtually a Catholic retreat.[4]

Loretto's spiritual leader was an ascetic priest, Demetrius Augustine Gallitzin, who had originally come to the village in answer to the call of a dying woman for the last rites of the Catholic Church. On Christmas Eve, 1799, Father Gallitzin celebrated his first mass in Loretto. He remained the village patriarch until his death in 1840. Loretto soon became noted for the fact that the only Catholic priest between Lancaster Pennsylvania, and St. Louis, Missouri, lived there. For those to whom strict religious observance and access to the holy Sacraments were matters of importance, Loretto seemed an ideal place to settle.

Although Father Gallitzin was self-effacing, he was known far beyond Loretto, for he was not an ordinary man. He became a naturalized American in 1802 under the name of Augustine Smith, but he had been born a prince of Russia. At the time of his birth in The Hague in 1770, his father, a cousin of Czar Paul, was the Russian ambassador to Holland. Tradition had destined the young prince for a military career, but he broke with his family, converted from the Greek Orthodox to the Roman Catholic Church, sailed to America, took Holy Orders in a seminary in Baltimore, and thereafter dedicated his life to the spiritual salvation of the handful of Catholic souls living in the midst of America's almost impenetrable forest. The magnitude of his sacrifice and the intensity of his religious devotion served as a beacon, attracting to Loretto other religious groups intent upon carrying on his mission. In 1834, Father Peter Henry Lemke received permission to assist Father Gallitzin in Loretto, and in 1843 the newly created Bishop of Pittsburgh assigned two priests to aid Lemke in carrying on Gallitzin's work after that patriarch's death. Then, in 1847, the Franciscan Brothers came to Loretto, and in 1848 the Sisters of Mercy established a convent there.[5]

5

In 1834, when Charles Schwab first settled in this religious sanctuary, he was able to earn enough to support himself and his wife. By 1857, however, he also had eight children to support, and the limited economic prospects in Loretto outweighed the religious advantages. Moreover, the winter of 1856 had been particularly bitter and brutal; over ninety-six inches of snow had fallen on Loretto, immobilizing the village and isolating Schwab from his sources of raw wool and from his customers.[6] In 1857 he moved his family to Williamsburg, thirty-two miles to the east. In Williamsburg he was able to lease a large woolen factory in the region known as Clover Creek, bordering the Juniata River.[7]

For Schwab's eldest son, John, born in 1839, the family's move from Loretto must have been a welcome change; it spared him further embarrassment at the hands of a neighbor and classmate, Pauline Farabaugh. The Schwabs' nearest neighbors in Loretto had been Michael and Genevieve Farabaugh, who were also German-born Catholics and also the parents of eight children. The first recorded encounter between John and Pauline was less than auspicious. John had trespassed on the Farabaugh property to pick apples; when Pauline spotted him, she unleashed her dog to chase him away. They were also classmates in the one-room country schoolhouse, where an itinerant teacher, known as Master Thomas, held classes for six weeks each year. By his own admission, John had been a poor student in "skule," but he was especially sensitive when Pauline chided him about his deficiencies as a speller, for she regularly triumphed over him in competitive spelling bees, even though she was four years his junior.

John Schwab learned his father's trade, and as he matured he was given a greater responsibility in the family business. Then in 1859, at a party in Munster, John met his old tormentor, Pauline. She had outgrown her boy-baiting days, and their earlier coolness gave way to cordiality and then to courtship. They were married in Loretto on April 23, 1861, in the remodeled chapel of Father Gallitzin's old church.

After the wedding festivities, John and Pauline left for their new home in Williamsburg. The young couple had their first disagreement then and there: John was eager to enlist in the Union Army. Only a week before their marriage President Lincoln had called for 75,000 volunteers to suppress the Southern rebellion. John wanted to join a number of his friends who were preparing to leave for Harrisburg to answer the President's call, but Pauline was able to dissuade him.[8] Their first child, Charles M. Schwab, was born ten months later.

On Tuesday, February 18, 1862, Dr. Ake, the only physician in Williamsburg, was summoned to the Schwab house to assist at the birth. Charles Michael Schwab, named in honor of both his grandfathers, weighed in at over ten pounds.

In later years Schwab would own the largest and most luxurious mansion in New York City, but he spent his infancy and boyhood in almost spartan surroundings. His first home was a one-story wood-frame house in Williamsburg, close to his grandfather's woolen mill. The house was modestly furnished; two of Pauline's brothers, like their father, were woodworkers, and they gave the couple a bed, a table and chairs, mixing bowls, and a washtub, among other things. Pauline's mother's gift was a cow. The cow proved to be a prized possession, for it was wartime and food prices were soaring. As an economy measure, Pauline made all the clothing for the family, a practice she was slow to abandon. Charlie was seventeen before home-made gave way to store-bought clothes.[9]

In 1861, after Charles Schwab the elder had obtained a government contract to produce blankets and overcoats for the Union Army, John was promoted to foreman of the woolen mill. He and his father worked unstintingly to meet the delivery dates specified in the contract. Since no record survives of the details of the contract, we do not know how profitable it was; however, it appears that even though the family lived modestly, they were not poor, even by the standards of that era. We do know that the profits from the government contract supported thirteen people—not only John and Pauline Schwab and their two sons, Charles and Joseph (who was born in 1864), but also John's father and mother and their seven other children. When the war ended, so did the contract.

All of the family income was earmarked for necessities, and any surplus went into savings, as a hedge against an uncertain future. As a child, Charlie neither expected nor received many toys or gifts. When he was seventy-three years old, he recalled the day when his Christmas stocking contained only a single marble.[10]

There was, however, one form of entertainment which did not require any outlay of cash. All the Schwabs loved music, and singing was their primary source of pleasure. On Saturday evenings Schwab's grandfather would borrow the small church organ and lead a choir of his children and grandchildren.[11] From the age of four, Charlie was a key participant. A photograph taken of him at that age reveals a chubby-cheeked, dark-haired child with large, intense eyes.[12]

Charlie entered school when he was five. The school year was only three months long, but the teacher, Miss Sadie Stevens, ran a supple-

7

mentary session each year for her brightest students. Charlie was one of them. His mother permitted him to attend despite the additional expense, because she was flattered that he was one of the few chosen.[13]

Charlie excelled not only as a singer and a student, but also as a showman, even at this early age. Whenever the class put on a pageant, play, or recital, he was usually the star.[14] The future orator of the steel industry, the man who never passed up an opportunity to regale his colleagues and customers with jokes and anecdotes and who freely acknowledged his pleasure in "playing the part of clown and entertainer for the edification of my delighted friends," [15] made his debut as a performer at the age of seven, when he won first prize in a poetry recital. Beyond good looks and a fine voice, he had another asset—a superb memory. It was a natural skill, but he developed and improved upon it until he had an almost photographic memory for names and numbers, faces and facts, short passages and whole pages—an ability he would capitalize on throughout his business career. He also learned various mnemonic devices and tricks of memory—paradoxically, from a book whose author and title he could not recall.[16]

At the age of eight Schwab earned the first installment on his multimillion-dollar fortune. He solicited neighbors who wanted their pathways cleared of snow, enlisting his brother Joe as his assistant. The financial rewards were meager—his share on each job was only five cents—but he did acquire a reputation as a "good boy." His first customer, the local Presbyterian minister, complimented Mrs. Schwab on the fact that Charlie was not raucous and rowdy like most boys his age.[17]

Charlie spent much of his time developing his musical skills. John Schwab bought a small parlor organ on an installment plan, and Pauline arranged for Charlie to receive lessons once a week. Since the family's budget was already strained to its limit, Pauline gave the music teacher a free dinner in lieu of payment. Charlie soon mastered the organ and was ready for a public performance. The best place to display his talent was in the local church, where he would have the largest audience. But there was one problem: although he could easily play the hymns on the church organ, his legs were too short to reach the pedals. The elder Charles Schwab came to the rescue; while Charlie sat in the church loft playing the organ in full view of his admiring neighbors and proud relatives, his grandfather, concealed from view, worked the pedals.[18]

In 1874, when Charlie was twelve, his family moved back to Loretto. For five years his father had been suffering from a mysterious malady which was becoming increasingly severe. He had bouts of dizziness and

8

fainting. Charlie made his first trip to a big city when he accompanied his father to Philadelphia, where John was examined by Samuel David Gross, the well-known surgeon and diagnostician, at the Jefferson Medical College. Doctor Gross suggested a change of locale and occupation; he recommended that John live in a town with a higher elevation than Williamsburg's and that he find work which would keep him out-of-doors most of the time. Although John Schwab had dug into his savings to pay for his visit to Philadelphia and the physician's fee, he had enough money left to purchase a small livery stable in Loretto.[19]

The move was beneficial for the whole family. For John and Pauline it meant a return to their birthplace; for Charlie and his brother it meant a chance for a better education. Pauline, who was deeply pious, also hoped that the schools in Loretto would provide her sons with a more intensive religious training than they had received thus far. Although Charlie did finish grade school in Loretto at an institution run by the Sisters of Mercy, the major benefit he derived from his schooling was a heightened skill as a musician. The Reverend Horace S. Bowen, who had been a pupil of Franz Liszt, was then serving as chaplain to the nuns, and Charlie asked him for special singing lessons. Bowen agreed, and for several years Charlie studied under him.[20]

When Charlie was graduated from grade school, he began classes at St. Francis College, which then offered the equivalent of a high school curriculum. The school had been founded by six Franciscans from Ireland, and courses in Christian doctrine were given first priority. But Schwab remained remarkably unaffected by his years with the Sisters of Mercy and the Franciscans; he was never influenced by the religious atmosphere of the community which so ardently embraced the teachings of Father Gallitzin. Schwab's teachers were men and women who had renounced the pleasures of this world in preparation for a life to come. They may by their example have intended to inspire him to do the same; if so, they were completely unsuccessful. They had eagerly embraced poverty, chastity, and obedience; Charles Schwab would forever find those three states intolerable.

The secular portion of the curriculum at St. Francis did introduce Schwab to subjects which were useful to him in his future career. In addition to being taught literature, history, and mathematics, he received instruction in public speaking, perspective drawing, bookkeeping by double and single entry, surveying, and engineering.[21]

Charles Schwab stood first in his class, not through effortless brilliance, but rather, as one of his teachers later explained, through bulldog tenacity—and bluffing. Brother Ambrose Laughlin later gave this description of his prize student:

> Charlie especially liked arithmetic. Generally it was easy for him, though sometimes it wasn't. But if it wasn't, Charlie would never let on that he didn't know his problems. Instead he'd go to the blackboard and mark away with might and main. And he wouldn't stop until he had solved the problem, or had convinced us that he knew how to get the right answer.
>
> In all things Charlie was a boy who never said "I don't know." He went on the principle of "pretend that you know and if you don't, find out mighty quick." [22]

Schwab relished any opportunity to be the center of attention and to make a favorable impression. His greatest pleasure was to find a responsive audience for his talents, and he did not disguise his pride in his accomplishments. His mother often told of a Sunday afternoon when a number of their relatives had come to visit. Charlie entertained them with songs, jokes, somersaults, and magic tricks. Reveling in their applause, he did not want the occasion to end. When he finished his basic repertoire, they rose to go home, but he held them with the announcement that "I can do something else yet!" [23]

From the beginning, Charlie demonstrated a strong sense of competitiveness and a drive to excel. His father, by contrast, was an easygoing, carefree, self-effacing man. Once, when recalling his early family life, Charles Schwab said:

> Father was always amiable. He would let us [Charlie and Joe] do anything. Mother was the disciplinarian because she had to be—despite her sense of humor. She would have to do whatever punishing there was to do. They always cooperated. They were a great team. But although Father would give his consent to what we wanted to do, it was always to Mother that we would go to get our money. [24]

During the months when Charlie was not attending school he helped his father in the livery stable. When John Schwab had purchased the stable he had also acquired a contract to carry the mail, and Charlie's major responsibility was to drive a wagon to the near-by town of Cresson, the local railroad stop, where he would pick up the mail and then deliver it to the people of the neighboring villages. [25]

On rare occasions, Charlie carried passengers in the wagon. He used these opportunities both to show off his fine voice and to display flashes of his developing entrepreneurial talent. At regular intervals a group of Franciscans, heading for the monastery at St. Francis College either for retreats or for commencement exercises, would arrive by train. Charlie worked to get them as passengers. According to his mother, "As soon as the train pulled in, Charlie seemed to acquire a dozen pair of hands. He had a grasp on the travelers' bags before they realized what was happening, and bundled them into the coach." [26] He also entertained his passengers with ballads, so that they would ride with him when they came to town again. He earned a nickname, "the singing cabby." [27] On one such occasion, a lady passenger, apparently impressed by the quick-witted boy, gave him a travel book. Schwab later said, "I suppose it wasn't a very good book but it opened my eyes to the glories of the outside world, and stimulated my imagination tremendously." [28]

When he was not in school or performing household chores or driving the mail wagon, he enjoyed wandering to secluded spots where he could be alone to read and daydream. Among the books he read were English novels, such as Fielding's *Tom Jones* and Richardson's *Pamela*. [29] The bond between Charlie and his grandfather included not only a love of music, but also an interest in history and biography. When Charlie was in his teens he discovered a personal hero, Napoleon Bonaparte. Years later he told a reporter that he had read just about every book in English on Napoleon. [30]

Another interest entered Schwab's life just a few months before his seventeenth birthday. Mary Russell, an attractive young actress from Pittsburgh, came to Loretto to visit her older sister, who had recently married a widower named Abernethy. As a relative of Charlie's said,

> Charlie got sweet on her and loafed around old man Abernethy's doorstep all day and pretty nearly all night. Seems as if the girl was gone on Charlie too. . . . Charlie wanted to marry her. She told Charlie that the stage was the place for a nice fellow like him who could play so well and sing so sweetly. But all Charlie's people were dead set against Charlie's marrying an actress and going on the stage, so after a good deal of hard work Charlie was kept from running away with the girl, as he'd raved he would do. Then the girl went away and after a little while Charlie went down to Braddock. [31]

Although this youthful episode was not revealed until 1902, it was soon authenticated by Mrs. Abernethy, who explained her own role in breaking up the romance. "When I learned they were engaged I had her sent to Pittsburgh, as I had been informed they contemplated eloping. I did not consider the match a judicious nor wise one, and also knew my sister was entirely too young, she not being 17 years old at the time." [32]

The breakup of Schwab's romance and the interference with his plans for a stage career left him dejected, restless, and distracted from his studies. His father, hoping that a change of scenery and the need to be self-supporting would help Charlie deflect his mind from the frustrations he had experienced, suggested that at the end of the school year he should "get out in the world." Presumably, the change was intended to be only temporary, since he had not finished his studies at St. Francis; he had one more year to complete before graduation.

Schwab's parents asked a friend, A. J. Spiegelmire, to give him a job. Actually, Spiegelmire was Charlie's friend, originally. He had first come to Loretto as a traveling salesman for McDevitt's, a grocery and dry goods emporium of which he was part owner. He was a regular passenger on Charlie's mail wagon, and it was Schwab who had introduced "A. J." to his parents.

Spiegelmire offered Schwab a job for $10.00 a month, clerking at Mc-Devitt's, which was located near the entrance of Andrew Carnegie's Edgar Thomson Steel Works in Braddock. The school term ended on July 1, 1879, and a few days later Schwab left Loretto. Years later he recalled that the scene of his departure had reminded him of a popular old engraving, "Breaking Home Ties." His parents had wept even as they had encouraged him. [33] The ambitious young man who had longed for stardom was going off to sell sugar and cigars.

2

A New World of Steel

When Charles Schwab reached Braddock he was thrust into an unfamiliar and unwholesome environment. Loretto was a sleeply hamlet whose 300 inhabitants shared a common life-style and religion; Braddock was a growing industrial town whose population of nearly 9000 was torn by tension and bitter resentments.

The "native stock" of Braddock, descendants of settlers from England, Ireland, and Scotland, feared the growing influx of "foreign" laborers, most of whom were Italians, Slavs, and Hungarians. These men, unskilled workers at the Edgar Thomson Steel Works, were regarded as vulgar, ignorant, unclean, ruthless, and degenerate "foreigners." Many of them had come to America to earn sufficient cash to return to the old country and buy a small farm, or to send home ticket money so that their wives and children could join them. Meanwhile, alone in an inhospitable town, they sought pleasure mainly at Braddock's bars and brothels—or so it seemed to the natives.[1]

Schwab was wearing his first store-bought suit when he arrived in Braddock. His anxious mother had pinned a five dollar banknote inside the lining of his new jacket.[2] He found McDevitt's store on the ground floor of an old three-story brick building; the second floor served both as a storage area and as sleeping quarters for the five clerks. Schwab rose at dawn each day, went to breakfast at a near-by boardinghouse, and hurried back to the store, which opened at 7:00 a.m. His workday was fourteen hours long, minimum, plus overtime when necessary at no additional pay.[3] He worked as a bookkeeper and as a clerk, but neither of the jobs was exciting or rewarding. Adding a column of figures, pushing a broom, selling cigars or shoes or sundry items—none of this challenged him. McDevitt finally decided that his newest clerk's performance was less than satisfactory, and he suggested to the boy that his talents might be better employed elsewhere.[4]

Fortunately for Schwab, he had made a friend in Braddock, a man whom he could ask for a job. Schwab's friend had fought in the Union Army during the Civil War, and had risen from the ranks. By 1864 he was captain of a regiment. Fifteen years later that veteran was commanding a far larger body of men; by then he had become the general superintendent of the Edgar Thomson Steel Works in Braddock, Pennsylvania. He was William R. Jones—"Captain Bill."

Jones was short, stocky, and clean-shaven, with closely cropped hair. In 1879 he was only forty years old, but, young as he was, he had had thirty years' experience in steel. Jones's father, a Welsh minister, had emigrated to America in 1832, and when he became ill his only son, Bill, was forced to leave school and become the family's breadwinner. Bill Jones was only ten years old when he was apprenticed to a foundry at Catasauqua, Pennsylvania. First he worked as a molder in the foundry, then as a machinist. By the time he was twenty he had acquired the skills—and was being paid the salary—of a master mechanic. He went to work for the Cambria Iron Company in Johnstown, Pennsylvania, but Cambria could not keep him. The owners of an iron company in Chattanooga, Tennessee, wanted a blast furnace built, and they hired him to construct it.

Then the Civil War broke out. Jones had just gotten married, and, late in April of 1861, fearing that they might be trapped behind Confederate lines, he and his wife fled north. In July Jones enlisted in the Union Army as a private. He was supposed to serve only nine months, but during the Gettysburg campaign he re-enlisted, and by 1864 he had become a captain.

In 1865, George Fritz, one of the early masters of steel-making, was the general superintendent of Cambria. When the war ended he asked Jones to come back to the company and work for him. Jones accepted, and by 1872 he was Fritz's chief assistant. Then, in 1873, Fritz died. Jones should have succeeded him, but Cambria promoted another man to the post. Jones resigned in anger. He went to New York, where he met with Alexander Holley, the leading design engineer of the steel industry. Jones and Holley knew each other well; they had worked together at Cambria. Holley had designed Cambria's Bessemer furnaces, and Jones had supervised their construction. Now Holley was working on new plans for Bessemer furnaces. They were to be built at Pittsburgh, the coming capital of steel. Andrew Carnegie had commissioned Holley to design the furnaces of the Edgar Thomson Steel Works. Jones and Holley talked about the job, and when Holley went to Pittsburgh he had a new chief assistant—Captain Bill Jones.

Again, it was Holley who designed the new furnaces, and Jones who supervised their construction. And over the months, Andrew Carnegie, who eagerly sought men of ability and restless energy, watched Jones at work. By the time the furnaces were completed, Carnegie had decided who should run his steel works. In 1875 he made Jones the general superintendent of Edgar Thomson.[5]

Bill Jones was the best-known man in Braddock. He was both brilliant and forceful. And he was the only man alive who dared to talk back to Carnegie. Jones had a volatile temper. If a subordinate defied him or did a bad job, Jones would fire him on the spot. And if the offending party was Carnegie himself, he would tender his resignation, effective immediately. (In fact, he kept such resignations in his pocket for just that purpose. One day his secretary wrote out a dozen letters of resignation so that Jones would always have one handy when he needed it.) Jones resigned every month or so, and each time he did Carnegie persuaded him to return.[6]

Captain Jones was a regular patron of McDevitt's; he bought his cigars there. Early one morning during the summer of 1879 he went into the store for his daily supply. Schwab waited on him that day, as he often did thereafter. The two hit it off at once. From the first, Jones enjoyed Schwab's self-confidence and quick wit. Then he learned that he and Schwab shared an interest in music. Jones had a piano in his home, and when they became friends he invited the young man to visit him.

Jones suggested to Schwab that his job as a grocery clerk was not demanding enough for a bright, self-confident boy, and he asked him if he had any interest or background which might suit him for a job at the steel works. Schwab showed him two diplomas from St. Francis College which certified that he had completed instruction in surveying and engineering.[7] (Schwab, of course, had not been graduated, but he had earned the diplomas. St. Francis awarded a diploma for each completed course. He did not have to misrepresent his qualifications; he could say, quite honestly, that he had attended a "college," bring out his diplomas, and leave any inferences to Captain Jones—which he did.) Jones offered him a job at the steel works; he began work on September 12, 1879.[8]

When Schwab first reported for work, Captain Jones sent him to Tom Cosgrove, the labor supervisor at the Edgar Thomson Works. Cosgrove assigned him to the engineering corps, which was headed by Pete Brendlinger. In his first job he worked as a rodman, carrying the leveling rod used by the surveyors who were laying out the plans for the construction of new furnaces. If he had not attempted to enlarge upon the job, it would have proved no more exciting or challenging than clerking had been. But by parlaying a little knowledge and a lot of bluff he was able to create the impression that he was vastly experienced in surveying and engineering. During the day he worked, and, whenever possible, he asked questions which would not betray how green and unseasoned he actually was. At night he read borrowed books, to increase his knowledge.

Schwab was fortunate in having Pete Brendlinger as his superior, for Brendlinger was quite willing to let the eager youth handle as much work as he successfully could undertake and complete. Schwab was a zealot for work, and Brendlinger did nothing to dampen his zeal. He did not view the young man's ambitiousness as any threat to himself.[9] Just six months after Schwab began work, Brendlinger was transferred temporarily to the Scotia Ore Mines in Center County, Pennsylvania, to handle a special engineering job. In Brendlinger's absence Schwab served as chief engineer; he was advanced over men older and far more experienced than he was.

Schwab subsequently became a draftsman at the Edgar Thomson Works. Perhaps Jones transferred him so that he would have an opportunity to explore a number of fields and determine the area of his greatest interest and aptitude. In any case, Schwab once again demonstrated his capacities; as always, whenever he felt sufficiently motivated, either by intellectual curiosity or by the hope of promotion, he would plunge into any subject, however unfamiliar, and master it. Years later, Schwab recalled:

> You would be surprised at the variety of things I studied in those days. I even took up shorthand, and later came to know Pitman very well. I didn't take up shorthand with any idea of becoming a professional at it. It merely appeared to me to be a good thing to know—something that might come in handy. . . . Similarly, I became interested, through reading the works of some novelist, in Egyptology and made a study of the pyramids. It was just a hobby, but I had a desire to know all I could about everything I could.[10]

When he first became a draftsman, Schwab joined a corps of young men all of whom were roughly equal in ability. Soon thereafter there was one opening, a promotion, for a diligent and dependable man. Jones decided to hold a competition for the post, but he did not announce it. He ordered the chief draftsman to tell his men that they were to work overtime one day, without additional pay, on a special project. The chief was to send Jones the names of the men who worked unremittingly, without clock-watching and without complaint. He sent only one name: that of Charles Schwab.[11]

Even after Schwab took the job in the steel works, he still maintained a friendship with McDevitt's senior clerk, Tom Wagner. Wagner had married the eldest daughter of Mrs. Mary Elizabeth Dinkey, a widow who ran a boarding house in Braddock. Mr. and Mrs. Wagner lived in the boarding house. When Schwab was invited to visit the house, he accepted; he had learned that Mrs. Dinkey owned a piano. He was then invited to come and use it as often as he liked. One evening as he sat playing, so absorbed in his music that he barely noticed that the room was almost dark, a young woman entered to light the kerosene lamp. When the light from the lamp cast a glow over her face, Schwab's interest was aroused.[12]

The young woman was Emma Eurania Dinkey, but everyone called her Rana (pronounced Raynee). She was just over twenty years old, two and a half years older that Schwab.

The first Dinkey (he spelled his name D'Inque) to come to America was a native of the region which is now Alsace-Lorraine. He landed in Philadelphia in 1743. By the middle of the next century the family had moved west, along with most of America's settlers. Rana's father, Reuben Dinkey, worked for the Lehigh and New England Railroad. He had purchased a small farm in Weatherly, Pennsylvania, where he and his family lived. Mrs. Dinkey was a widow with two daughters when she married Reuben, and Rana was the eldest of their children. She was born on September 12, 1859, in West Penn Township, Carbon County, Pennsylvania. In addition to her two half-sisters, Rana had a full sister, Jeanette, and two brothers, Alva and Charles. In 1875, when Mrs. Dinkey was again pregnant, Reuben was killed in a railroad accident. His posthumous child was a girl, Minnie. Rana, who was then sixteen, assumed the responsibility of raising the baby while her mother worked to support the family.[13]

In 1879, Mrs. Dinkey, acting on a friend's suggestion, moved the family to Braddock, where her sons might find jobs at the Edgar Thomson

17

Steel Works. Alva, then thirteen years old, became a water boy there. He began work in May 1879, three months before Schwab took his first job in the mill. Mrs. Dinkey used what money she had left from her husband's insurance to buy an ugly, dilapidated, but spacious house in Braddock. It was adjacent to the railroad tracks. Her daughter Minnie vividly recalled the day they left their farmhouse in Weatherly and arrived by train in Braddock. Seeing the grotesque house at the side of the tracks, the little girl exclaimed, "I hope it's any house except that one." It *was* "that one." Mrs. Dinkey rented the extra rooms in the house to boarders, and each day the younger girls were sent out to pick up bits of coal which had fallen from passing trains. The trainmen were so pleased at the sight of the little girls that they often threw them extra coal.

Shortly after he met Rana, Schwab became one of the boarders at the Dinkey house. He helped to reduce the household's expenses by getting up at 3 a.m. several mornings a week to keep the ledger books for a local butcher shop in exchange for free meat. He was a proficient bookkeeper; he and his brother Joe had kept the books for his father's livery stable in Loretto.

The friendship and then courtship of Schwab and Rana lasted for three and a half years. There are no surviving love letters or personal reminiscences; one can only guess what drew them together. Rana was not beautiful; even when she was young she was quite heavy, having inherited a disposition to obesity which became more pronounced over the years. She did have a flawless complexion, and she never lost it. Like Schwab, Rana was a voracious reader, enjoyed music, and loved dancing. She expressed her likes and dislikes in a manner which some considered sharp-witted and others thought bluntly outspoken. She was quite self-assertive and ambitious; she started a small business of her own, making and selling hats—no small feat for a woman of that day. From the start, Rana admired Schwab's vitality and self-confidence. He, in turn, felt warmed by her encouragement of his ambitions. When he first broached the subject of marriage, she said, "Why do you want to marry an old lady like me, lad?" For the next fifty-five years they called each other Old Lady and Lad. The names capture the essence of their relationship—they were close companions, not passionate lovers.[14]

There was one major obstacle to their marriage—Rana was a Presbyterian. Although Schwab's ties to Catholicism were never strong, his mother's were; she opposed his decision to marry outside of his family's faith. Since Rana had no intention of converting to Catholicism, and since a marriage between a Catholic and a non-Catholic cannot be consecrated in a Catholic church, they decided to hold the wedding in her home. But even this required a special dispensation from Father Hickey in Braddock.[15] Charles and Rana were married on May 1, 1883, and they spent their honeymoon in Atlantic City. Photographs of Charles and Rana were taken at that time: Schwab was a tall, barrel-chested, athletic-looking man then, and his clean-shaven face was almost adolescent in its vigor. (He never did succumb to the fashion of wearing a beard or a moustache, so he always looked somewhat younger than he actually was.) In his honeymoon photograph his expression is confident and serene; he seldom looked serene in later years.

Some months after he got his first job with Carnegie Steel, Schwab developed another source of income. He began giving organ and piano lessons. Among his first pupils were Bill and Cora Jones, the children of Captain Jones. He continued to teach after he got married. He also continued to send money to his parents, which he had been doing ever since he had worked for McDevitt; it was his way of thanking them for providing him with music lessons and a parlor organ.[16]

The money Schwab sent his parents helped them to raise their youngest children. Charles Schwab was the eldest of eight. His brother Joe, two years younger, was born in 1864. The next three children died in infancy, of spinal meningitis. Then in 1876 Mary Jane was born, and she was followed by Gertrude in 1879 and Edward in 1885. John Schwab was, as his son Edward later recalled, a jack-of-all-trades; he could sell or fix almost anything. John was eager to increase his income to meet his family's needs, so he sold his livery stable in Loretto and became a traveling salesman for the D. M. Osborne Company, which manufactured farm machinery. But in 1890, though he was only fifty-one years old and still had small children to support, he retired. He could afford to: in 1889 Charles Schwab had bought his parents an annuity which gave them each $300 a month for life[17]—a sizable sum in an era when a dollar had eight to ten times the purchasing power it has today.

Charles Schwab had worked out a strategy for getting ahead at Edgar Thomson: he would demonstrate his initiative and reliability and make himself indispensable to Captain Jones. Jones demanded a high level of performance from his workers, and the men who met his standards won promotions and salary increases. Describing his early days under Jones, Schwab said:

When I first went to work . . . I had over me an impetuous, hustling man. It was necessary for me to be up to the top notch to give satisfaction. I worked faster than I otherwise would have done, and to him I attribute the impetus that I acquired. My whole object in life then was to show him my worth and to prove it.[18]

Jones and Schwab apparently never quarreled, but the Captain could—and did—explode. In later years Schwab still vividly recalled times when Jones had been furious. Jones had a slight speech impediment, and it rendered him incoherent when he became enraged. Schwab witnessed that fury fairly often. He described one scene:

He [Jones] was in a white suit, covered with grease, making some repairs under an engine. From beneath the engine, he shouted unintelligible gibberish at an apprentice, who said, "What's that?" Bill's rage increased and with it his inability to talk coherently. Each time the apprentice asked him to repeat it his anger mounted higher. Finally he stopped, looked at the youth and said, in clear, one-syllable words, "I want —a—mon—key—wrench—is—that—god—dam—plain—enough—for—you." [19]

Schwab also remembered Jones's chief superstition—that it was necessary to work on January 1 to ensure that the company would have a profitable year.

On one New Year's day I was sick in bed. Along in the morning a wagon drew up to the house. New drawings for the mill had to be started. Jones had taken a cart, loaded it with some of the office furnishings, including my drawing board, and had brought it home to me, so that I could get to work, New Year's day, on the new rail mill.[20]

Wisely, Schwab offered no resistance to Jones's demand, and he carefully avoided provoking Jones's wrath. He never permitted himself to lose sight of his objective: to prove that he was indispensable. He did his best to study and master every aspect of steel-making, and whenever possible, to come up with ideas which would increase output or reduce production costs.

Jones rewarded him. He made Schwab his messenger to Andrew Carnegie. As Schwab later said,

Mr. Carnegie was living in Eighth St., Pittsburgh, ten miles away. Every night he got a report from Captain Bill and frequently asked the Captain himself to bring it. This was a trying experience for Captain Bill. He grumbled about it no little and one night said to me, "You go, Charlie. If he asks any questions, tell him all about things here. You know them as well as I do." [21]

On one of these occasions, Carnegie was delayed, and Schwab sat down at the piano and began to play. The sound of the music immediately attracted Carnegie. As he entered the parlor Schwab stopped playing, but Carnegie invited him to continue. Andrew Carnegie loved all music, but his particular passion was Scottish tunes and ballads. When Carnegie asked Schwab if he could play and sing some Scottish tunes at a party Carnegie was giving for some friends in a few days, Schwab said he would be happy to oblige. Then he returned to Braddock; in the next three days he learned the music and lyrics of Carnegie's favorite songs. Schwab was a great success at the party, and the next time Carnegie saw Jones he congratulated him on his choice of a messenger.[22]

Early in his career, Andrew Carnegie, the steel tycoon to whom Schwab now regularly carried messages, had been a messenger himself. Carnegie had come to America in 1848 from Dunfermline, Scotland. His father, William Carnegie, had been a handloom linen weaver, and when the new steam-driven looms came in, Will Carnegie, fine damask weaver though he was, could not compete. Young Andrew's parents decided to emigrate rather than face grinding poverty. The family settled near Pittsburgh, in Allegheny, which was then a community of Scottish immigrants.

Although young Carnegie was eager to pursue his education, his family was poor, and he had to get a job. Nevertheless, he borrowed books and read avidly. He was as much self-taught in the realm of ideas as he was self-made in the world of business. He had a fierce determination to improve himself—to acquire wealth, knowledge, and prestige.[23]

At his first job, as a bobbin boy in a cotton factory, Carnegie earned $1.20 per week; a half-century later he was earning more than $1.20 per second. At his second job he made $2.50 a week; he was a messenger boy in the main telegraph office in Pittsburgh. He quickly mastered a rare skill; he learned how to decipher messages by ear rather than by writing down the code, letter by letter. This won him a promotion to telegrapher and an increase in wages to $4.00 a week. Thomas A. Scott, superintendent of the Western division of the Pennsylvania Railroad, met Carnegie at the telegraph office, and in 1853 he hired him to be his personal telegrapher and secretary. In 1859, when Scott was promoted, Carnegie succeeded him as superintendent. During the Civil War Scott served as Assistant Secretary of War, and he chose Carnegie, then twenty-six, to command and coordinate the military railroads and telegraph lines in the eastern United States. Scott gave his young protégé several opportunities to invest in profitable businesses. On the first of these occasions, in 1856, Scott lent Carnegie $600 so that he could purchase a few shares of stock in the Adams Express Company. When Carnegie received his first dividend, a check for $10, he exclaimed, "Eureka! Here's the goose that lays the golden eggs." He got far greater returns from a later investment in a company which produced sleeping cars used by the Pennsylvania Railroad; Carnegie's monthly dividends exceeded the monthly payments for his shares, and once the stock was fully paid for his annual income from the investment was nearly $5000. Small wonder Carnegie said, "Blessed be the man who invented sleep."

Carnegie as a young man never followed his most widely publicized maxim of later years: "Put all your eggs in one basket, and then watch that basket." His biographer, Joseph Frazier Wall, has observed that between 1865 and 1875 Carnegie "was busy depositing eggs in many different kinds of nests. . . . it was a mark of his genius that he could keep careful watch over so many nests, any one of which could have provided him with a full-time career." He invested in a wide variety of ventures, including some companies which drilled for oil and others which operated telegraph lines.

Carnegie had had more than a decade of experience with bridges and railroads before he began to concentrate on steel-making. While working for the Pennsylvania Railroad at the outbreak of the Civil War he had helped to start a company which produced iron bridges for railroads— wooden bridges had been used before that time. He had also invested in a company which manufactured cast-iron rails with a new coating which made the rails less brittle and more resistant to severe changes of temperature and the heavy traffic which passed over them. But not until 1872, when he started construction on two blast furnaces in Pittsburgh, did Carnegie become directly involved in producing the steel from which rails and bridges were made.[24] Seven years later, Schwab entered Carnegie's employ.

In the 1880's, the steel industry was still in its infancy. Although products had been made of iron since at least as far back as 1000 B.C., it was not until the nineteenth century that steel went into mass production.[25] Steel is not an element, but an alloy. It is pure iron which is mixed with a carefully controlled amount of carbon to increase its strength. But iron is never found in a pure state; it always contains impurities—oxygen, carbon, manganese, silicon, phosphorus—which cause it to be brittle or soft. For centuries, through a slow and costly process, men made iron products. They began by heating iron ore in a furnace until it melted, using coke as a fuel (coke is soft coal which has been baked to eliminate the sulphur and phosphorus in it). After the molten contents of the furnace was poured into molds, it began to cool and harden into pig iron. Then the manufacturer had two options: he could produce wrought iron or cast iron. Wrought iron was made by placing pig iron in a puddling furnace. There it was

> reduced to liquid form and boiled and stirred about until most of the impurities were driven off. When the bubbling mass thickened and assumed a pasty consistency, the puddler passed a long bar through a small opening in the furnace door, and rolled the paste into a ball. This ball was then withdrawn and carried, dripping with liquid fire, to a queer arrangement of big wheels which crushed and rolled the ball over and over, squeezing out all sorts of useless stuff and further solidifying the mass. . . . The ball was then re-heated, and passed under hammers and through rollers; and the kneading it thus repeatedly underwent gave it the fibrous quality of wrought-iron. When it had been finished into bars it was ready for the market.[26]

Cast iron was cheaper to produce than wrought iron was. It did not require the services of puddlers, who were both highly skilled and highly paid. To make cast iron, the manufacturer remelted the pig iron, mixed it with limestone to remove some of its impurities, and then poured it into molds shaped like the end products desired. But cast iron is limited in its usefulness; it is very brittle. Wrought iron is far superior because it is so malleable—it can be hammered or drawn without rupturing.

What was needed—and what Bessemer discovered in 1856—was a low-cost process which would purge iron of its impurities, leaving an end product which was ductile and malleable rather than brittle and subject to fracture under strain. The Bessemer process inaugurated the age of steel. It permitted mass production; it allowed the manufacturer to use inexpensive, low-grade ore and cheap fuel; and it required few skilled workmen.

The Bessemer process was actually discovered by two men—Henry Bessemer in England and, earlier, William Kelly in America—working independently and unaware of each other's efforts. They both found that pig iron can be purged of its remaining impurities by placing it, molten, into a huge vessel (fifteen-ton capacity) now known as a Bessemer converter, and then introducing a blast of cold air into the liquid mass. The carbon in the iron unites with oxygen, producing a waste gas, carbon monoxide, while two other impurities, manganese and silicon, also combine with oxygen, forming a slag. The sudden appearance of a white flame at the top of the converter is the dramatic signal that the impurities have been purged. Finally, a carefully controlled amount of carbon is reintroduced. In James Bridge's words:

> a few shovelfuls of spiegeleisen or ferro-manganese are thrown into the mass, which is then poured into moulds, to solidify into ingots of steel. When taken out of the moulds the steel is passed under heavy rollers to give it the shapes needed for its intended use as rails, beams, or plates. . . . The first rolling thus makes blooms; and these cut into lengths make billets, which again are shaped into a hundred and one things as needed." [27]

Soon after Bessemer made his discovery, the American steelmakers found that the process had a serious limitation: it could not be used successfully with ore which contained a high percentage of phosphorus and sulphur, ore such as that in the huge, newly discovered Mesabi range in the Lake Superior region. That ore, however superabundant it was, could not be utilized until someone could discover a way to eliminate its vexatious impurities. Meanwhile, Carnegie and the other American steelmakers adopted the Bessemer process, using low-phosphorus ore. But they eagerly awaited a better method of making steel.

In 1861, five years after Bessemer's discovery, a young German mechanic, Charles Siemens, discovered a new process of making steel. Although Siemens's new method—the open hearth process—was considerably slower than Bessemer's, it had many advantages. The steelworkers could take samples from the open hearth furnace at various stages and determine its precise chemical composition; then, if necessary, they could add ingredients to the mix which would improve it. Also, open hearth furnaces were enormous—while a Bessemer converter could hold only fifteen tons, each open hearth furnace held fifty. Further, the steelmen could add scrap metal to the furnace mix; that resulted in a considerable saving. And the contents of the huge, open-topped furnaces boiled slowly, so the furnaces did not need many highly paid workers to attend them, which the Bessemer furnaces did.

Unfortunately, the open hearth method, like the Bessemer process, had one drawback: it could not handle ore containing a high percentage of phosphorus. But then a young Englishman, Sidney Gilchrist-Thomas, who was an amateur chemist, became interested in the problem. And in 1878 he discovered that if the mix in a furnace is heated 500 degrees above the temperature the steelmen ordinarily used—that is, up to 2500°F.—and if limestone is then added to the mix, the phosphorus will chemically unite with the limestone to form a slag, and the slag will float to the surface. It seemed that the last great problem of steel-making had been solved.

The prestigious Iron and Steel Institute in London spurned Thomas's discovery and cast doubt on its validity, but Andrew Carnegie did not accept that opinion. He purchased exclusive American rights to the Thomas process. A short time later, an American inventor, Jacob Reese, claimed that he, not Thomas, had made the discovery first, and he brought suit. Carnegie then offered to share the rights to the Thomas patent with other American steel companies if they would support him in the suit. The case was settled out of court; in 1881 Reese sold his own patent to the Bessemer Steel Association, an organization led by Carnegie.

Carnegie and the other American steelmakers first tried the Thomas process in their Bessemer furnaces. They were disappointed. It failed. Even so, Carnegie would not drop the idea; he kept it in mind. Then, in 1885, while he was traveling in England, he visited a steel mill, as he did so often. And there he saw the Thomas process working perfectly—in an open hearth furnace. Once again Carnegie took the lead in America: he ordered the steelmen at his newly built open hearth furnaces at Homestead to use the Thomas process.[28]

While Carnegie was eagerly embracing each new technological advance, he was at the same time trying to reduce the economic uncertainties which confronted his business. The American steel industry was highly vulnerable to every economic crisis, and that situation disturbed Carnegie. The steelmakers' primary customers were the railroads, which purchased rails either to lay new lines or to replace old ones. But track expansion or replacement depended on factors beyond the control of the railroads—namely the volume of rail shipments of both agricultural products and manufactured goods. During periods of depression, the demand for agricultural and industrial products declined, prices fell, and profits shrank—all of which affected the railroads, either through a decrease in freight traffic or by defaults on payments owed for past shipments. The impact upon the railroads was then quickly transmitted to the steel mills. When orders for steel declined, the mills' suppliers faced retrenchment or ruin. Those suppliers were the owners of coal fields and coke ovens, iron mines, and ore ships whose greatest customers—sometimes, whose only customers—were the steel mills. Any failure in the economy's general prosperity spread like a contagious disease, and the steel industry enjoyed no immunity.[29]

In periods of recession, the steel mills received reduced orders for rails, and other orders were canceled. A mill owner who had expanded his production facilities to meet a rising demand for rails in one year might find himself with an idle plant or a substantial overcapacity the next. Carnegie, however, refused to accept the situation. He sought to minimize the impact of the erratic swings in demand which plagued the industry. To the dismay of his partners, in good times he did not pay out profits in the form of dividends; rather, he retained the profits in order to ride out periods of national economic crisis—and, in fact, to use those periods to expand. Carnegie's two Bessemer furnaces in Pittsburgh became operational shortly before the onset of the Panic of 1873. Many businesses went under during the crisis, but Carnegie was able to survive it by drawing upon the profits he had retained over the good years. He also created a steady market for the output of his furnaces: he used the entire tonnage in his own rail rolling mill, the new steel works at Braddock which was begun in 1873 and became operational in 1875, in the middle of the depression.

Carnegie was able to enlarge his share of the market for rails during the 1873–78 depression because his production costs for Bessemer rails were lower than those of his competitors'. He had a dual advantage; he obtained the pig iron needed for the rails at cost from his own Pittsburgh furnaces, and he had built his rail mill at a time when the costs of labor and raw materials were abnormally low. To Carnegie, expansion during depressions was a matter of policy; he acquired the Homestead works in 1883 and the Duquesne works in 1890. He regarded Pittsburgh as his bastion, the center of his growing industrial empire, and he dealt with rivals in his home territory quite simply: he bought them out during periods of economic decline.[30]

Carnegie never hesitated to spend money to improve the productive efficiency of his enterprises and thereby reduce his operating costs. In 1885 he decided to phase out Bessemer production at the Homestead plant and to produce open hearth steel there instead. As Wall has noted, ". . . Carnegie had hundreds of thousands of dollars invested in Bessemer converters . . . [even so,] he was ready to scrap them in favor of a better and more economical process. Construction costs never bothered

Carnegie. It was operational costs that mattered, and that simple truth was a major reason for his success." [31]

No one was less complacent or more willing to embrace each new business or technological opportunity than Carnegie. One of his many innovations was a system of "promotion within the ranks," which was designed to attract ambitious young men to manage his mills and to implement his over-all business strategy. He was especially eager to encourage those who displayed initiative and who shared his passion for innovation.

Charles Schwab was one such man.

During the late 1870's and early 1880's, the Pennsylvania Railroad was the largest customer for Carnegie's steel rails. J. Edgar Thomson, was then president of the railroad, and he became interested in the chemistry of steel-making. In his orders to the steel mills he gave precise specifications on the quality and composition of the rails to be produced. On one occasion the railroad rejected a large shipment produced at Braddock under Captain Jones because the rails had failed to meet the specifications. When the defective rails were returned to Braddock, Jones told Schwab, "This damned chemistry business will be the ruination of us yet." [32]

This problem was not Jones's alone; it was common throughout the steel industry. Most furnace masters, Captain Jones included, worked on a trial-and-error basis, judging the richness or impurities of iron ore or coke by the single test of visible appearance. This method sometimes resulted in costly and calamitous errors: if the ore used were of a low grade, the end product might be a brittle steel which would flake and fracture under pressure. On the other hand, if the ore used were too rich, it might cause a furnace explosion. [33]

Schwab perceived the growing importance of chemistry in steel-making and resolved to master this arcane subject. He later recalled:

> In my own house I rigged up a laboratory and studied chemistry in the evenings, determined that there should be nothing in the manufacture of steel that I would not know. Although I had received no technical education I made myself master of chemistry and of the laboratory, which proved of lasting value. [34]

In the small cottage to which Schwab brought his bride in May 1883 there was a room which she originally intended to use as a sewing area; instead, it became his laboratory. He brought home specimens of steel and subjected them to rudimentary analysis, testing them to find out how much carbon and phosphorus they contained. He also tested the corrosive effects of acids upon steel. A co-worker once described these early experiments:

> I was the "carbon boy" in the laboratory, and CMS used to get filings from me to analyse at home. His analyses were good, as I remember, and most of the variation from the official analysis would come from the fact that he had a pair of worn-out scales that were never accurate. Scales would have been one of the most expensive pieces of equipment for his laboratory.[35]

At first Schwab's experiments were limited in their range because his laboratory equipment was so makeshift. But then Henry Phipps, one of Carnegie's major partners, heard of his studies. Phipps gave Schwab a check for $1000; he told Schwab to buy any equipment he needed.[36] If Schwab had made only a single minor discovery about the chemistry of steel it would have more than repaid the investment. In fact, he never discovered anything new, but he did acquire a first-hand knowledge of chemistry and metallurgy, and widened his mastery of steel-making.

Again and again Captain Jones offered Schwab new and challenging assignments. In 1885, when Schwab was twenty-three, Jones ordered him to design and build a bridge over the Baltimore and Ohio Railroad tracks; hot metal was to be carried over it from the furnace area to one of the finishing departments. Schwab completed the job sooner and at less cost than originally estimated, and Jones gave him a reward. It was a pin in the form of a spider, with a diamond set in it. Jones also handed him "a package from Mr. Carnegie"; in it were ten $20 gold pieces.[37] But John Schwab's response to his son's achievement apparently meant even more to Charles than the gifts from Carnegie and Jones did. John, as his son once said of him, was "never lavish in his praise." Schwab recalled that when he invited his father to Braddock to view the bridge he had built, "he could not comprehend that it was true—that his son had done this—but as hard as he tried to conceal his pleasure, I knew it was there."[38]

The hideous spider pin from Jones and the gold coins from Carnegie were acknowledgments of the strides Schwab had made and indications that he would be given greater challenges and responsibilities in the future.

The Homestead Works, which Carnegie had purchased in 1883 from the financially weak Pittsburgh Bessemer Steel Company, had originally been designed to produce rails by the Bessemer process. But Carnegie was not interested in expanding his own rail-producing capacity. In 1887, he ordered Homestead to be rebuilt; instead of rails, it would specialize in structural shapes. Carnegie saw a growing market for structural steel because more and more new office buildings were multi-story structures, and such structures could best be built with steel beams. Homestead also made semi-finished products, such as bars, blooms, and billets, which were sold to companies making cables, wire, nails, and tubes. Carnegie had planned shrewdly; when rail orders fell during depression years, the demand for other products kept the mills in operation.[39]

In 1886, on Jones's recommendation, Carnegie appointed Schwab general superintendent of the Homestead works.[40] But before Schwab began his new job, Carnegie sent him to Europe, where he visited the best foreign steel mills, especially the Krupp Iron Works in Essen, Germany, and the Schneider Works at Le Creusot, France. There he studied the blast furnace and rolling mill methods of the European masters, so that the Carnegie mills could duplicate or adapt their methods at home.

His promotion had one drawback: he was required to live in a company-owned house on company land. It was one of five "managerial houses" which Carnegie had had constructed for his key personnel so that they would be available at any hour of the day or night. The house was small, it had no bathroom, and its only sources of heat were coal grates and the kitchen stove. As Schwab later said, "We lived there, right on the cinder dump." [41]

Schwab's increased salary, $10,000, may have compensated him for the discomforts of his new home, yet it hardly compared to the amount Carnegie paid Captain Jones. Carnegie hoped to keep Jones permanently, so he had offered him a partnership. But Jones had declined; it might create a gulf between him and his men and impair his management of the works. When Jones refused the partnership, Carnegie told him that he would increase his salary, and substantially. He did just that. Carnegie suggested, and Jones accepted, $25,000—the same salary as that paid to the President of the United States.[42]

Schwab was now general superintendent of Homestead, and under his direction the new open hearth facilities were built quickly and smoothly. In late 1887, Sir Arthur Keene, the British steelmaker, visited the works, and he was struck by the men's efficiency and *esprit de corps*. He was so impressed by Schwab that he offered him $50,000 a year to run the steel works in Birmingham. Schwab later told of his own reaction: "I was surprised, flattered, and asked him about the contract. He told me he would give me any contract I liked; in fact, [he] proposed a five years' engagement with an annual increase in salary." Schwab tentatively accepted, but the deal fell through when Keene would not agree to Schwab's proposal that his assistants also receive large salaries.[43]

Carnegie was pleased; now his two major steel works were under the direction of the most capable steelmen in America: Jones at Braddock and Schwab at Homestead. But this arrangement ended abruptly in September 1889. One evening a furnace at Braddock jammed, and its contents could not be released. Jones took charge. As he was probing the furnace it suddenly exploded, and the blast threw him back several feet. He hit his head sharply on a metal surface and his body was badly burned. He never regained consciousness; he died of a skull fracture.[44]

Captain Jones was given a funeral befitting a reigning head of state. Every Carnegie employee was given the day off to attend the memorial services, and nearly 10,000 men and their families stood along the sidewalks, their heads bared, as the funeral procession passed through the streets of Braddock.

Carnegie had lost his ablest superintendent and Schwab had lost his first friend and mentor in the steel industry.

Carnegie now had to choose a man who could succeed Jones as head of the Braddock works. He was not inclined to transfer Schwab to Braddock; he did not want to jeopardize the smooth operations Schwab had achieved at Homestead. But Schwab was eager for the job, and he urged Carnegie to appoint him to it. Carnegie finally relented, saying, "Very well, if you wish to have the honor of succeeding to Capt. Jones' position, I will oblige you." At twenty-seven, only a decade after he had began work at Edgar Thomson as a dollar-a-day laborer, Schwab became general superintendent of the largest steel works in America.[45]

3

Success and Scandal

Schwab's view of other men was an extrapolation of his own attitudes. He assumed that working men had the same desires and ambitions he had, and that they could be spurred to greater efforts by the same inducements and rewards which appealed to him. He was keenly aware that he would do his best work if he knew that he would be well paid for it, so, as superintendent of Braddock, he offered his workers rewards for maximum effort. Throughout his life he gave his subordinates positive incentives, bonuses and promotions, rather than threatening to fine or fire them if their efforts fell short. In his early days at Braddock he had to deal with the problem of "seconds"—substandard steel, which was one of the greatest drains on the company's profits. He initiated a policy of paying cash bonuses in return for first-class results. The rail workers at Braddock were made eligible for bonuses; each month a prize of $20 in cash would go to the scraper, pourer, blower, and heater on whose shift the fewest second-class rails were produced. The number of "seconds" went down dramatically, and the savings which resulted more than compensated for the bonuses paid.[1]

Schwab also believed that most men would strive for excellence even when no monetary reward was at stake. Competitive himself, he believed that there was an inherent competitiveness in most men and that they could be stirred to prodigious feats if they were given a sufficient challenge.

In one mill at Braddock production was below average, even though Schwab had a capable manager there.

31

"How is it that a man as able as you," I asked him one day, "cannot make this mill turn out what it should?"

"I don't know," he replied; "I have coaxed the men; I have pushed them; I have sworn at them. I have done everything in my power. Yet they will not produce."

It was near the end of the day; in a few minutes the night force would come on duty. I turned to a workman who was standing beside one of the red-mouthed furnaces and asked him for a piece of chalk.

"How many heats has your shift made today?" I queried.

"Six," he replied.

I chalked a big "6" on the floor, and then passed along without another word. When the night shift came in they saw the "6" and asked about it.

"The big boss was in here today," said the day men. "He asked us how many heats we had made, and we told him six. He chalked it down."

The next morning I passed through the same mill. I saw that the "6" had been rubbed out and a big "7" written instead. The night shift had announced itself. That night I went back. The "7" had been erased, and a "10" swaggered in its place. The day force recognized no superiors. Thus a fine competition was started, and it went on until this mill, formerly the poorest producer, was turning out more than any other mill in the plant.[2]

By using such incentives, Schwab was able to maintain the high level of production and general harmony which Jones had achieved at Braddock.

But Carnegie's industrial empire did not enjoy good will in all its plants. Homestead was the scene of a brief strike in 1889, before Schwab left to succeed Captain Jones. In 1892 another strike, one of the most violent in American industrial history, erupted there, and Schwab had to go back to the plant to restore peace.

When Carnegie had purchased Homestead in 1883, he had accomplished two objectives: he had eliminated a competitor for Bessemer rail sales in the Pittsburgh area and had acquired a site along the Monongahela River where he could produce structural steel. The purchase, however, had also brought him what he most feared: an entrenched union. Homestead was the stronghold of one of the most powerful unions in America, the Amalgamated Association of Iron, Steel, and Tin Workers.[3]

The strike of 1889 was precipitated by the skilled workers' refusal to accept a new system of calculating their wages. Under the existing system they were paid a flat sum per ton produced. Beginning in 1886, however, Carnegie had spent millions of dollars to improve the Homestead plant, and those improvements had raised the productivity of the skilled workers, increasing their income. Carnegie was always suspicious that his skilled workers were earning too much. Now, having spent large sums to improve his plant, he was doubly disturbed. The improvements had increased rather than decreased his fixed labor costs. For that reason he favored a sliding-scale wage plan, rather than the flat per-ton system. He had outlined just such a sliding-scale plan in an article published in April 1886:

> What we must seek is a plan by which the men will receive high wages when the employers are receiving high prices for the product, and hence are making large profits; and *per contra*, when the employers are receiving low prices for products, and therefore small if any profits, the men will receive low wages. If this plan can be found, employers and employed will be "in the same boat," rejoicing together in their prosperity, and calling into play their fortitude together in adversity. There will be no room for quarrels, and instead of a feeling of antagonism there will be a feeling of partnership between employers and employed.[4]

But the skilled workers at Homestead had no interest in Carnegie's proposal; obviously, it would reduce their income.

On July 1, 1889, the Homestead workers went out on strike. They had refused to accept the sliding-scale system and they had balked at Carnegie's proposal that each man should sign an individual contract with the company—which would have been one way to oust the union. William L. Abbott, president of Carnegie, Phipps & Company, was handling negotiations with the strikers. (Carnegie was then in Great Britain, where he spent his summers.) Abbott ignored Carnegie's written advice to keep the Homestead plant shut down until the strikers agreed to the company's terms. Instead, acting on his own authority, he hired new workers to replace the strikers. When the new recruits, escorted by the sheriff and his deputies, arrived at the plant, they found themselves vastly outnumbered by a menacing crowd of strikers. They made no effort to force their way through the ranks of the strikers, and violence and bloodshed were averted. Abbott's impatience mounted when he heard that the men at the Edgar Thomson plant in Braddock were threatening a strike to show their solidarity with the strikers at Homestead.

However, it was Abbott's unwillingness to risk violence or a widening of the strike which ultimately defused the explosive situation. In a meeting with the strike leaders, he proposed a compromise settlement. The

33

union accepted the sliding-scale wage system in a new three-year contract, and in return Abbott granted the union what it had long sought but never before had been able to obtain: formal recognition and exclusive jurisdiction to bargain for the Homestead workers. As Wall has observed, "The men had reason to cheer. For, in spite of the sliding scale, the union was more strongly entrenched than ever before. Because it was now accepted as the only bargaining agent with management, not a man could be hired or fired at Homestead without the union's approval." [5]

While Carnegie was pleased that the Homestead workers had accepted the sliding scale, he was deeply disturbed that Abbott had granted the union formal recognition. Even worse, as he wrote Abbott, "the great objection to the compromise is of course that it was made under intimidation—our men in other works now know that we will 'confer' with law breakers." Even so, Carnegie was somewhat consoled by the fact that there would be three years of peace under the new contract, and that Schwab, acting as superintendent of Homestead, had been so skillful in restoring the plant to normal operation after the settlement. Writing to Abbott on August 7, Carnegie said, "So glad Schwab proved so able. If we have a real manager of men there Homestead will come out right now. Everything is in the man." [6]

Carnegie's irritation at the terms of the compromise was shared by his chief partner, Henry Clay Frick, chairman of Carnegie Brothers & Company, the other major division of Carnegie's industrial empire. Frick, born in 1849, had built a business worth more than a million dollars before he reached his thirtieth birthday. He produced coke, one of the essential ingredients used in making steel. Frick purchased control of vast reserves of coal in the Connellsville region of Pennsylvania, north of Pittsburgh. In 1871, when he organized the Frick Coke Company, he owned 300 acres of coal lands and 50 ovens; by 1880 he held nearly 3000 acres of coal fields and close to 1000 coke ovens. Frick was eager to continue expanding his business, but he lacked the necessary capital, so when Carnegie offered to buy a share of his business, Frick agreed.

Early in 1882, Carnegie invested $225,000 in Frick's newly reorganized company, for which he received 11.25 per cent of the stock. Over the next months Carnegie increased his investment, and by the end of 1883 Frick was no longer the majority owner of his coke business; Carnegie was. But Carnegie had no intention of ousting Frick or challenging his control over the operation of the company; he knew that Frick's untiring ambition and ability matched his own. In 1889 Frick proposed that the two divisions of the Carnegie business, Carnegie, Phipps & Company and Carnegie Brothers & Company, be formally united. Carnegie agreed, and the two companies were reorganized. In 1892, Frick became president of the new entity, the Carnegie Steel Company.[7]

In that same year, 1892, the Homestead contract came up for renewal. Carnegie and Frick had agreed in advance not to capitulate to the union; in fact, they had decided to oust it. In April 1892 Carnegie announced that, since a majority of his employees were non-union, the union must go. "This action," he stated disingenuously, "is not taken in any spirit of hostility to labor organizations, but every man will see that the firm cannot run Union and Non-Union. It must be one or the other." [8] Frick, long noted for his hostility to unions, supported this policy.

Carnegie and Frick had also agreed to retain the sliding scale system; it enabled them to reduce wage rates whenever steel prices fell. In 1889, when the sliding scale was adopted at Homestead, the market for steel was excellent; selling prices were highly profitable and the minimum wage rate per ton on the sliding scale seemed tolerable to the company. But by 1892 the economic situation had changed substantially. Steel prices had fallen during the preceding three years, and Carnegie and Frick insisted that in the new contract the minimum rate on the sliding scale—that is, the base rate below which skilled workers' wages could not fall—would have to be reduced.[9]

Carnegie Steel, as a matter of policy, paid wages no higher than the current state of the labor market required. In periods of rising sales and profits, when labor was scarce, a wage increase would be necessary and justifiable: good workers would stay on and new ones would be attracted. The company was also willing to match wage increases offered by competing firms in order to keep the unions out. But in periods of falling demand, when men who were looking for work could easily be hired for less than the company was currently paying, Carnegie and Frick believed they should reduce wages. They thought it both reasonable and necessary.

The Homestead contract was due to expire on July 1, 1892, but representatives of the union tried to extend the existing contract for another three years, until June 30, 1895. The company, however, did not intend to renew the contract on the same terms. Because the market price of steel billets had declined, Carnegie and Frick wanted to lower the minimum rate on the sliding scale. The union was unwilling to accept such a reduction. Frick then announced the company's final offer: a 15 per cent

35

reduction in pay for those skilled workers who still were paid on a ton-nage basis, a lowering of the lower limit of the sliding scale, and a con-tract expiration date of December 31, 1894.[10] (In mid-winter it would be easier for the company to withstand a disruption of production and harder for the workers to sustain a strike.) Frick intended to discredit the union no matter what option it chose, "either a strike or an abject surren-der of all its recent gains. Either way, he stood to gain." [11]

On June 24, when the union refused the company's offer, Frick closed down the plant and announced that it would reopen on July 6. The Amalgamated Association had 1800 members at Homestead, out of a total work force of 3800. Only 325 of them went out on strike. Frick was confident that the plant could operate effectively until the small band of holdouts capitulated, and Carnegie shared his view.

When Carnegie left for his summer vacation in Scotland, he was cer-tain that Frick would not give in to the strikers' demands and that the overwhelming majority of workers would return to work when the plant reopened. Carnegie knew that Frick had been in contact with Robert Pinkerton, founder of the Pinkerton National Detective Agency. Frick intended to enlist the services of 300 Pinkerton men to protect the plant and those of its workers who were not on strike. Carnegie felt that the workers whose contracts had been settled would neither side with the strikers nor be intimidated by them.

The 325 union members who refused Frick's final offer believed that the company was bluffing—that it would compromise again, as it had in 1889. They thought there were several factors in their favor. Carnegie had often made public statements about the sanctity of a striker's job. Then, too, it was an election year; neither the State of Pennsylvania nor the government would risk sending troops or militiamen to quell a labor disturbance. Finally, the strikers expected that Carnegie's highly placed friends within the Republican party would prevail upon him to yield to the union's demands; the company's resistance might antagonize the voters and throw the election to the Democrats, who would then launch a new assault on the protective tariff on foreign-made steel.

The 300 Pinkertons whom Frick had hired were to guard the steel mill when it reopened. They were brought to Homestead on July 5, under cover of night, on two large barges drawn by a tugboat. But when the strikers learned that the guards were coming, they marched to the dock-ing area to prevent their landing. As the Pinkertons attempted to come ashore, someone—it was never determined who did it—fired a shot.

That started a bloody battle, and several men on both sides were killed. But the outcome was never in doubt. The 300 Pinkertons trapped on their barges were no match for several thousand enraged workers and their families on shore. The rout of the Pinkerton guards gave the strikers and their newly won supporters control of the steel mill, for the sheriff was unable to raise an adequate posse to dispossess them. He finally appealed to the governor of Pennsylvania for state militiamen, and the governor sent 7000 to Homestead, where they remained for nearly four months.[12]

Order was quickly restored, and the managers took control of the mills again. On July 16, the company announced that during the following week strikers could reapply for their jobs and return to work. If they did not, they would be replaced by nonunion workers transferred from other mills and by unskilled laborers recruited from other towns. But the strikers did not return to work, and some of the nonunion workers at other Carnegie mills went out in sympathy.

Overnight, the eruption of violence at Homestead turned the wage dispute into a major news event. Carnegie had always said that he was concerned for the welfare of workingmen; he was famous for his statements on the subject. He was deeply troubled when he learned of the outbreak. He was also disturbed to find himself the target of bitter attacks in the British and American press. In private letters to his friends he absolved himself from any responsibility for the events at Homestead, placing the blame upon Frick, "my young and rather too rash partner." But he issued no public statement and refused all requests for interviews, saying that "silence is best." He preferred that "Homestead and its deplorable troubles . . . be allowed to rest." But over a month later Homestead was still front-page news and Carnegie was still being censured in the press. He was in despair. He told Prime Minister Gladstone, "This is the trial of my life (death's hand excepted)." [13]

Two of Carnegie's senior partners, Henry Phipps and George Lauder, managed to persuade him not to return to America and personally direct the company's response to the strike. They feared that if he intervened Frick would resign, and the company would then have lost its chief executive officer. Carnegie could not afford to do anything which might reflect unfavorably upon Frick, either directly through public criticism, or indirectly by removing John A. Potter, Frick's lieutenant, who was superintendent of Homestead.

On July 17, Carnegie wrote to Lauder:

. . . must keep quiet and do all we can to support Frick and those at the seat of war. . . . We shall win of course, but may have to shut down for months. . . . Potter should be sent abroad and Schwab sent back to Homestead. He manages men well and would soon draw around him good men from E.T. [Edgar Thomson] and other Works. Have suggested this to Frick.[14]

Although it was mid-July when Carnegie made the suggestion to Frick, Schwab was not sent in to replace Potter as superintendent of Homestead until mid-October.[15] Frick delayed the action because the strikers might have seen any major shift in personnel as a sign of weakening determination on the part of the company, or they might think that Carnegie was repudiating the policies of Frick and his lieutenants. Schwab's move to Homestead had to wait until the strike was virtually over, when the tide of victory clearly had turned in the company's favor. By early October the union's strike funds were almost exhausted and contingents of the state militia were being withdrawn.

When Potter was ousted, his removal was elaborately disguised. The company said that it was a promotion. Potter had gotten his first job with Carnegie at the age of fourteen, working as a greaser boy at the Lower Union Mills. He won rapid advancement because he had great ability in engineering and metallurgy, but he never acquired any skill in dealing with men, particularly with subordinates. When Schwab took over the superintendency of Homestead Potter was named "chief mechanical engineer" for the entire company. It was a newly created consulting post. Carnegie Steel, in its official announcement, stated that "all the changes are promotions, and were not brought about by reason of any dissatisfaction with Mr. Potter or with anybody else." A year later, Potter, still bewildered, resigned. He wrote to Carnegie, "About the time the strike was ended and our victory won, my career of usefulness also ended, and I was removed to a new field of labor. The cause of the sudden change is still as great a mystery to me as it was at the time it was made." [16]

On October 18, Schwab left Braddock for Homestead. He had tried to decline the assignment because he enjoyed his position at Braddock, but Frick induced him to go by persuading Carnegie to give Schwab his first ownership interest in the Carnegie Steel Company, one-third of 1 per cent. "I hope," wrote Frick, "it will prove as profitable to you in the future as such interests have been to others in the past. It largely rests with you, however, and I have no doubt that it will not be your fault if it does not prove profitable [sic]." [17]

In the fall of 1892 Carnegie had one great desire—that "Homestead" be erased from public consciousness. If peace and normal production could be restored, then the memory of past violence might quickly recede. Carnegie's hopes lay with Schwab, the new superintendent. Schwab certainly realized the significance of his new responsibility; the fact that he had been awarded a partnership interest made it clear that this was far more than a recognition of past services. It was an important opportunity to win Carnegie's gratitude and esteem, and Schwab acted on it.

Unfortunately, the record of Schwab's actions at Homestead is meager. The company avoided issuing any statements which compared Schwab's policies with those of Frick or Potter; he could not even be mentioned as a peacemaker. Any such comment would, by implication, have cast aspersions upon his predecessors. Reporters were excluded from the Homestead plant, and the company's personnel at every level were ordered to make no public statements and to grant no interviews. Schwab's accomplishments can only be pieced together from fragmentary evidence.

At first Schwab thought that the only problem at Homestead was labor discontent. But he soon discovered that the plant was in a general state of decay. As he later testified before a congressional committee:

> During the great strike of 1892 I was asked, much against my wishes, to reorganize and take charge of the Homestead works. I finally consented to do so. . . . When I went to Homestead I found it in a thoroughly disorganized condition. The works were badly run down, and the men were unsuited to their work, and they did not have competent foremen. Now, the first four months of my time at Homestead were devoted entirely to reorganization.[18]

However, before Schwab could begin the necessary physical improvements, the strike had to be settled.

The first break in the strikers' ranks occurred on November 17, when two hundred of them, recognizing the impossibility of winning their demands, decided to go back to work. Once the strikers' unity was broken, the end was inevitable. The Amalgamated Association released its members from their strike pledges, telling them that they were free to return to their old jobs if they wanted to, and if they could obtain them.

Schwab himself greeted the returning strikers, not *en masse*, but individually—calling the many old-timers he remembered by their first names. His approach, so unlike Superintendent Potter's dour formalism, made him all the more popular with the workmen. Schwab thought the strike could have been settled and further ill will and violence prevented if Potter had been able to meet with the striking workmen, win their confidence, explain the company's position, and invite them to return to work, assuring them that a full and fair hearing would be given to their demands and grievances. But, according to Schwab, Potter did not have the temperament for the job. He could never be a conciliator or a counselor. As Schwab later said:

> Johnny Potter was not the proper man to have been in charge at Homestead. He was a handshaker but there was no sincerity back of his interest in the workingman. You can tell a workingman you like him, but he knows whether you are sincere or not. You can't make him believe you are interested in his welfare unless you are.[19]

No smile, no word of greeting, however, could erase the workers' memories of Homestead's bitter strife. The strikers had lost their union, their savings, and, perhaps they thought, by surrendering, their self-respect. If Homestead ever again was to operate efficiently and harmoniously, the discord would have to be resolved, and that would take time and a tactful manager. Schwab held a series of informal meetings with small groups of the leading workmen. He believed that if the heads of the various work crews could be won over then they, in turn, would win over their assistants and helpers.[20]

During his first four months at Homestead, Schwab did not set foot outside of the plant; he lived in one of the five executive houses on the grounds. Drawing upon an uncommon ability (one which he shared with his hero, Napoleon), Schwab was able to work for seventy-two hours at a stretch, taking only short snatches of sleep. He decided that in his first weeks he should go everywhere, greet and speak to everyone, on both the day and night shifts, and compile an inventory of physical and human assets on which he could base the reconstruction of Homestead.

It is tempting to speculate whether violence could have been avoided if Schwab had been general superintendent of Homestead at the outset of the strike. He seems to have thought so. In 1935 he said:

> If the Homestead strike had been mine . . . I would have called the men in and talked to them, and shown them exactly where the company stood on the issues. Then I would have told them I was sorry, but that it would be necessary to close the plant. But I never would have brought in other workers to take their places. There is nothing a worker resents more than to see

some man taking his job. A factory can be closed down, its chimneys smokeless, waiting for the worker to come back to his job, and all will be peaceful. But the moment workers are imported, and the striker sees his own place usurped, there is bound to be trouble.[21]

Whatever the general wisdom of this plan might be, it would not have applied at Homestead. Violence was not precipitated there when new workers were hired to replace the strikers, but earlier, when the Pinkerton guards were brought in. The new workers first came to Homestead more than two weeks after the night the strikers fought off the Pinkertons. And, since Carnegie and Frick were agreed on the use of the Pinkertons, it is improbable that they would have yielded to Schwab if he had counseled a more conciliatory policy.

In the same 1935 interview Schwab said that

Frick had a fight on his hands before he started because he didn't understand men. Frick was, by nature, a fighter, and when he saw it was to be war, he imported 300 men from New York "to make war." If Carnegie had been here that wouldn't have happened. Carnegie and I would have agreed on the same policy—that of closing the mills until an adjustment could have been made.[22]

That argument is simply untenable. Carnegie himself had agreed to hire the Pinkertons, and Schwab probably knew that he had. Furthermore, Schwab was confronted by a similar strike threat in 1910, and at that time he did not adhere to the policy which in 1935 he claimed he would have pursued at Homestead. In fact, in 1910 he adopted policies similar to Frick's. Even if Schwab had been superintendent at the outset of the strike, it is unlikely that he would have been able to alter the course of events—neither side was willing to compromise, and Schwab would not have challenged or refused to implement the strategy of Carnegie and Frick.

Although Schwab wanted to rehire all of the old workers, he did not want to fire any of the new men who had come to Homestead during the strike. In his talks with the returning strikers, he offered them a vision of a new and greater Homestead, one which would surpass every other plant, both within and beyond Carnegie's realm. There would be jobs enough for everyone, he said, but preference would be given to the older, more experienced, returning strikers. He promised them that their grievances would be given a fair hearing and assured them that they would not be worse off financially, even though the company was resolutely determined to be non-union.[23] " 'That,' I told them, 'is a situation that cannot and will not be changed. Otherwise we should have to close down completely and dismantle our plants.' When I said that they knew I was telling them the truth because I had always told them the truth." [24]

Schwab's "truth" was only a ploy. Threatening to close down and go out of business was one of his favorite weapons. He used it on at least four occasions: to settle strikes at Homestead in 1892 and at Bethlehem in 1910, to try to prevent tariff reductions in 1913, and to forestall construction of a government armor plant in 1916. But he managed to persuade the workers to give up their union, and so he was able to proceed with the reconstruction of Homestead.

After four months of hard work, Schwab needed a vacation. He also had to report to Carnegie, so he rested aboard a steamer headed for the British Isles. When he arrived in Scotland, he told Carnegie that Homestead was now at peace and that production at the mill was soaring. He also suggested that the company should spend millions of dollars there on new plant and equipment. Carnegie was delighted that his Homestead troubles were over, and he agreed with Schwab—now was the time to spend money on new facilities. A fortune, if necessary. He sent Schwab home with the authorization.

To Carnegie, Schwab was a hero; he had settled Homestead. And to the workers, he was a great man; he had saved them their jobs.

Less than a year later, however, Schwab was under attack, and seriously so. He had been accused of masterminding a conspiracy to defraud the United States Navy.

Ever since the early 1880's, a growing number of Congressmen had become convinced that the government should enlarge the Navy and, at the same time, that it should reduce American dependence on foreign steel producers for its defense needs.[25] President Chester Alan Arthur agreed, and his Secretary of the Navy, William E. Chandler, made the government's first overture to the American steelmakers. He urged them to construct plants for the production of armor plate, to be used to fortify warships against cannon fire, and gun forgings. His arguments did not move the steelmakers. However, his successor, William C. Whitney, was more realistic. He knew that no American steel company would be willing to spend several million dollars on a plant for armor or gun forgings unless it received enough orders to justify the investment. He went to Congress and obtained authorization to stop buying armor and gun forgings abroad; then he grouped the Navy's estimated requirements for several years, making them into one large order. He hoped that that would encourage several American steel companies to submit bids.[26]

In the fall of 1886 Carnegie ordered the construction of an armor plate mill at Homestead, so that he might submit bids to the Navy early in 1887. He sent Secretary Whitney a message: "You need not be afraid that you will have to go abroad for armour plate. I am now fully satisfied that the mill we are building will roll the heaviest sizes you will require, with the greatest care."[27] Because Carnegie and the heads of some other companies seemed interested and confident, Whitney assumed that armor plate would soon be produced domestically.[28]

But Secretary Whitney was wrong. Most American steel producers were not interested in obtaining government contracts. By March 22, 1887, Whitney had spent a year conferring with steelmakers throughout the country, yet he had received only three bids. The Midvale Steel Company entered a bid to produce gun forgings and the Cleveland Rolling Mills submitted one for armor. But Whitney's prospectus had stated that preference would be given to firms bidding on both armor and gun forgings. The third bid did not come from Carnegie, but from the Bethlehem Iron Company. Since Bethlehem was the sole bidder on both items, forgings and armor, it was awarded the entire contract, which was valued in excess of $4,000,000.[29]

In 1886, when Schwab had visited Krupp and Schneider, the foremost European armor-producing plants, he had learned what he later reported to Carnegie—that the production of armor plate was not at all simple. The main problem was that the producers could not achieve uniform results: no two plates in any group were identical. Even different sections of the same plate varied markedly in quality, thickness, tensile strength, and resistance to penetration when fired upon. These variations were inevitable, because no one could fully control the chemical reactions which took place within the furnace, nor the rate at which the various parts of an ingot would cool, nor the congealing of impurities, which made some areas of the plate thinner and less resilient than others.

Despite Carnegie's initial interest in obtaining the potentially lucrative contracts, he developed serious misgivings, and not just about production. He was having difficulties with the Navy Department. He found himself in sharp disagreement with the Department over the procedure for producing armor. The Navy insisted on prescribing the process by

43

which the armor was to be produced. But Carnegie believed that his company would have to have both the flexibility and the discretion to do whatever it believed necessary if it were to produce armor of consistently high quality. In a letter to Secretary Whitney Carnegie complained about the Navy's rigorous production specifications:

> Your "specifications" are the serious point, as our experience with Government officials of Army and Navy—is that they are martinets only and insist upon technical points to an absurd degree. Practical men know that "tests" are necessarily approximate[,] no two can result alike—that we can give you plates equal to any made in the world is true but we believe "inspectors" abroad know that variations exist and allow for them. I do hope we shall have to deal with a practical experienced Inspector should we contract.[30]

When the Navy refused to permit more flexibility in production techniques, Carnegie decided not to enter a bid and the contract went to Bethlehem.

When Bethlehem began construction of its armor plant it ran into unanticipated complications and costly delays. The contract had been signed on June 1, 1887, and under its terms Bethlehem was to begin deliveries of armor in two and a half years—that is, in December 1889. But when the delivery date came, Bethlehem's armor plant had not even been completed. In January 1890, the Navy Department granted Bethlehem a six-month extension, but the company still failed to meet its deadline.[31] In May, Secretary of the Navy Benjamin F. Tracy ordered an investigation of Bethlehem's progress. He learned that the plant was only partly built and that the company would need an additional fifteen months to complete it. Only then could it begin to cast ingots and to experiment in the production of armor plates according to the precise specifications of the Navy Department.[32] Tracy granted Bethlehem a second extension, twelve to fifteen months to complete the plant.

These delays cost Bethlehem the position it would have held as the sole domestic producer of armor plate. In 1890, when Secretary Tracy learned that Bethlehem would not be able to make deliveries in 1891, he wrote to Carnegie, urging him to complete the plant he had begun in 1886. But Carnegie again refused; only after President Benjamin Harrison asked him to reconsider did he relent.[33]

Carnegie's armor plant at Homestead was completed in 1892, a few months before the Homestead strike erupted. The armor department was headed by William Ellis Corey, an ambitious and competent young man who had begun working for Carnegie in 1881, and Corey's chief assistant, W. A. Cline. When Schwab returned to Homestead as general superintendent that fall, he paid little attention to the department; he knew that Corey was an able manager, and he was busy trying to reestablish good working relations with the returning strikers. He was also supervising the design and construction of new furnaces, forges, and rolling mills in an effort to increase efficiency at Homestead. Armor plate was only one of twelve major departments under his jurisdiction; and he had to supervise fifteen sub-departments.

Schwab's involvement in the operation of the armor department was minimal; the plant seemed to be running smoothly. The Navy Department had assigned resident inspectors to supervise the entire armor operation and to ensure that production would strictly conform to Navy specifications. Schwab did make brief, unannounced visits to the armor plant to watch the work in progress, and he received periodic progress reports from Corey. Also on frequent occasions, he asked the chief resident inspector whether he was satisfied that the work was proceeding according to contract specifications. He never received a word of complaint.[34]

Then how could Schwab have been accused of perpetrating a deliberate fraud against the Navy Department? What was the evidence of fraud? And what explanations did Schwab offer in his own defense?

In September 1893, Secretary of the Navy Hilary A. Herbert was approached by a Pittsburgh attorney who represented four employees of the armor plant at Homestead. They had information to sell about acts of fraud which they alleged were being committed by the Carnegie Steel Company. When Secretary Herbert told them that he had no funds for buying information from civilians, the informants made another proposal: in the event that their charges of fraud were substantiated, they were willing to accept as their reward a percentage of any penalties charged against the company. At first they asked for 40 per cent of the penalties. After clearing the legality of such an arrangement with Attorney General Richard Olney, Herbert made a take-it-or-leave-it offer of 25 per cent. The informants accepted.[35]

Secretary Herbert appointed a three-man board of inquiry, headed by the Chief of the Bureau of Naval Ordnance, Captain William T. Sampson. The board conducted a speedy *sub rosa* investigation, which it based on data furnished by the informants. Without ever having informed the company of the accusations made against it—that is, without offering the accused a chance for reply or defense—the board submitted a report to Secretary Herbert which declared that the company was guilty. Herbert then telegraphed Henry Frick to come to Washington immediately. When Frick arrived, he was confronted by the indictment, the evidence, and the verdict. In effect, after being tried *in absentia*, the company was summoned for sentencing: the Navy was proposing to levy a fine equal to 15 per cent of the value of all armor delivered to date.[36] After hearing the charges and being informed that the company could appeal Herbert's decision to the President, Frick withdrew—both to seek counsel and to prepare a reply.

Frick was not totally surprised by the Navy's charges; the attorney for the four informants had offered to sell their "evidence" to him before they had gotten in touch with Secretary Herbert. Frick later explained that, ever since the bloody strike at Homestead, hardly a week passed when he did not hear of some "vermilion-hued" plot or conspiracy thought up by the disgruntled and defeated strikers. He believed this latest "revelation" was another such plot.[37]

When the attorney had approached him, Frick had been concerned that there might just possibly be some truth in the allegations, so he had written to Schwab, warning him to double his vigil. He had also cautioned Schwab to carry out the work for the Navy in strict conformity with the specifications and procedures enumerated in the contract. Schwab, in turn, had sent similar messages to his foremen and supervisors. As an added precaution, Frick had urged the Navy's resident inspectors to be especially vigilant so that no improprieties could escape their detection.[38]

Carnegie was enraged by the charges leveled against his company and his partners. He wrote an impassioned plea for justice, which he sent to the White House. On the envelope he wrote:

> This is a personal
> letter to Mr. Cleveland
> not to the President
> and I ask the Secretary to
> hand it to him unopened.

Carnegie's letter was a strictly private message between two men of honor and integrity. "No one, not even Mr. Frick knows of this letter, it is between you and me alone—I keep no copy—."[39]

Carnegie did not ask Cleveland to rescind the fine or suppress the government report, but rather that he have a new board of inquiry appointed. He gave his version of how the first board had operated:

> We have been accused, tried, found guilty and sentenced without ever having been heard—The vilest criminal has always the right to be heard in his defence—The Secretary of the Navy even condemned us and after notifying Mr. Frick that he had approved the finding of the so called Board (which was not a Board but only one man with two assistants upon whom he might call to aid him if necessary) and then allowing us to say what we had to offer in defence—monstrous this—After we had been sentenced we were asked to state our side of the case, not till then—But this is not the worst of it—
>
> The so called Board which should have been our Judges were not allowed to judge. They were *instructed practically* what to find. The Secretary called them together only once I think.—*Instructed them* as to the law, gave them the *rule of damages*,—gave them a long lecture as to the enormity of the offence &c, Instead of acting in the capacity of an Impartial Judge[.]

He accused the Secretary of the Navy of behaving as though he were the "Attorney for these informers." But Herbert's motives were not dishonest, said Carnegie; instead, his behavior was "chargeable to over zeal. . . . He did not intend injustice—."

Carnegie also reminded Cleveland that the company had worked hard and well to fulfill the Navy's needs:

> Four years the first Contractor [Bethlehem] tried to make armor—had delivered none—ships stood on the stocks.—In *one year* we delivered the best armor ever made and won three premiums. . . . Spent millions, subordinated every other branch of our business to the Government's needs, succeeded—and then upon the testimony of spies we are charged with irregularities and our men with fraud—I cannot stand this . . . we must ask to be tried by a Court who will at least visit our Works, *listen to explanations upon the ground*, and see for themselves before they judge—

A week later, Carnegie wrote to Cleveland again, lamenting the "over-zeal of an inexperienced Secretary who charges 'fraud' upon people (Mr. Schwab & others) quite as incapable of attempting to defraud the Govt as the Hon. Sec'y himself." [40]

Carnegie placed the President in a difficult position. Cleveland could uphold the decision and fine levied by Secretary Herbert; that would end the matter. Or he could order a new board of inquiry to investigate the charges—but if he did, he would in effect be saying that he had no confidence in the judgment of the Secretary of the Navy and the Chief of the Bureau of Naval Ordnance. Also, if Cleveland chose to appoint a new board and it did not absolve the company and rescind the fine, the whole problem might be thrown back to him again.

Cleveland carefully reviewed the controversy; then he rendered his verdict. The central point at issue was whether the company had defrauded the Navy by delivering plates which were inferior in quality to those it could have produced with greater care and effort. In fact, all but three of the plates had met the minimal standard of quality specified in the contract. The fine was not intended to be a penalty against the company for the three substandard plates; those plates would simply be rejected. Rather, the company was to be penalized because most of the armor it had produced was only 5 per cent better than the quality specified in the contract, whereas a few of the plates were 20 per cent better.[41] Since one provision of the contract required that the company produce the best possible quality of armor, the question was whether or not the company had in fact produced plates of consistent excellence.

Although President Cleveland admitted that he felt unqualified to judge all of the technical intricacies involved, he nevertheless reached a conclusion:

> I am satisfied that a large portion of the armor supplied was not of the quality which would have been produced if all possible care and skill had been exercised in its construction. . . . I am of the opinion that under the terms of the contract between the Government and the Company this constituted a default entitling the Government to damages.[42]

Cleveland did not directly challenge the verdict of the Navy's board of inquiry; he expressed confidence that it had shown "an honest desire to meet the case fairly." Nevertheless, in view of what he cryptically called "the indefiniteness of the proofs obtained," he ruled that the fine should be reduced from 15 per cent to 10 per cent. The reduced fine still amounted to $140,484.94.

By this decision Cleveland accepted Captain Sampson's rationale for fining the company. But he totally ignored Frick's rebuttal. Frick had objected to Sampson's recommendation, saying that it was unjust. He argued that if the company had been able to produce a few specimens of armor which were 100 per cent better than the majority of the plates delivered, then by the same logic the Navy would have been justified in fining the company for the 100 per cent discrepancy. What could be better calculated to suppress any efforts to experiment or to make improvements? To the extent that the steelmakers succeeded in raising the quality of some, but not all, of the plates delivered, they would be penalized for the discrepancy between the standard and the exceptionally good. Thus the reward for successful experimentation might be a penalty.[43]

The company had one final recourse: it could initiate a civil suit, asking the courts to reverse the President's decision. Frick declined to do so. He realized that delay in paying the fine would give the Navy grounds for canceling its order for armor, most of which was still undelivered. He also feared that if the company chose to prolong the controversy the Navy might retaliate by refusing to accept any future bids on armor from the Carnegie Steel Company.[44]

Predictably, President Cleveland's decision pleased no one. Some of the Republican members of Congress shared the opinion held by Carnegie and Frick—that although the fine had been reduced, it was still tantamount to a guilty verdict. An even larger contingent of Democrats in Congress believed that the President had reached a cowardly compromise with the men responsible for the bloodshed at Homestead in 1892. Some of the Democrats—those who favored inflating the money supply—welcomed an opportunity to discredit the President; although Cleveland was himself a Democrat, he was a "gold-bug," a man who feared that the dollar would be endangered by an expansion of the money supply through the free coinage of silver. Many of those same Democrats also saw an opportunity to discredit the protective tariff by exposing Carnegie's "villainy" to full public scrutiny. (The Carnegie Company was a beneficiary of the tariff which restricted the sale of foreign steel in the United States.) [45]

Hostility to Cleveland's decision was not confined to members of Congress. One of Carnegie's severest critics was his former friend and defender, Whitelaw Reid, publisher of the *New-York Tribune*. Reid, a Republican, was convinced that Carnegie's handling of the 1892 Homestead strike had caused President Benjamin Harrison's defeat when he ran for re-election that year. Reid's ire was not based on party loyalty alone; he had been Harrison's running mate, and he blamed Carnegie for ruining his own chance to obtain high office.[46] Reid's *Tribune* joined the call for a full-scale investigation of the armor plate controversy.

The two-pronged campaign fought by Carnegie's friends and enemies was successful: a resolution authorizing the Committee on Naval Affairs to conduct hearings on the case was passed by the House of Representatives.[47] A special subcommittee was then created, with Amos J. Cummings, a Democrat from New York, as chairman. The subcommittee's investigation extended beyond the question of armor quality, which was what the board of inquiry had studied. The subcommittee was investigating a series of charges against Schwab; he was accused of either authorizing or condoning fraudulent irregularities.

For the first five weeks, a series of witnesses, including the four informants, described the alleged irregularities. None of them offered any evidence, either direct or circumstantial, showing that Carnegie or Frick had been aware of the deceptive acts in question. But when William E. Corey was called to testify, he admitted that his superior, Charles M. Schwab, had been aware of and had not objected to the irregularities. Consequently, it was Schwab, and Schwab alone, who bore the ultimate burden of establishing his and the company's innocence.

There were a dozen charges, but the four most serious were the failure to follow specified procedures in the manufacturing of armor plate; the falsification of the results of preliminary tests of quality; the concealment or disguising of defects; and, most important, the secret re-treating of plates which had been selected for the crucial ballistic test.

Each of these irregularities was a violation of the contract. Since it was impossible for Schwab to deny that they had occurred, the best he could hope to do was to convince the subcommittee that no criminal fraud had been intended. He could claim that he and the company were guiltless only if he could establish that the alleged acts of fraud were merely innocent deceptions, carried out for the benefit of the Navy. He attempted to do so, but with only mixed success.

Schwab began his testimony on July 6. He tried to absolve himself of any wrongdoing, denying that he had ever authorized any deviations from the contract specifications, and claiming that he had no personal knowledge of any irregularities.[48] It was a lame defense; even if it were true, it would merely have shifted the charge against him from fraud to negligence.

The first of the four charges concerned the violation of an explicit contract provision that all plates were to receive identical treatment. In his testimony Schwab challenged the wisdom of this requirement. He explained that plates varied widely in quality within the same group, and that even different areas of the same plate were not of the same quality.

It was technically impossible to achieve uniform thickness in a plate and just as impossible to prevent impurities in the molten ore from congealing together, thereby making the plate weakest at the point where they did. And, finally, it was impossible to prevent an ingot from cooling at an uneven rate, which further prevented all areas of the same plate from being of uniform quality. Therefore, Schwab claimed, American producers should have had the same freedom that foreign producers did; they should have been allowed to subject the plates to unequal treatment in order to bring about uniform results. And he said, "I believe everything has been done that would give the best possible results. There are some things . . . which, if literally interpreted, would have given the Government a very bad lot of armor plate." [49]

Schwab argued that these were indisputable facts, that technical experts and "practical men" would agree with him, and that the Navy's refusal to concede or acknowledge the intrinsic unevenness in quality simply reflected the inexperience of the Navy's "experts." Schwab thought that the Navy should not set any procedural requirements. He recommended that the United States adopt the same policy that the German government employed on its armor plate contracts with Krupp: resident inspectors should not be assigned to oversee production, nor should there be any specifications about process. The only test of quality for an armor plate should be its capacity to withstand the impact of a high-speed projectile. The Navy, said Schwab, should be concerned exclusively with the results as demonstrated in the ballistic test; it should not interfere with the methods by which those results were achieved by the armor makers. [50]

The second major charge of fraud was that the company had falsified the results of the preliminary test of quality.

Each group of steel plates weighed about 300 tons, worth approximately $175,000. Before the Navy would accept and pay for any group, a sample from that group had to be submitted to the ballistic test. Since the Navy could not test every plate, it selected the worst plate in the group for testing, on the theory that if the worst plate passed then all the rest would as well.

When the Navy inspector chose what seemed to him to be the worst plate in a group, it was subjected to a preliminary test before being shipped to the Naval Testing Grounds. The purpose of this test was to determine the tensile strength of the plate. This was done by cutting off a small strip of the plate, rolling it into a cylinder, and then subjecting it to a massive weight load. If the strip did not rupture or shatter when subjected to a strain of 90,000 pounds of pressure per square inch, the plate was approved for the ballistic test.

If, however, a strip did not meet the prescribed minimum resistance level, it was possible to jockey or juggle the speeds on the testing machine so that a plate which would have been rejected for insufficient tensile strength could be made to pass. This, the accusers claimed, had been done by the company.

Schwab replied to this charge by saying that the preliminary test was unnecessary and potentially misleading, since there was no direct correlation between a plate's performance on the preliminary test and on the ballistic test. At most, the preliminary test might be indicative of good quality—but not necessarily so. He testified, ". . . where you are making a mistake is in assuming that these tests are an exact science or an exact method. They are not. Two tests from the same plate, I will guarantee, will show a greater variation than the most skillful manipulator can show on any machine." [51]

He pointed out that Krupp of Germany was not required to use any other test except the ballistic. To clinch his argument, he noted that the other American firm producing armor plate, the Bethlehem Iron Company, was not required to submit plates to any preliminary test. Only the Carnegie Company had this test in its contract. The test had been introduced in 1890, when the first contract was being negotiated. At that time William L. Abbott, then the chairman of Carnegie, Phipps, & Co., was confident that the company could produce armor of high quality, and he had offered to submit the plates to an extra test before they were shipped to the Navy for the ballistic test. [52]

Schwab did not deny that surface defects in the armor had been secretly disguised and that test results had been falsified. He explained, however, that the steelmakers did so in order to please or placate the resident inspectors. The inspectors, Schwab said, were the real source of the trouble. They were novices, filled with book knowledge, perhaps, but lacking the experience and expertise which only a professional steelman could possess. The presence of these tyros at the armor plant was positively harmful, Schwab believed, since they had the power to reject plates which to them seemed damaged or inferior but which the steelmaker knew was of high quality despite superficial flaws or failure to pass the preliminary test.

Schwab conceded that the practice of tampering with test results and concealing surface defects was a bad policy, but one which he could understand. He said:

> . . . some of the superintendents have made a great mistake in the past in endeavoring and in trying to please the inspectors. They were men ordinarily who would not and could not understand these things. I can easily understand how some of our superintendents might have made a statement which would please them where the value of the plate was not concerned. For instance, Lieutenant Ackerman was recently at the works and he did not understand absolutely anything about it; he was worse than a good smart schoolboy would have been, but he reported to myself and Mr. Frick on the day of his leaving the Homestead works that he was entirely satisfied with everything; entirely." [53]

The steelmen had also deceived the inspectors by plugging up "blowholes"—visible fissures and holes in the plates caused by uneven cooling of the ingots. Inspectors sometimes rejected plates because of blowholes, even though, according to Schwab, the holes in no way detracted from the strength of the plate. It was impossible to reason with the inspectors about this issue, Schwab claimed. This left the company only two alternatives. First, it could have added silicon to the molten steel, thereby preventing blowholes completely. The proof of the company's good faith, said Schwab, was that silicon was *not* used—even though it was not barred by the contract and was inexpensive. If the aim had been to palm off an inferior product, Schwab argued, the company would have used silicon, but that would have weakened the plates. [54] The other alternative was to plug up the blowholes when the inspectors were not present—and that is exactly what the company did. The inspectors were only on duty sixty hours each week, while the plant operated twenty-four hours a day, seven days a week, so it was easy enough for the armor plate workers to plug the blowholes without being detected.

Schwab testified that he had not initiated or authorized this policy, but he admitted that he did not suppress it when he had reason to believe it was being carried on.

Congressman Cummings questioned him:

Q. Do you know whether the company really did conceal the fact of blowholes in the plate?
A. I think likely that was done.
Q. Was it done with your knowledge?

A. Well, the concealment was not, no; but I had knowledge of this fact, that they did not make any plates that did not have blowholes.

Q. You had knowledge that these blowholes were plugged in many instances?

A. No, sir; I did not say that. I said I had knowledge that the plates had blowholes, and that I should not be surprised if they were concealed.[55]

Once again, Schwab cited the armor produced by Krupp. At the recent Chicago World's Fair, Krupp had exhibited examples of its best work. They were massive plates of armor, and they were filled with blowholes.[56]

Even if the subcommittee had accepted Schwab's opinion of the unimportance of the preliminary test and of blowholes, and even if his opinion had been corroborated by the testimony of impartial experts, two major problems remained. He had to account for the fact that three plates had failed the ballistic test and had been declared "inferior and defective." And he had to explain why six plates which had been selected for ballistic testing had been re-treated secretly, at night, without the knowledge of the inspectors and in blatant defiance of the contract.

Schwab explained. There was a controversy, he said, between the armor plate experts at Bethlehem and those at Homestead over whether re-treating a plate could actually strengthen it. Each of the six re-treated plates had been given a different treatment, and therefore the re-treatment was really an experiment. The reason this was done to the plates which had been selected for testing by the Navy was that only in this way could the various methods of re-treatment undergo the one definitive test of their worth—the ballistic test.[57]

But Schwab's explanation was unconvincing. He knew quite well that the Navy would have been willing, as it had been in the past, to give a ballistic test, free of charge to the company, to any experimental plate which might show an improvement. And, as Captain Sampson, Chief of the Bureau of Naval Ordnance, had argued previously, whether retreatment improved or worsened the quality of the plate was irrelevant: in either case, re-treatment would make that plate unrepresentative of its group.[58]

Schwab was questioned on this by Democratic Congressman Hernando D. Money of Mississippi.

Q. . . . how do you explain that that plate, which was selected for a ballistic test, afterwards received a treatment; would not that take it out of the class of uniformity and make it singular in its qualities?

A. Unfortunately, it did make it very much singular, and made it very much worse. It was but an experiment, as I think Mr. Corey explained to you fully. . . .

Q. Then in fact they expected and the intention was to improve the plate?

A. The intention was to discover a method of improving the plate.

Q. Right there I want to ask a question. I want to know why you could not tell the Government about that?

A. That was a mistake; I admit that. They should have done it; but, on the other hand, as you know these Government people, if you had been Mr. Cline or Mr. Corey I could readily understand why they did not tell them because as I say to you they [the inspectors] were not practical men and they would not have understood this thing.[59]

At the time of his original investigation, Captain Sampson had concluded that re-treating was obviously intended to improve the quality of the plate. But since three of the re-treated plates had not passed the ballistic test, Sampson had assumed that the rest of the plates in the three groups were of even poorer quality and also would not have passed. Re-treating, in Sampson's opinion, was a method of raising a sample from an inferior group to a passable level.[60]

Schwab emphatically denied this, and he offered convincing proof. Owing to the massive size and weight of each plate, when one was chosen for ballistic testing it was cut in half. Three of the six plates which had been re-treated had failed the ballistic test. Several months earlier, at Schwab's suggestion, the Navy had agreed to test the other halves of the three defective plates. When this was done, the plates were found to be of passable quality. Therefore, contrary to Captain Sampson's original opinion, the groups of plates which they represented were not of inferior quality.[61]

Captain Sampson then offered another explanation. He claimed that Schwab had ordered the plates to be re-treated so that Carnegie Steel would win the cash premiums which the Navy awarded for plates of outstanding quality.[62]

Each armor plate tested was fired at three times, and the caliber and speed of the projectile was increased proportionally with the thickness of the plate being tested. The plate was bolted to a thirty-six-inch-thick slab of oak. If the plate could withstand three shots without being shattered, then it was accepted. Whenever the company thought it had produced plates of especially good ballistic quality, it could try to win an extra premium of $30 per ton. Then the caliber and speed of the projectiles were further increased, and if the plate successfully withstood the heightened assault the premium was awarded. Since Schwab was a partner in the Carnegie Steel Company, he personally would share in any added profits made by winning the cash premiums. Schwab's reply to this was, ". . . my stock in the company is not only at Homestead but in the Carnegie Steel Company, and Homestead is only one department of the Carnegie Steel Company, and the armor-plate department is only one department of Homestead; so you can readily see what a slight effect this tonnage would have upon the profits." [63]

If a 300-ton group won a premium of $30 a ton, Schwab, as a partner holding one-third of 1 per cent of the shares, would only make a profit of $30—scarcely enough money to tempt him into acts of fraud and deceit which might cost him his job or his partnership, or possibly even a term in prison, even if he were inclined to be dishonest.

Then why did he allow the secret re-treatment of plates and the other irregularities? It was quite simple; Schwab had nothing but contempt for the Navy's inspectors—and Corey and Cline felt the same way. Once, when he was mocking a series of suggestions made by Lieutenant A. A. Ackerman, the chief inspector, Schwab said,

> They were so ridiculous that it would be almost laughable to explain what he did try to prove to me, or say to me. For instance, he said a plate going into the furnace ought to be encased with firebrick, so that the flame would not get to it. It was the most ridiculous thing I ever listened to, and I did not have the patience to listen.

Earlier, Congressman Cummings asked Schwab, "Didn't you consider it the duty of the man at the [preliminary testing] machine to give the Government the correct figures?" Schwab answered, "Yes sir; but I was about saying [sic] what Cline said to you: 'You have never worked for a Government inspector.' You have no idea of the loss of time involved in their requirements." [64] Schwab (and Corey) were not willing to deal with or obtain permission from men whom they considered to be meddlesome, ignorant novices.

Because he was confident that the plates being produced were of the highest quality and that experimental re-treating was a valuable means of discovering methods to strengthen armor plate, Schwab ignored the fact that secret re-treating constituted misrepresentation. Even if he had made technological improvements by re-treating the plates, it was still wholly improper for him to have acted without the Navy's knowledge and consent. He believed that the Navy officials and inspectors "were not practical men and they would not have understood this thing." And once he had adopted and condoned a policy of secret defiance, it was difficult for him to admit to it. He could not offer the one explanation which probably would have made his behavior understandable, though hardly commendable. He could not openly say, "The Navy be damned, I will do things my own way." By his own silence he reinforced the Navy's conviction that he had intended to perpetrate a malicious fraud out of motives of personal financial gain.

Schwab's defense was wholly unsuccessful. Even if the subcommittee had accepted his claims that the Navy's specifications were arbitrarily rigid, that its inspectors were ignorant intruders, and that the irregularities had been committed for the benefit of the Navy, it still could not condone the secret violation of the requirements of the contract.

The outcome of the hearings was predictable. The Committee on Naval Affairs unanimously affirmed the original findings of the Secretary of the Navy, and the reduced fine was not rescinded.[65]

In its report, the subcommittee completely ignored the testimony and explanations offered by Schwab, Corey, and Cline. It did so on the grounds that "the unblushing character of the frauds to which these men have been parties, and the disregard of truth and honesty which they have shown in testifying before your committee, render them unworthy of credence." Yet, the subcommittee offered no explanation of its charge of perjury, and no indictment was suggested or initiated.

The subcommittee also claimed, without explanation or evidence, that the irregularities were "acts whose natural and probable consequence would be the sacrifice of the lives of our seamen in time of war, and with them, perhaps, the dearest interests of our nation." This statement contradicted Secretary Herbert's finding—that all the irregularities had been found in the light armor used to make enclosures for the guns on warships, and that none had been discovered in the heavy armor used to fortify the sides of the ships.[66]

Finally, despite the fact that both Secretary Herbert and Captain Sampson had stated that no defective armor had been produced and that all of the armor was above the minimal standard specified in the contract, and despite the fact that the three plates which were rendered defective by secret re-treatment were simply rejected by the Navy, the subcommittee drew the conclusion that Carnegie had deliberately sold defective plates to the Navy.[67]

These three unproven but lurid conclusions were seized upon by editorial writers and political cartoonists. Whitelaw Reid's *New-York Tribune*, in an editorial published on August 31, 1894, concentrated its fire against Carnegie and Frick: "In palming off those defective and inadequate armor plates upon the government they were imperiling the lives of thousands of our seamen and jeopardizing the nation's honor and welfare, but they were making money. It is an appalling conclusion. One shrinks from believing a thing so monstrous. And yet there is the record in all its hideous *simplicity and clearness*." [68] (Italics added.)

In point of fact, there has seldom been an episode less clear or less simple. The subcommittee's report to Congress was brief, but the Congressional hearings were not; they alone occupy nearly one thousand closely printed pages, and they are filled with technical explanations concerning the production and quality-rating of armor plates, as well as the detailed accusations against Schwab and Carnegie Steel and the explanations offered by the accused.

Newspapermen are not scholars. They work under the pressure of daily deadlines, and they never have time to make a complete investigation of any story. So it is understandable that editorial writers in 1894 took the subcommittee's report at face value, using its dramatic, though erroneous, conclusions, and quoting its most lurid accusations.

Although the *New-York Tribune*'s editorial concluded that "Messrs. Carnegie and Frick will go down in history covered with the odium of the Cummings report," in fact it was Schwab whose reputation was most sullied by the episode. His role in the armor plate scandal was resurrected again and again during the next four and a half decades of his career. Writers of the muckraker tradition revived this scandal as "evidence" of Schwab's boundless capacity for criminal activity. Whenever Schwab was accused of new acts of fraud, or of negligence, or of fomenting wars in order to increase munitions profits, his detractors cited the armor plate scandal as proof of his guilt. On at least eight different occasions Schwab was denounced by writers whose knowledge of the episode went no deeper than a reading of the subcommittee's brief report. A few did not even go that far. They merely read earlier, secondary reports of the scandal and then liberally embellished their own accounts with imaginative inaccuracies. [69] These writers helped to sustain the myth of Schwab's criminality. Yet it must be said that Schwab himself, by his actions in 1892–93, placed in the hands of his detractors a weapon which they would use against him many times.*

* See Appendix B.

4

Promotion to the Presidency

Schwab's position at Homestead was secure despite the armor scandal; Carnegie refused to sacrifice his ablest young partner on the altar of public opinion. He fully shared Schwab's low estimate of government inspectors; he, too, viewed them as martinets and meddlers. Carnegie believed that he, Frick, and Schwab had been victims of a vendetta instigated by members of a defeated union, that they had been persecuted by over-zealous government officials, and that they had been condemned by journalists and Congressmen who, for a variety of motives, sought to vilify him and his company.

Just as he had remained loyal to Frick when Frick had been condemned by Congress and the press for his handling of the Homestead strike, so Carnegie now shielded Schwab. To have replaced him would have been tantamount to admitting that the charges of fraud were true, and that Carnegie strongly denied. Perhaps Carnegie's determination to keep Schwab and Corey was reinforced by President Cleveland, however unwittingly. At the time Cleveland had agreed to reduce the fine and thereby settle the dispute, he had said to Carnegie: "I'll take no more plates from Schwab or Corey." Carnegie then told Cleveland, and later told Schwab, "Even the President of the United States can't tell me how to run my business." [1]

But even if Carnegie had been tempted to dismiss Schwab or merely to demote him, there was no one within the company who could take his place. He was precisely the kind of man Carnegie needed: he was a seasoned expert in managing men, in devising and implementing technological improvements, and, above all, in reducing costs.

Schwab recognized that sometimes large sums must be spent in order to achieve small savings, and that even a savings of a few pennies per ton of steel could justify spending many thousands of dollars to buy new equipment or to rebuild existing facilities. One day, late in 1889, Frick was touring the Braddock plant with Schwab, the new superintendent. While they were in the rail finishing department, Schwab said that if he had designed the facility he would have done it somewhat differently; he would have improved the flow of materials and reduced the need for labor. He told Frick, "I could save ten cents a ton on the cost of steel." Frick asked how much it would cost to redesign and rebuild the department. Schwab answered, "About $150,000," and Frick authorized the expenditure.[2]

On another occasion, however, when Schwab wanted to spend far less money to achieve a cost reduction, he failed to get Frick's permission. The company owned its own supply of natural gas, which was used as fuel in the furnaces. Schwab proposed that meters be installed, so that he could measure the quantities of gas being consumed. Frick disagreed: the supply of natural gas was "limitless." Nonetheless, Schwab installed the meters; the cost of their purchase and installation was small enough that he did not require the authorization of any senior official. His decision was justified, Schwab later claimed, because the presence of meters caused the furnacemen to monitor their consumption of gas and to stop wasting it.[3] Like Carnegie, Schwab believed that every man in the mills should be intensely cost-conscious, and that every source of waste and inefficiency should be eliminated. Less than three years after the armor scandal, Carnegie named Schwab president of Carnegie Steel. Schwab was then only thirty-five.

Carnegie liked Schwab personally, and he respected his ability to manage men and his knowledge of the industry's newest technological developments. But he would never have made Schwab president of the company if they had not been of one mind about cost-cutting. Carnegie believed that the company's growth depended upon its ability to undersell its competitors. He stressed the urgency of reducing costs—and Schwab was the man best able to carry out this policy.

At the beginning of the nineteenth century, before the nation's railroad network was built, ironworks were small-scale operations, comparable in size to paper mills, breweries, tanneries, and brickyards. They were essentially neighborhood industries, which drew upon local ore and fuel supplies to make products largely intended for local consumption. In the pre-canal and pre-railroad era, overland transportation of raw materials and finished goods was prohibitively expensive. Consequently, ironworks were located in river valleys where there were ore sites, and the rivers could be used for transporting products to local customers.[4] But the development of the railroad network in the middle of the century completely altered the character of American industry, for the railroads made almost all areas of the country readily accessible.[5] The creation of the railroad network also eroded the local sales monopolies which many manufacturers had enjoyed; now their sales areas could be penetrated by distant producers whose products were superior or whose prices were lower, or both. Now there were few "zones of safety" in which producers could stagnate, immune from competitors.

Carnegie recognized the revolutionary changes that were taking place, and he realized that Carnegie Steel was threatened just as much as any other company was. So from time to time he joined with his fellow industrialists in their attempts to deal with the problem—even though his own solution to it was quite different from theirs, and, in the end, far more successful.

As competition intensified, many businessmen, Carnegie included, sought to stabilize prices or to protect their markets from rival producers. In years of depression and declining demand, they tried to prevent prices from falling to a level which left them no profits. And during years of prosperity and rising demand, they worked to obtain the highest prices they could get without being undersold by more aggressive firms. Instead of "competition"—a term many businessmen prefaced with such words as "wasteful," "pernicious," or "cut-throat"—they generally preferred "cooperation," "collaboration," or "confraternity." This cooperation often took the form of agreements to fix prices, to divide markets according to a specified percentage for each company, or both.[6]

During the 1870's and early 1880's, price-fixing and market-sharing agreements were casual and short-lived. They were based on oral promises, so-called "Gentlemen's Agreements," which broke down—as they always did—when one or more parties to the agreement secretly cut prices in order to win a larger share of the market.

In the mid-1880's, Gentlemen's Agreements were replaced by "pools," written agreements between potential competitors which stated that they would respect each other's sales territory. If a product was already being sold by various firms in a national or regional market, the competitors agreed to divide the market according to fixed percentages, depending upon the relative productive capacity of each firm involved, and to sell that product at a fixed price. In an effort to enforce compliance, the members of the pool set up a policing arrangement: each company deposited a sum of money which it would forfeit if it were discovered to be violating the agreement, and each consented to open its sales records for inspection by all members of the pool.

These pooling agreements had very limited effectiveness. Firms could secretly violate the terms of the agreement and escape detection simply by supplying falsified sales records to the other members of the pool. Even if a firm were caught cheating, the sum to be forfeited rarely exceeded $25,000, a small penalty to pay for weeks or months of increased sales.

Another flaw became evident during depressions. Pooling agreements seemed to work during periods of prosperity and rising sales when there was little financial incentive for any firm to defy the agreement, but they proved ineffective when demand slackened and new orders became increasingly difficult to obtain. At that point, a firm would openly withdraw from the pool, announcing its determination to cut prices, and the pool would collapse. Pooling agreements were like broken umbrellas—they seemed to be of value on sunny days, when they were not needed, but when a downpour began, it became clear that they were useless. If pools had been enforceable in courts of law (they were not, because they violated common law prohibitions against price-fixing), then many of them would never have been formed. The most competitive firms, like Carnegie's in steel, would not have agreed to a legally enforceable moratorium on their ability to excel. Carnegie was only willing to enter into pools, occasionally, because they were so fragile and unenforceable.

Carnegie believed that pools and Gentlemen's Agreements were of dubious value, but he entered into them when he saw an advantage in doing so. He wanted to gain access to competitors' sales records—the right to such access was one of the provisions of the pooling agreements.[7] But he saw no reason for making any permanent market-sharing agreements with rival firms. In 1893 Carnegie told steelman Abram S. Hewitt, "I can make steel cheaper than any of you and undersell you. The market is mine when I want to take it. I see no reason why I should present you all my profits." [8] And, in an interview with *The Iron Trade Review*, Carnegie said, "I never believe in combinations. They are only for weak people. They give a little temporary strength to these weak persons, but they are not good for the strong and healthy." [9] He counseled his partners, "Put your trust in the policy of attending to your own business in your own way and running your mills full regardless of prices and very little trust in the efficacy of artificial arrangements with your competitors, which have the serious result of strengthening them if they strengthen you." [10]

Carnegie fully intended to remain among the "strong and healthy." And he knew that in order to do so he would have to be in a position to undersell his competitors, no matter how severe any economic downturn might be. For that reason, he did not rely on pooling agreements.

He cut costs instead. He reduced his costs to the barest minimum so that he could cut his prices and "scoop the market"—that is, get so many orders that even in bad times Carnegie Steel would be able to run at full capacity and still show a profit. [11]

This was exactly the policy he followed during the severe depression of 1893–97. [12] In 1890, Baring Brothers, a famous British investment bank whose clients held large investments in American securities, went bankrupt. The failure of Baring Brothers triggered a heavy sell-off of American stocks owned by European investors and a drying up of new sources of investment capital from abroad. At the same time, a balance of payments deficit produced a drain on United States gold reserves, and that drain was accelerated by declining European confidence in the stability of the American monetary system. By 1893 credit had become so tight that a panic had spread throughout the American economy. When the Philadelphia & Reading Railroad, a long-established and seemingly solid line, went into insolvency, anxiety within the business community increased, as it did at the demise of the National Cordage Company, one of the largest mergers formed between 1889 and 1892.

During the four years that followed, hundreds of national and state banks, as well as loan, trust, and mortgage companies fell into receivership. The same fate overtook 156 railroads, including three of the nation's largest. Retrenchment became the byword, unemployment soared, and gold reserves continued to decline.

Steel prices fell drastically during the depression. [13] As the railroads went bankrupt, the rail manufacturers' sales declined. Even the roads which remained solvent could not afford to make new purchases. Those steel companies whose primary product was rails were hardest hit, as were the firms with above-average costs for ore or coke or labor.

But Carnegie had already anticipated that there was a declining market for rails and a growing demand for structural steel. And, unlike the other manufacturers, he was now using the open hearth method, which gave Carnegie Steel a further competitive advantage—it enabled the company to use low-cost iron ore from the Mesabi Range. When the depression came, Carnegie Steel, having concentrated on diversification and cost-cutting, was better able to weather it out than its competitors were; when good times returned, the company was immeasurably stronger than they.[14]

In 1890, Pennsylvania Steel, which was located in Harrisburg, tried to strengthen its competitive position by increasing its capacity. It formed a new subsidiary, the Maryland Steel Company, which built a rail-producing plant at Sparrows Point on Chesapeake Bay. But in 1893 Pennsylvania Steel passed into receivership, the victim of over-extended debt brought on by building the Sparrows Point plant. The company was reorganized and reopened shortly thereafter, but for the next four years it paid no dividends and was unable to accumulate any cash surplus. It did not operate at full capacity and it lost money in 1896 and 1897.[15]

In 1889, three of Carnegie's major rail competitors merged. The new company, Illinois Steel, included the North Chicago Rolling Mills, Joliet Steel, and Union Steel. Its rail-making capacity was then twice as large as Carnegie's, and its total output made it the world's largest steel company. It owned Lake Superior ore lands and Connellsville coal lands and coke ovens, as well as railroads and rolling stock to transport its raw materials. It could produce rails, rods, plates, and structural shapes. Illinois Steel seemed an adversary worthy of Carnegie. Yet in 1893 the company ran at a deficit, with much of its plant idle. The Union works were closed down completely, while the Joliet works were open only six weeks during the year. In 1893, 1894, and 1895, the new merger was unable to pay any dividends, and when it resumed payments in 1896, the amount was greatly reduced. It operated at only 60 per cent of capacity in 1896.[16] In 1894, Illinois Steel lost nearly $1,000,000, while Carnegie Steel's profits exceeded $4,000,000; and in the following year, Illinois earned only $360,000, while Carnegie earned over $5,000,000.[17]

Carnegie's cost-cutting policy was effective: because his costs were always lower than those of his competitors, he could match their prices or undersell them no matter what the state of the economy was. Carnegie Steel made profits every year, whether there was depression or prosperity.[18]

In 1889, Carnegie Steel made $3,540,000, and in the next year it earned $5,350,000. Then, in 1891, as the nation began to react to the economic stresses which had been set off by the collapse of Baring Brothers, profits began to decrease, and the company earned only $4,300,000. By 1892 profits were even less, $4,000,000, and in 1893, when the entire country was in the grip of depression, Carnegie made only $3,000,000—but that was still clear profit. Between 1894 and 1897 thousands of businesses went under; Carnegie's profits went up. The company earned $4,000,000 in 1894, $5,000,000 in 1895, $6,000,000 in 1896, and $7,000,000 in 1897. Then, as the depression ended, Carnegie Steel's profits began to skyrocket—to $11,000,000 in 1898, $21,000,000 in 1899, and an incredible $40,000,000 in 1900. From $3,540,000 to $40,000,000 in twelve years, and five of those years depression times, is nothing short of phenomenal.

While Carnegie himself always determined the company's over-all strategy, that strategy was implemented by three men—first by Henry Clay Frick, then by John G. A. Leishman, and finally by Charles M. Schwab. Frick served as chairman, the equivalent of chief executive officer, until late 1894. At that time Carnegie proposed that they invite a certain man in the coke business into partnership with them. The coke manufacturer was one of Frick's arch-enemies, and Frick was furious at the suggestion. He resigned. When he regained his temper he tried to retract his resignation, but the Board of Managers had already elected a president to exercise the powers he had formerly held as chairman. Frick then asked that he be given a nominal office in the company, and, in a conciliatory mood, Carnegie agreed. The office carried an impressive title—Chairman of the Board—but the man who held it was to have no power. Frick's only official responsibility was to serve as chairman at the meetings of the Board of Managers.[19]

Carnegie agreed to name Frick chairman because he wanted to appease Henry Phipps, his partner. Schwab later described Phipps as "a little, dandified man, who was not assertive. He had no vigor or initiative, and was of the bookkeeper type."[20] One would never think of Phipps as a man who could be Carnegie's friend or business partner. But Carnegie was fond of him; they had known each other as boys, and John Phipps, Henry's older brother, had been one of Carnegie's closest childhood friends. These ties of sentiment bound them together.

Phipps felt greater admiration for Frick than for Carnegie. He considered Frick to be a conservative businessman, as he himself was—a man who looked upon every new spending proposition with skepticism, and one who reached decisions by a carefully reasoned, dispassionate process. They were natural partners.

Phipps and Carnegie were not. Time after time, Carnegie horrified Phipps. He would make rash decisions, and he had a penchant for authorizing extravagant expenditures on new technology. That is, rash and extravagant to Phipps's mind; Carnegie thought they were not only reasonable, but forward-looking.

Phipps was convinced that Frick had been a valuable counterweight to Carnegie. So when Frick resigned as chief executive officer and the new president was appointed to take his place, Phipps wrote Carnegie in near hysteria: "Mr. Frick is first, and there's no second, nor fit successor. With him gone a perfect Pandora's box of cares and troubles would be on our shoulders. . . . The Herculean labors that would confront us, troubles unnumbered—unending, life too short, the game not worth the candle. . . ." Phipps sought "a haven of enjoyment and rest, instead of what may beset us any day, a sea of trouble, cares and anxieties," and he was sure that Carnegie would be unable, or, more likely, unwilling, to provide that haven. He was desperate, so much so that he proposed that he, Carnegie, and Lauder consider "some plan of sale and security." [21] But Carnegie had no desire to withdraw from active business; he decided to allay Phipps's anxieties by naming Frick chairman.

The minute Frick became chairman, he immediately sought to regain his lost authority—and Carnegie would have been extremely naïve if he had ever believed that Frick would content himself with a purely titular post. Frick was temperamentally incapable of playing a passive role or of being a ceremonial figurehead. With Phipps's encouragement, he steadily expanded his power and prestige within the company by actively supervising and criticizing the policies of the presidents. Since he encountered neither overt resistance nor secret defiance from Leishman and Schwab, there was no crisis of authority. For the next five years, until he had a new fight with Carnegie, which precipitated his expulsion from the company, Frick wielded vast power—and he must be credited with a major role in building Carnegie Steel.

The man chosen to succeed Frick was John G. A. Leishman. Leishman had been born in Allegheny, Pennsylvania, in 1859. He had entered Carnegie's service in 1886 as a salesman and was promoted to vice-chairman, under Frick, in 1887. Carnegie regarded Leishman as Frick's protégé, and he kept him under close surveillance after Frick resigned in 1894. When Carnegie discovered in January 1896 that Leishman was speculating in pig iron and was mounting up personal debts, he sent him a stern warning:

> The Carnegie Steel Co. is daily compromised by its President owing private debts. I am told of an instance where you borrowed from a subordinate. This is madness, and proves to me and to others that you do not realize what the Presidency means, just as little as you admit you do in regard to your gamble in pig. . . . The President must have no private affairs which the Board does not know. He must sell out all investments carried on margin, or by loans; *he must make no contracts of importance, except as the Board after full discusssion as a Board, directs him to make.* . . . We all appreciate your ability in some directions, but you have much to do before you regain the confidence of your partners as a safe man to be the Executive head of the Company. (Italics added.) [22]

By that letter alone Carnegie told Leishman exactly what kinds of actions he would not tolerate. Yet only a month later he had to scold Leishman again—this time for making an unauthorized purchase of two bankrupt furnaces:

> I beg to enter my protest against what I consider a departure from business principles and a stretch of authority. I leave you to imagine how unsatisfactory it is to have such surprises sprung upon one. . . . I am sorry to have to write this, but I cannot live and have the Carnegie Steel Company degraded to the level of speculators and Jim Cracks, men who pass as manufacturers, but who look to the market and not to manufacturing, and who buy up bankrupt concerns only to show their incapacity." [23]

A year later, Leishman was no longer president. Carnegie wrote a statement and released it to the press. In it he claimed that "Mr. Leishman has been urging his partners for some time to relieve him from the harassing duties of the presidency of the company, owing to the state of his health." [24] The resignation took effect on March 31, whereupon Leishman "recovered" and accepted an appointment to serve as American ambassador to Switzerland. [25]

Once Carnegie had decided to remove Leishman from office, he was faced with the problem of choosing a new president. Two of his chief partners, George Lauder and Henry Phipps, were middle aged; they were not inclined to undertake the strenuous pace of work which the job would require. Henry Frick was ruled out because of his previous disagreements with Carnegie.

A few weeks before Leishman was ousted from the presidency, Carnegie had offered to make Schwab the vice president of Carnegie Steel. Apparently Carnegie believed that Schwab, then superintendent of Homestead, was not yet ready for the presidency because his technological expertise was not matched by administrative experience. But Schwab declined Carnegie's offer; he preferred to remain the ruler of a principality rather than become heir apparent to a kingdom. He told Carnegie, "I am no good carrying out other men's orders, and I should have to do that as vice-president. As superintendent I am boss of the plants I manage." [26] It is likely that Schwab knew of Carnegie's growing dissatisfaction with Leishman, and equally likely that he saw no clear candidate to succeed Leishman. If Schwab's refusal was a gamble, it paid off. On April 17, 1897, he became president of America's largest and most profitable steel company.

Carnegie's decision to offer Schwab the presidency was justified, as was Schwab's belief that he was qualified for the job. At thirty-five, he possessed boundless energy and ambition; his eighteen years in the steel business had given him widely diversified experience. He was an expert in the design and construction of production facilities; he had successfully managed Carnegie's two largest plants; and he enjoyed the respect and admiration of his partners, foremen, and workers. The two skills he lacked—negotiating with customers and administrating a multi-plant business—he could acquire quickly, just as he had acquired his other skills.

Once Schwab became president, cutting costs became his foremost objective. Appropriately enough, his first act as president was to end a wasteful practice which he himself had condoned when he had been superintendent at Homestead.

Every superintendent and every department head worked to please Carnegie, in the hope of being invited into the select circle of partners. The road to preferment was paved with statistics—records of steadily rising tonnage output and declining production costs. Since everyone knew that Carnegie carefully studied the monthly reports on output and costs, those seeking recognition and promotion tried to increase tonnage each month and to outproduce comparable departments in other Carnegie plants. This rivalry between department heads sometimes led to deceptive practices.

One of the men most eager to win a partnership share was William E. Corey, now superintendent of the 119-inch plate mill at Homestead. Several years later one of the Carnegie veterans described Corey's production policy:

> Calling for the order books, he [Corey] went over them, tearing out the orders on which the greatest tonnage could be made, entirely disregarding the promised delivery dates. Gathering them up in his arms he handed them to the Assistant Superintendent with instructions to run the mill on them, regardless of "hell or high water."
>
> His orders were obeyed with the result that the shipping facilities, inadequate even under normal conditions, were swamped and the plates had to be piled up on both sides of the tracks until they resembled an unroofed tunnel. Then came chaos. A customer telegraphed asking the cause of delay. In response to our inquiry, the mill said it had been rolled on such a day. We wired back: "But when will you ship?" The mill said the plates were at the bottom of the pile and it would not pay to handle the whole pile to get at them. The result was that in many cases, the same order had to be rolled several times as there was no let-up in the attempt to make a record for the month.[27]

While Schwab was general superintendent at Homestead he did not order Corey to stop this practice, but once he became president his attitude changed. He instructed William B. Dickson, the assistant head of the Entry Department (which had responsibility for assigning orders to the appropriate mills for production), to visit each mill weekly to establish rolling schedules which would meet shipping deadlines. Thereafter, according to Dickson, "the weekly rolling schedules were arranged by a compromise between mill efficiency and contract obligations."

Schwab's second innovation as president was to establish Saturday meetings of all the mill superintendents. There they would discuss common problems and exchange ideas about ways to cut costs and increase output. The meetings were intended to heighten competition between the plants by openly comparing the tonnage records and operating costs of each. When Schwab wrote to Carnegie to tell him about the meetings, he did not conceal his pride: "To my mind they are the best thing inaugurated for some time, because it stimulates healthy rivalry between the works without creating undue friction."[28] Carnegie agreed, calling the meetings "one of the best ideas that has ever entered your fertile brain. I see how many points of trouble are met and solved, giving unity of action among the various works." [29]

Schwab knew from his own experience that the superintendents played a crucial role within the company. Their foremost responsibility was to fulfill the tonnage orders sent to them, but they were also permitted to experiment with any new production methods which might lead to reduced costs and to offer financial incentives to stimulate extra effort by the workers. They had authority over their own personnel; they could hire, fire, or promote workmen and foremen, so long as production schedules were maintained without interruptions. And finally, they were in charge of the design and construction of new furnaces and rolling mills.

These superintendents were paid salaries which Carnegie and Schwab said were "low" or "rather modest." But any superintendent might be awarded a partnership interest in the company and thus share in its profits—a major incentive. As president, Schwab was responsible for nominating superintendents who warranted a partnership interest, and he made an annual review of each man's performance for this purpose. At the Saturday meetings Schwab regularly reminded the superintendents that a share in the partnership awaited the man who could demonstrate his indispensability to the company. [30]

When Schwab first became a partner in 1892, the company was capitalized at $25,000,000, and his share was one-third of 1 per cent. By 1897, the minimum share for new partners had been reduced to one-ninth of 1 per cent, which was drawn out of the 3.66 per cent of the shares held in trust for new partners. The success of the company, Carnegie declared, "flows from having interested exceptional men in our service; thus only can we *develop ability* and hold it in our service." [31] The system of awarding fractional partnership shares was intended to entice ambitious young men into permanent careers with the company. The longer they stayed, the more valuable their share would become. But if a partner voluntarily left the company, or if he were expelled, he was unable to sell his share to any outsider, and he received only the book value of his share. The Carnegie Steel Company was intentionally structured as a closed partnership so that its shares would not fall into the hands of outsiders. A partner was entitled to his fractional share of the total dividends declared in any year, but he did not receive the money directly. Instead, the dividends were credited to paying for his partnership share;

only when it was fully paid off did a partner begin to receive cash dividends.[32]

Carnegie, who owned 55.33 per cent of the shares, deliberately pursued a policy of low dividends, seldom paying out more than 25 per cent of the profits in any year. The balance was retained for modernization and expansion of the company. After the first nine months of 1897, Carnegie declared a dividend of $750,000 on earnings of $4,000,000. One of the junior partners who received the dividend was a man whose progress deeply interested Schwab—his own brother, Joe, the superintendent of the Duquesne Works.

Joe Schwab had followed closely in his older brother's footsteps. He entered Carnegie's service in March 1883, after being graduated from St. Francis College in Loretto. In his first job, which his brother helped him to obtain, he was a draftsman in the Edgar Thomson Works at Braddock. The brothers bore a striking physical resemblance, except that Joe had a more angular face and trimmer physique. And Joe, like his brother, was quick-witted, competent, and undisguisedly ambitious.

After working for a brief period as a draftsman, Joe Schwab served as a civil engineer. Then he was promoted to run the blooming mill at Braddock. When his brother became general superintendent at Homestead in 1886, Joe was named superintendent of the structural steel department, a post he held until 1896, when he was promoted to run the Upper Union Mills, another Carnegie plant. Six months later, when Charles Schwab became president of Carnegie Steel, Joe was promoted again; he became general superintendent of the Duquesne Works. He was also awarded a small partnership interest in the company, one-third of 1 per cent.[33]

Joe's share of the 1897 dividend was $2500, or the equivalent of $50 a week. It was not paid to him, but credited against the total amount he owed for his partnership interest. At that rate, it would have taken him thirty-one years to pay off his interest. Even the upsurge in profits during the last three months of 1897 did not substantially alter the situation. When the statement of year-end earnings revealed profits of $7,000,000, Carnegie increased the total dividend for the year from $750,000 to $2,500,000, but even at that accelerated rate, it still would have taken Joe more than a decade before he could fully own his interest.

Carnegie believed that the young partners' riches should be deferred, that that would serve as the greatest incentive for their continued pro-

71

ductive effort. Schwab disagreed. He argued that they would have even greater motivation to excel if they had some hint of what their future prosperity might mean to them. He urged Carnegie to allow them to receive some of the dividends directly, before their partnership interests were fully paid off. When he proposed this to Carnegie in April 1900, Carnegie made an unprecedented concession: he agreed to pay out 25 per cent of the dividends directly to the young partners, while the remaining 75 per cent would be credited against their shares. But he gave his consent grudgingly. He was not fully convinced that it was a good idea to give cash to young men; it might dilute their dedication. "Still I ask myself," Carnegie wrote to Schwab, "what would I as a young man desire to do if I owed a Company for an interest which it gave me without risking a cent—I would pay every cent off, live modestly on my salary, until I really owned my interest. . . . It is best to keep young debtors economical I think you must conclude." [34] And, in a second letter to Schwab, he added, "All I wish to submit to you as the Manager of men is that you will have better service from a body of young men who have to 'labor and to wait' until they are out of debt—Truly the best use any young man can put his money to—or rather our money, is to pay for his interest. [35]

Just as Schwab closely supervised the superintendents, so he in turn was watched by Carnegie. Although Carnegie claimed that he was only "a shareholder" and that his role was merely advisory, [36] as the majority shareholder he exercised ultimate control over every aspect of the business. The bylaws of the company required that all important changes in equipment or operating methods be approved by owners of two-thirds of the stock. Since Carnegie owned more than 50 per cent, no major changes could be authorized without his consent. Even during his four-month summer vacations in Scotland he maintained continuous contact with his Pittsburgh headquarters. He used trans-Atlantic cables, either coded or cryptically terse, offering advice to his partners and giving his opinion of urgent pending proposals. When the pace was more leisurely, he dictated long letters to elaborate the reasoning behind his decisions.

In 1895, Carnegie instructed the company's secretary to begin keeping a full record of everything that went on at the weekly Board of Managers meetings. The minutes would not only serve as an *aide memoire* for the participants; Carnegie, though he might be absent, would still know what had been discussed, and he could send back detailed comments. Carnegie's instructions read:

> We want every motion to show who voted for it, and who against it. . . . ,
> The Minutes should, in addition to this, record every reason or explanation
> which a member desires to give for his vote. If this were properly done,
> then any of us looking over the Minutes would be able to judge of the judg-
> ment displayed by the voter, which of course would affect his standing with
> his colleagues. It would bring responsibility home to him direct, and I do
> not see any other way that will enable us to judge whether any of our
> partners has good judgment or not . . .[37]

The first known occasion when Carnegie doubted Schwab's "good judgment" occurred in 1897, when Schwab proposed some costly changes at Homestead and Duquesne without offering a detailed justification for the expense or a precise estimate of the profits which could be expected if the changes were made. While Carnegie was in Scotland, Schwab persuaded the Board of Managers to authorize the spending of $1,000,000 to build new furnaces and a blooming mill at Homestead and to remodel the Duquesne Works. When Carnegie read the minutes of that meeting, he complained, "sweeping changes which have never before been heard of, the board instantly approves; no word of criticism made." Carnegie objected that "A very large question is presented to us which should receive *months of study* and especially should time be given for the principal owners to hear all sides." [38]

To Schwab himself Carnegie sent a sharp note, criticizing the hastiness and incompleteness of his proposal:

> What we have a right to ask is figures and not an expression of your
> opinion. Show us why the blooming mill at Homestead is one of the best ex-
> penditures we could possibly make. You must have arrived at this opinion
> from some data. You have not founded it upon nothing. Why do you keep
> this data from your Board and partners?
> My dear young friend, you say you will answer any list of questions I
> may ask, but why should a report of a President, recommending an invest-
> ment, render questions necessary? The report should be full, accurate; fact,
> not opinions, or impressions: really you do not do yourself justice in the ap-
> parent "slapdash" manner in which you recommend this to the Board.[39]

Schwab, always highly sensitive to any criticism, bristled at Carnegie's first letter. He sent him two long letters justifying his proposal; in

73

the second he wrote, "I appeal to you, in all the years connected with your firm, and having recommended the principal improvements that we have made in these years, can you point to a single error?" [40] But Schwab's first letter, containing a detailed explanation of the proposed changes, had already convinced Carnegie, who wrote, ". . . yours of the 1st October puts a new face upon the matter and I am delighted to know that the additions are going forward. You are a hustler!" [41] To Schwab's second letter, Carnegie replied:

> Believe me, I am rejoicing equally with yourself at your brilliant success and at the improvements which I am sure you are going to make.
> Here is a letter from Lauder which I adopt as my own sentiments: "Everything about the business is in first-class shape. I do not hear a single word of adverse criticism. There has never been a time on my return that I can recall when everything seemed moving so smoothly." [42]

When Schwab was elected to the presidency his partnership interest was increased to a full 1 per cent, and he was promised that he would get an additional 1 per cent if Carnegie was pleased with his work after the first six months. Carnegie wrote to Frick about Schwab's interest after less than five months. "I vote for giving Schwab 2% [more] instead of 1 promised—cheapest man we have at 3%—President should have that and not be on level only with vice presidents—I hope Schwab will have a big interest some day—gives every promise of being the man we have long desired—." [43]

Frick agreed: ". . . there is no question about Mr. Schwab. Our business was never handled so well, and he certainly is entitled to 3%." [44]

Just as Carnegie, Lauder, and Frick were pleased with Schwab's performance as president, so, too, Schwab was finding that job increasingly satisfying. In a letter to Carnegie he said:

> I had felt on coming into the City Office in general charge of the business, that the faculty which I possessed of devising new methods, etc., and which was given such free play at the works would be narrower by reason of being away from the works. On the contrary, I find that I have many more ideas, better ones, and much broader ideas with reference to improvements in machinery and methods than I ever had in the works, and I find great delight in having the details of such schemes and improvements worked out and solved by the Superintendents of the works. . .[45]

Carnegie, Frick and Schwab were in full agreement on how to deal with the threat of nascent unionism. Schwab was an intransigent opponent of labor unions because they would interfere with management's ability to cut costs and thus threaten the company's competitive position. He wanted to be free to reduce the tonnage rates paid to skilled workmen and to lower the rate paid to those who worked on a flat daily basis if prices fell. Schwab's objective was to obtain enough orders so that the mills could continue to run at full capacity. This, in turn, would prevent shutdowns and layoffs. Schwab was convinced that cutting wages would redound to the long-range benefit of the workers; if Carnegie Steel remained foremost in the industry, then the workers' jobs would be secure. It is a tragedy in human terms that the two parties—the workers and the owner-managers—appeared to each other as adversaries, as rivals for a static sum of wealth, rather than as allies in a common cause: increasing production and sales, thereby increasing job security for all, at every level of the employment hierarchy.

Cutting wages was the fastest way to reduce operating costs. Since Carnegie Steel had no unions, it could immediately retaliate in kind if a rival firm sought to gain a temporary competitive advantage by reducing wages. In 1897 Carnegie learned that the Pennsylvania Steel Company had reduced wages by 10 per cent. He told Frick, "I think this forces us to take off the bonus to teach our competitors that they cannot take any advantage over us by the reduction of wages.[46] Later that year, Schwab acted on this same principle when he recommended that the company eliminate the 10 per cent bonus then being paid to tonnage men. His reasoning was that even though the cut was not strictly necessary, despite the decline in business, it was desirable; "When business improves, we will be expected to raise wages, which we cannot do unless we end the bonus now.[47]

When Schwab made this proposal to the Board of Managers, one member suggested keeping the bonus on but refusing a wage increase later when business improved. Schwab replied that although that policy would give the men greater total income, they could not be made to see the point. Therefore it was necessary to cut the bonus now so that later, when other companies granted increases, Carnegie Steel could follow suit. The alternative policy, Schwab commented, "might satisfy a thinking man, but most of the men do not think. Argument doesn't count for as much to the average workman as an advance." [48]

Schwab was talking about some of the skilled workers at the time he made that remark, but he held an even lower opinion of the unskilled Hungarian laborers, whom he considered to be ignorant and intractable.

In March 1899, when disgruntled workers were demanding an increase in wages, Schwab felt it was futile to try to explain the company's policy and the reason for its refusal. He voiced his impatience to Carnegie: "it is exceedingly hard to reach the class of men to whom we wish this particularly to apply. I refer especially to the Hungarians at our furnaces. You cannot make them understand why we should not advance wages when everybody else is doing it although their rates have been higher during the past year than any other firms have been paying." [49]

These comments, which were written in confidence, directly contradict the optimistic, benevolent view he expressed in his 1935 reminiscences: "When reductions have to be made the proper thing to do is to call the men in and take them into your confidence. Show them your cost and profit sheets. Demonstrate your position, frankly and honestly, and you'll have no trouble." [50]

Each time the company introduced new equipment which increased the output per day of workers who were paid on a tonnage basis, Schwab proposed to reduce the pay rate. Late in 1897, after one of his periodic reviews of wages, Schwab decided on a substantial reduction in the rates paid to the tonnage men at Homestead and the Union Mills, where new equipment had been installed. He estimated that the savings would be at least $20,000 to $25,000 a month—more than $250,000 a year.[51]

Schwab realized that the reduction would be unpopular, but that the men at Homestead and Union Mills would be even more angry if they alone among the workers at the various Carnegie plants received a rate cut. He decided to make a small, token reduction in the wage rates at the Duquesne and Edgar Thomson works as well. When Carnegie objected on the ground that this savings was so slight that the cut hardly seemed warranted, Schwab explained:

> It might look to you that the reduction of Duquesne and Edgar Thomson is trifling and should not be disturbed, but it should be remembered that the objection at Homestead and Union Mills will be much greater if it is understood that Duquesne and Braddock [Edgar Thomson] are excepted entirely. . . . I think the adjustment absolutely necessary because the wages being earned by some of our tonnage men are beyond all reason. The Plate rollers at Union Mills earn about $15 per day and our beam rollers at Homestead, about $12 per day. This is too much. I have based the rollers on an average of about $6 per day which I think is very liberal. . . . This, as you know, is much above what we have paid in this position in recent years.

76

Rest assured I would not take any steps in this direction for the amount involved if I thought we were going to be led into any difficulty, but I really anticipate none.[52]

He was correct; the rate cut was not resisted. Without unions, the men found it difficult to organize any resistance—and as individuals they had almost no bargaining power.

Schwab was determined to keep the unions out. In May 1899 he learned that the Amalgamated Association of Iron and Steel Workers was again attempting to unionize the workers at Homestead. William E. Corey, Schwab's successor as general superintendent at Homestead, reported that he had discharged four of the leaders of the unionization movement. Schwab called Corey's action "admirable," and he informed Carnegie of it: "Labor seems to be giving us some little trouble in all directions. This week at Homestead there was an effort on the part of some men to reorganize a union. We promptly took action and discharged a half dozen of them yesterday and will do the same today. I feel this will nip the move in the bud." [53]

Schwab sent a warning to all superintendents to "act promptly and vigorously, and stand by the consequences, no matter what they be." Aware that they might interpret his remark as an incitement to harsh or repressive measures, Schwab added that "it is advisable to treat the men in a broad and liberal spirit, where requests were reasonable." He undercut this statement, however, by warning them "to make no concessions in principle which would only lead to further demands." [54]

Schwab was adamant: "we will have no labor organization in our works." He believed that the unionization attempt would fail because the majority of workers did not want the union and because the company could discharge the tiny minority who did. In his view the men were "being well taken care of" and would not be "so foolish as to obey the commands of a few dissatisfied men." [55] He opposed any conciliatory policy; as he told the Board of Managers:

We should not permit the Amalgamated Association to get into Homestead again—we should keep them out at any cost, even if it should result in a strike. We should not adopt any shifting policy, especially at Homestead. If they get in there, they would be able to get into the other Works much more easily.[56]

A month after renewing its organizational drive, the Amalgamated Association sent a delegation to Superintendent Corey to demand that the

77

company recognize the union. Schwab described the events that followed in a letter to Carnegie:

> We promptly discharged the Committee and any workmen who were in sympathy with the movement; they then declared a strike on June 30, and did their best to bring out our men but failed completely. A few, of course, came out, but not enough to cripple us. On July 3, they gave up the attempt and advised all men to save their places that could do so. We maintained a very firm policy from the start and it won.[57]

According to Schwab, only young workers, newcomers to the company, had attempted to unionize. Older men who had participated in the 1892 Homestead Strike "stood firmly with us."[58] Perhaps the older men realized that any attempt to force a union upon the company would be futile and probably would cost them their jobs.

Schwab was fortunate that he began his presidency of the Carnegie Steel Company just as the Depression of 1893 was drawing to an end. By August 1897 there were unmistakable signs of recovery.[59] For Schwab the return to prosperity meant there would be fewer financial objections to his plans for modernizing production techniques, introducing new machinery, and building new plants.

Henry Phipps did not share Schwab's optimism. Only a year away from his sixtieth birthday, Phipps had grave misgivings about incurring new debts for further expansion. After years of struggle his fortune was secure, he told Carnegie, and now he would be quite content if the company were to pursue a cautious, conservative policy which would earn an unspectacular but safe 5 per cent interest on his stock.[60] But Schwab had a lifetime ahead of him, and he was impatient to make his fortune.

Schwab also found that some of his proposals for expansion were unacceptable to Carnegie—that there were factors other than potential profits which influenced Carnegie's decisions. He suggested that the company build a mill in Chicago to take advantage of the growing midwestern market for steel products. Carnegie opposed the recommendation; he stated that the company's best markets were in the East and overseas, and that the "West is not growing in a way that will create a better market for Steel—it is an agricultural district."[61] Schwab accepted this, but he named one major exception, steel rails, which could be produced more cheaply in Chicago than in the Pittsburgh area; from

78

Pittsburgh they had to be freighted westward. Carnegie answered that he "would rather manufacture Rails in Pittsburgh at $15.00 and pay $1.00 freight to Chicago, than make them at Chicago for $15.00. Our attitude should be that we do not need to go into any parts of the Country to compete in its local markets. This is our headquarters; let us grow here." [62] Carnegie was determined to concentrate his energy and his empire in Pittsburgh and its immediate vicinity; he wanted the entire area to be his monument. It was futile for Schwab to suggest expansion elsewhere, and he made no such suggestions again.

Schwab was not always able to persuade Carnegie about where to expand, but he and Carnegie generally agreed about when to do so. Carnegie Steel's greatest expansion had always occurred during periods of economic depression, when bankrupt steel works could be bought up at greatly reduced prices and new facilities could be built cheaply. But this was too slow and uncertain for Schwab—his ambitions could not hang on the vagaries of unpredictable depressions. He persuaded Carnegie that the company should expand during times of prosperity and high profits as well as during periods of depression. He said, "while we are making good profits, we should not hesitate to spend money freely for improvements, and in keeping our Plants up to date in every respect." [63] He argued that it was a mistake to forego expensive improvements during prosperous times; "when times are dull, we do not feel inclined to spend money for betterments." [64]

But when hard times did come, Schwab remained equally zealous about the necessity for expansion and improvements. In 1900 there was a brief but severe slump in business. The price of Bessemer pig iron fell nearly $8 a ton between May and July, from $24.90 down to $17.00, and the price of steel billets fell $6.25 in a single month. [65] It was a precipitous decline, following a year of stable, high prices. In reaction, George Lauder proposed a moratorium on all improvements until the economic downturn was over. He argued that the company should conserve its dwindling financial reserves, that funds would be needed for working capital just to keep the company operating. Lauder urged that work be canceled on the new wire and nail mill which was being built at the Duquesne Steel Works. Schwab countered with a proposal that the company should spend an additional $1,400,000 immediately, to hasten the completion of the plant. He challenged the premise of Lauder's argument: ". . . if we had adopted the policy of holding back because we

79

thought we were to meet a panic we would never have developed our business to its present state. We would have no Structural Steel Departments, no Open Hearth furnaces."

Schwab contended that expansion was crucial just then, when the sales outlook was most dismal. Unless the company moved into a new area of production, it would not be able to consume its full production of billets, which meant that its steel producing mills would have to shut down. He added that it would be desirable to have a completed wire and nail mill ready for the return of prosperity, but such a mill would be valuable even in the depths of depression, since the production of wire and nails would consume part of the furnace output and would generate some sales revenue. Lauder asked the Board of Managers to postpone a decision until Carnegie's return to America. But Schwab refused to accept delay; in a rare burst of anger he told Lauder, "If we want to drop back into an old fashioned way of doing business I want to be counted out of it." [66]

If Schwab's comment to Lauder was an oblique threat to resign, he was undoubtedly confident that his resignation would not be accepted. Throughout his presidency, he had repeatedly proven his value to the company. He continued to be "the man we have long desired," the man whose "fertile brain" Carnegie considered irreplaceable, the man who would navigate the company safely through the rapids of internal conflicts and external challenges.

5

Troubleshooter
and
Conciliator

Carnegie and Schwab had the same basic objectives: to run the works at
full capacity, with costs as low and profits as high as possible. But on occasion they differed about how those objectives were to be achieved.
Whereas Carnegie was a fierce competitor, determined to maintain his
industrial supremacy, Schwab tended to be more conciliatory, preferring
compromise and "cooperation" to achieve the same ends.

One area of disagreement between them concerned the company's
pricing policy during periods of prosperity. Carnegie believed it would
be a mistake to raise prices substantially during periods when rising
demand enabled the nation's steel companies to increase their prices
above depression levels. Unless the company undersold its competitors
by keeping prices below the highest level that could be obtained, its
competitors would earn substantial profits which they could then use to
expand and improve their own plant capacity. Should that occur, Carnegie feared, they might some day be able to engage in price competition
with his company during periods of recession.

But Schwab resisted this line of reasoning. Why, in times of prosperity, forego the highest possible prices, and therefore even higher profits,
just because high profits also would permit one's competitors to expand?
And why, in periods of declining sales, should prices have to go spiraling
downward just so the company could undersell competitors? He believed that a better and more profitable solution was possible. Rather
than getting involved in price competition, Schwab preferred to negotiate pooling agreements under which all firms could charge uniform

prices without fear of being undersold. On this basis, Carnegie Steel, by virtue of its superior efficiency, still would make a higher rate of profit than its competitors and thus would have surplus revenue with which to build additional plant capacity. Then, by using the threat of breaking away from the pooling agreement, the company could demand and expect to receive a larger percentage of sales.

Schwab knew, however, that pools were of little long-range value. The competing companies distrusted each other, and with reason: companies secretly produced more than their specified share of the anticipated market, and then secretly cut prices, luring customers from their rivals by charging less than the uniform price agreed upon by the pool. Yet Schwab believed pools were necessary, even if only for short periods.[1]

Whenever possible, Schwab preferred to avoid confrontations with competitors. And although Carnegie was opposed to pools on principle, he did permit Schwab to negotiate short-term pooling agreements for the company, particularly for the sale of steel rails. When Schwab took over the presidency of Carnegie Steel in 1897, the market for rails was declining; few new roads were being built, and the major demand for rails was to replace worn-out track. To arrange a pool, Schwab had to come to terms with John W. Gates, the head of the Illinois Steel Company. Gates was universally distrusted; Carnegie considered him an unscrupulous buccaneer with whom it was shameful to have any dealings.[2] Nonetheless, Schwab formed a tacit alliance with Gates. At the periodic meetings of the member firms of the rail pool, Gates would proclaim that he was entitled to a larger percentage of sales because of his company's proximity to the largest sales area. Schwab, in turn, would shout back that it was Carnegie Steel which deserved a larger percentage of sales because it had the greatest rail-producing capacity. As Gates and Schwab roared their demands for larger shares of the market, they were able, according to Schwab's account, to intimidate the "smaller fellows." He and Gates, separately or jointly, would threaten to withdraw from the pool unless they were granted the share they demanded. The other firms would usually relent and a settlement would be reached. Then Schwab and Gates would argue in private over the division of the share allotted to their companies.[3]

By 1898, however, Gates no longer headed Illinois Steel. Together with several smaller companies, Illinois had been merged into the new Federal Steel Company, with former Judge Elbert H. Gary as president.[4] Gary had no patience with unstable pooling arrangements. He proposed to Schwab that they should form a "Rail Association"—a more durable form of pool, one which could not be so easily violated because there would be strict policing and heavy fines or forfeitures levied against firms caught cheating. The two men negotiated such an arrangement.

But in order for it to come into being, the Board of Managers of Carnegie Steel would have to ratify it, and, in November 1898, the Board, with Carnegie present, discussed it. Since rail prices had taken a sharp dip that year, all of the partners agreed that "some form of association between and concert of action among the Rail Manufacturers" was necessary in order to prevent rail prices from falling to an unprofitably low level, which might have a generally depressing effect on all steel prices.[5]

But there was sharp dissension over what type of agreement should be made. Schwab and Frick led one faction. Schwab stressed the urgency of persuading all rail producers to join the pool, and he argued for the necessity of levying high fines against any firm which violated the agreement, stating that "any plan that did not properly punish a breach of faith would be evaded." His major argument was that even though Carnegie Steel would be allotted a smaller share of the market than it could get by cutting prices, it would be more profitable to have a smaller share of sales at higher prices.

Schwab did not convince Carnegie. Such a Rail Association had two dangers: it would be savagely attacked in the press and it would violate various state laws which forbade price-fixing. Even worse, railroad purchasing agents would be furiously opposed to the association if it raised rail prices.[6]

Despite Carnegie's objections Schwab's proposal was approved by a majority of the Board of Managers, which then authorized Schwab to sign a five-year agreement under which Carnegie Steel and Federal Steel each would be allotted 29.5 per cent of total rail sales. Carnegie seemed to give in. But Schwab's satisfaction was short-lived. Out-voted on November 14, Carnegie had "Second Thoughts" the following day. He told Schwab that the company would have nothing to do with a "trust." He ordered Schwab to withdraw, and to offer as the reason, if any were needed, that Carnegie was unwilling to approve the venture.[7] As an alternative to the vetoed scheme, Carnegie announced that he personally would negotiate with the representative of Federal Steel to arrange a sales and marketing agreement between the two firms.

Although Carnegie and Gary reached an agreement,[8] it proved ineffective because it applied to only two of the many firms producing steel rails. The following year, when Schwab proposed that the original Rail Association scheme be instituted, Carnegie withdrew his objections.[9] It seemed that Schwab had carried the day, but he soon had grounds for misgivings. As Carnegie had long maintained, the members of a pool are always at the mercy of an outsider. Early in 1900, the National Steel Company began producing rails, and it expected to be allotted a sizable share of the market by the pool members. The entry of National Steel, said Schwab, meant that "we will not be able to get the same percentage that we got before. National Steel will want 10 or 12%, thus reducing us from 30 to say 26%. We could not run full on such a percentage." [10] By June 1900, the survival of the Rail Association was gravely imperiled, and by the end of the year it had collapsed.

National Steel had used the same strategy to force its way into the Rail Association as the one Carnegie himself had used in 1899 to gain favorable concessions from the members of the Structural Steel pool. At that time Carnegie had counseled:

> . . . if our President steps forward at the right time and in the right way, informs these people that we do not propose to be injured, on the contrary we expect to reap great gains from it, that we will observe an "*armed neutrality*" as long as it is made to our interest to do so, but that we require this arrangement: — then specify what % is advantageous for us, very advantageous, more advantageous than existed before the combination, and he will get it. If they decline to give us what we want, then there must be no bluff. We must accept the situation and prove that if it is a fight they want, here we are "always ready." Here is a historic situation for the Managers to study—Richelieu's advice: — "First, all means to conciliate; failing that, all means to crush." [11]

Schwab followed Carnegie's advice about the Structural Steel pool and was awarded a handsome share of the market as the reward for not challenging the pool's pricing policies. Nonetheless, within eighteen months it was Schwab who was counseling withdrawal. In mid-1900, as business slackened, Schwab advised, "There is no question, of course, that if the business is left open and Associations done away with, that we can get our full share of business and sufficient to keep our mills going at full speed. We should all think seriously upon this point." [12]

Carnegie agreed with Schwab, adding that he himself favored withdrawal from all pools. He cited the company's past policy during depressions:

In the former depressions we announced our policy, viz., take all the orders going and run full. Our competitors believed that we meant what we said and this no doubt operated to clear the field. One after the other dropped out; finally Pennsylvania Steel dropped out & only a few remained who could meet the lowest prices. We averaged $4 a ton profit on all our product in the worst of times.[13]

Though Schwab actively favored the pooling arrangement as a means of avoiding price competition, it was essentially unsuccessful. In prosperous times a high price set by the pool would attract new entrants into production, thereby cutting into the percentages Schwab had negotiated for Carnegie Steel. And in times of depression, it was Schwab himself who advised withdrawal so that Carnegie Steel, by slashing prices, could win a large enough share of the market to be able to run at full capacity.

Carnegie and Schwab shared the view that the company should continually expand its operations, but, again, they disagreed on how this should be achieved. Schwab preferred to concentrate on the most profitable of the company's existing product lines—that is, to increase output of those products they already made and to enlarge profits primarily by reducing costs. Carnegie, on the other hand, was eager to expand into new lines of production, even though this would bring the company into direct conflict with firms already in the field. While Carnegie relished confrontations with other companies and was confident that he would ultimately triumph over them, Schwab did not share the Scotsman's enjoyment of conflict. He was content to stay out of those fields where well-entrenched companies were dominant, and he even favored withdrawal from certain of Carnegie's existing lines of production, those where future prospects for profit were dimmest. In late 1898, they were on opposite sides concerning two such ventures.

Their first disagreement concerned the disposition of the Keystone Bridge works, the least profitable division of the Carnegie Steel Company. Since Schwab preferred to concentrate the company's efforts in the most lucrative areas, he was delighted when, in December 1898, an offer was made to buy Keystone. Companies representing 95 per cent of the bridge-making capacity in America were attempting to consolidate into a single firm, and they offered to buy the works if Carnegie Steel would sign an agreement promising not to re-enter the bridge or structural fitting business. Schwab recommended the sale not only because it would eliminate an unprofitable division, but also because, in return for a promise not to compete, the newly consolidated bridge-makers were promising that Carnegie would receive a "satisfactory percentage" of the orders for all types of bridge building materials.[14]

The Board of Managers agreed with Schwab that this was a very good deal. Pleased with their agreement, Schwab wrote to Carnegie, asking for his reaction. Carnegie vetoed the proposal, saying, "Our policy should be to make finished articles, Bridges among them." And he offered Schwab this advice: "The concern that sells articles finished will be able to run all weathers and make some money, while others are half idle and losing money." Carnegie continued: "We hope our competitors will combine, for an independent concern always has the 'Trust' at its mercy." [15]

A more serious and more revealing disagreement between the two men concerned the production of railway freight cars. Late in 1898 Carnegie suggested that the company begin producing steel freight cars. It was a fast-growing and increasingly profitable line of business, for the railroads needed freight cars. That need could be met either by the expansion of the existing firms in the field or by the entry of a new firm, Carnegie Steel. Schwab saw one great obstacle to the latter alternative: the firms which were already producing cars purchased their steel plate from Carnegie. If Carnegie entered into competition with them, Schwab warned, they might turn elsewhere for their source of supply, or even build their own steel mills.

The leading producer of steel cars was the Schoen Pressed Car Company, headed by Charles T. Schoen, the pioneer in the field.[16] Schoen was trying to raise capital to build a new and larger plant for his company; he was also aware of Carnegie's interest in entering the field. Schoen was willing to make any agreement which would allow him to expand his business.

Schwab warned that Carnegie Steel would lose its largest customer for steel plate if it went into competition with Schoen. Schwab recommended that, rather than competing with Schoen, Carnegie Steel should buy an interest in Schoen's company; with the resulting increase in capital, Schoen could build a new and larger plant. But Carnegie refused to invest money in any company which he could not control. He doubted whether they would lose Schoen as a customer even if Carnegie Steel built its own car works, provided that Schoen was able to buy steel plate at an attractive price. Carnegie said, "either by price or by ability to make deliveries we can command his order. We ought to sell him Steel, if his credit is good, but should not buy his business. We never have bought a business; we have bought Works, and made the business." [17]

At the next meeting of the Board of Managers, Schwab, who was unconvinced by Carnegie's reasoning, reintroduced his plan. The Board rejected it decisively after Carnegie read a detailed report on the shaky financial status of Schoen's company. Frick proposed another alternative: Carnegie Steel should build its own steel car works, but give a small share of it to Schoen "in return for his experience in building and selling cars." Schoen would serve as a consultant and would share in the profits of his new competitor. Frick wanted the two companies to work in harmony and to "divide business in proportion to their capacities." Carnegie also rejected this compromise proposal, on the ground that Schoen would certainly demand a larger share in Carnegie's new company than he was actually worth. Carnegie then proposed a less costly scheme: Carnegie Steel would built its own plant, and then make an agreement with Schoen that as long as he purchased all of his steel from Carnegie, the two companies could divide the car business between them. On Carnegie's instructions, a committee headed by Schwab began planning the construction of a steel car works. [18]

Schoen tried to dissuade Carnegie from going into competition with him; he arranged a personal meeting with Carnegie but found him adamant and unyielding. Their talk ended inconclusively. Thereafter, Schoen dealt only with Schwab; he sensed that Schwab and Carnegie were not in agreement and that Schwab's attitude was more conciliatory. On December 9 Schoen offered Schwab a deal: he claimed that he had raised the capital needed to build a new plant and that no other car works presently were necessary; he proposed a long-term contract to purchase steel exclusively from Carnegie. Anticipating a requirement of 360,000 tons a year, Schoen said that Carnegie Steel would receive nearly $80,000,000 in business over the next ten years, and "without any effort or any competition." Schoen added, "Surely this would seem to any reasonable person a magnificent compensation for refraining from engaging in this industry." [19]

Although Schoen's proposal was attractive to Schwab, the final decision was not his to make. Schoen's letter was forwarded immediately to Carnegie, who fired back an angry reply. Since Schoen has made arrangements for capital, Carnegie wrote, "that ends all negotiations with us." He ordered Schwab to build the new car works without delay. [20]

Schwab still remained uneasy, but the matter was now out of his hands. When he reported Carnegie's final decision to the Board of Managers, George Lauder tried to reassure him that Schoen would not cease buying from Carnegie Steel. Lauder offered two reasons: no other firm could supply Schoen with the vast quantity of steel plate he needed, and Schoen knew that if he broke with Carnegie Steel the company could undermine him by selling freight cars at cost, thus disposing of the surplus steel plate.[21] Carnegie agreed with Lauder. He told Schwab that it was a mistake to overrate the importance of long-term contracts to buy steel, that if the long-term contract with Schoen had been signed it probably would have ended in a quarrel since the price eventually would have proved too favorable for one party or the other. As long as Carnegie Steel could undersell the other firms which produced steel plate, Schoen would not buy elsewhere.[22]

Schwab then met with Schoen to tell him of Carnegie's decision to build a steel car works and to assure him that the decision was irrevocable; it in no way hinged on whether he continued to buy from Carnegie Steel. Two days later, to Schwab's surprise, Schoen offered to sell his entire business to Carnegie. He asked for $3,000,000 for his firm, plus a $75,000 interest in either the new company which Carnegie would form or in the enlarged version of the Schoen company.[23]

Schwab was mildly receptive to the proposal, but Schoen's offer to sell was actually only a stall for time; simultaneously, Schoen was making a solidarity pact with other steel car manufacturers in order to strengthen his bargaining position with Carnegie Steel. The new alliance—Schoen's company, the Fox Pressed Steel Company, and the Michigan Peninsula Car Company—informed Schwab that it would be willing to buy as much as 1000 tons of steel per day, provided that Carnegie stayed out of the steel car business. The Board of Managers remained divided about what course of action to pursue, but Schwab had decided that any further opposition to Carnegie on the issue would prove futile, so he argued that they should proceed with the construction of the car works. The majority of the Board agreed.[24]

Since Schwab was leaving for a business trip to England, the burden of informing Schoen of the decision fell upon Alexander Peacock, head of Carnegie's Sales Department. Schoen offered Carnegie Steel one final threat and inducement: he told Peacock that he had taken an option on all of the wooden railroad freight car plants in the country. If Carnegie would agree to stay out of the steel car business Schoen would buy 400,000 tons of plate and 75,000 tons of axles from him each year; but if

Carnegie went ahead Schoen would exercise his options and thereafter he would make no further purchases from Carnegie. The threatened loss of business for Carnegie did not merely involve steel plate used in steel cars, but also steel axles used in wooden cars—a sizable segment of Carnegie's lucrative axle sales.[25]

Henry Phipps was convinced that Schoen was not bluffing; he feared that Schoen would build his new plant away from Pittsburgh and would stop buying from Carnegie. Because of Phipps's misgivings, Carnegie proceeded to make a slow and graceful retreat from his earlier decision. "We have decided to build Steel Cars, that settles it," Carnegie said, but he added a significant qualifying phrase: "unless contrary action be taken." If Schoen agreed to purchase all his plate and axles from Carnegie in large amounts and at high prices, Carnegie might be willing to postpone going into competition—but any agreement with Schoen would be subject to cancellation on one year's notice.[26]

Schwab had advised Carnegie that they could produce a steel car $200 cheaper than Schoen's. If the savings were passed along in the form of lower prices, Carnegie expected this would at least double Carnegie Steel's market for cars. As Carnegie told Frick, "to give up this business is pretty bad. I should be sorry indeed, and want a pretty big reward." As his compensation for postponing competition, Carnegie proposed that Schoen pay him a $1,000,000 bonus; "I believe we can get that." Frick agreed, with the slight modification that he was willing to accept $100,000 a year over ten years instead of the total amount outright.[27] Schoen agreed to the proposal. An agreement was concluded, subject to termination on one year's notice, by which Carnegie Steel was to be paid $100,000 per year as long as it did not produce steel cars.[28] This agreement, however, did not halt Carnegie's construction of a steel car works. When he returned to America Schwab learned that it was necessary for Carnegie Steel to have such a plant, to force Schoen to live up to the new agreement; meanwhile, Schwab was authorized to lease the Carnegie plant to Schoen, subject to cancellation on one year's notice. Schwab was now free to direct the search for new cost-cutting improvements in anticipation of the inevitable day when Carnegie would repudiate the agreement with Schoen and begin the no-holds-barred, price-slashing competition for which he was so famous and so feared.

In this episode Carnegie revealed a less savory side of his personality: a love of conflict for its own sake. Instead of attempting to surpass Schoen by producing and selling steel freight cars at lower prices, Carnegie settled for outmanuevering him. He seemed more concerned with the fact of his victory than with the means by which he won. Schwab was placed in an awkward position, having to carry out a policy which he found distasteful and imprudent; but he felt he had to defer to Carnegie, since Carnegie was the major owner and chief strategist of the company.

Although Carnegie sometimes overruled Schwab on competitive strategy, there was one area in which Schwab held undisputed preeminence. He was the company's foremost salesman, negotiator, and, where necessary, conciliator—the last a role for which both Carnegie and Frick were temperamentally unsuited. They both possessed explosive tempers and were easily offended; they were accustomed to command, not to negotiate and eventually to compromise. They were particularly inept at working on protracted negotiations, being impatient for quick solutions to complex problems.

Carnegie was wise enough to recognize that his vocabulary, laden with images of war and conflict, and his low threshold of anger disqualified him from trying to deal directly with the company's most difficult customer—the United States Navy. This was a job Carnegie gladly delegated to Schwab.

Although contract negotiations between Carnegie Steel and the Navy covered a span of four years, from 1896 to mid-1900, they were barely noticed in the press at that time. More than a decade later, however, they were resurrected and widely publicized by members of Congress who were seeking to discredit the Wilson administration's military preparedness program. Congressional opponents of mobilization feared the possibility of America's entry into the World War and therefore sought to discourage "militarism" by characterizing the munitions-makers as unscrupulous profiteers. The critics—Senators Benjamin R. Tillman and Robert M. La Follette and Congressmen Clyde H. Tavenner, Charles Lindbergh, and James M. Graham—charged that, during the late 1890's, the steel companies had eagerly sought government contracts to produce armor and munitions; that these contracts were especially lucrative because Navy officials failed to insist on reasonable prices; that Congress had virtually rubber-stamped the "exorbitant" prices agreed to by the Navy; and that the armor-makers had encouraged America's involvement in overseas wars in order to increase the demand for their products.

A search through Carnegie's confidential correspondence yields no evidence to substantiate such accusations. On the contrary, the evidence reveals that armor and munitions contracts were abnormally difficult to negotiate; that such business was not as lucrative as non-military production; that both the Navy and the Congress made arbitrary and erratic demands upon the companies; and that procurement of orders was, in fact, jeopardized by Carnegie's strong opposition to any American involvement in overseas conquests or colonization.

But the evidence is equally revealing about Schwab personally; he struggled to be patient and ingenious throughout the protracted negotiations, but he also was devious when he tried to manipulate a larger share of the market at the expense of other firms.

In the fall of 1896, almost a year before his promotion to the presidency, Schwab, then superintendent of Homestead, was placed in charge of armor sales. Thereafter, armor negotiations with the Navy occupied a disproportionate amount of his time and attention, even though the armor plant at Homestead represented only 2 per cent of Carnegie Steel's total investment and armor production comprised less than 1 per cent of the company's total tonnage output. Nevertheless, Carnegie had a $3,300,000 investment to protect in a plant with only one customer, the Navy, which was increasingly suspicious that it was being overcharged for armor. For that reason Schwab was assigned the sensitive role of negotiating sales to the Navy and of reassuring it that the prices charged by Carnegie Steel and its only competitor, the Bethlehem Iron Company, were reasonable.

In the aftermath of the 1893–94 scandal Secretary of the Navy Hilary Herbert had launched a general review of the Navy's armor procurement program. He had concluded that the two American producers were not really competitors; they presented the Navy with virtually identical prices on contract bids and subsequently divided the orders. During the summer of 1895 Herbert conferred with representatives of Carnegie Steel and Bethlehem about armor prices, telling them that the prices had to be reduced substantially, that they already had made sufficient profits to reimburse them for the initial cost of their plant and equipment, and that, consequently, the continuation of present prices would be "exorbitant." [29]

In responding to this, the companies were conciliatory. They agreed to reduce the price of armor from the prevailing $600 a ton, but Herbert considered their proposed cut—$59.54 per ton—grossly inadequate. He then sought and obtained authorization from Congress to investigate the actual cost of production and to decide what would constitute a fair profit. When he asked the companies to show him their cost data, they refused. He was told directly that "the Government had no right to pry into the secrets of [our] business affairs."

Determined to obtain the necessary information, Herbert decided to make a secret trip to England to investigate the production costs of British armor-makers, but news of his coming trip was leaked to Carnegie Steel, which then sent a representative to England—aboard the same ship as Herbert—to urge non-cooperation by the English armor-makers. According to Herbert, the ploy proved successful; he was given greatly exaggerated cost figures and heard identically phrased protestations about the unprofitability of armor production. Herbert was able to obtain data only by compiling fragmentary information from a variety of sources: the French Minister of Marine in Paris, an iron and steel broker in London, and the Navy Department's inspectors at Bethlehem's and Carnegie's armor works. Using this information, Secretary Herbert recommended that the current price of $600 a ton be cut to $400. Convinced that persuasion alone would not bring about the reduction, he recommended that the government threaten to build its own armor plant if bids were not forthcoming at the new price.

In 1895 Herbert's proposal won powerful support from three prominent Senators: Eugene Hale, Republican of Maine; former Secretary of the Navy William E. Chandler, Republican of New Hampshire; and Herbert's fellow Southern Democrat, Benjamin R. Tillman of South Carolina. But it was not enough. When support for a government armor plant was limited to a few Senators, there was no serious threat to Carnegie Steel.

Then, in 1896, Congress learned that Bethlehem was charging the United States Navy $310 more per ton than it had charged on a recent Russian order. Many Senators and Representatives were angry, and they backed a proposal that either the government set a statutory upper limit on the price of armor or that it build and operate its own plant and thereby end its dependence on private, profit-seeking companies.[30]

Both alternatives were distasteful to Carnegie, yet one or the other was a strong possibility. He therefore decided to withdraw from armor production, to propose to the Navy that it buy the Homestead armor plant. In July 1896 he told his Board of Managers, "There is no use in mincing matters. Let us press the Government to take our works, which I think will settle the agitation. If they do conclude to take them, allright. Let us get out of it." [31]

Carnegie was outraged by the report of a Navy board of inquiry which claimed that the government could build its own plant and produce armor for substantially less per ton than the price charged by Carnegie Steel and Bethlehem. But he was unable to challenge the cost estimates of the Navy board without publicly revealing his company's actual costs and profits—the former were higher and the latter lower than the Navy's estimates. Carnegie would not reveal his actual financial data; he had a long-standing policy against disclosure to any customer or potential competitor regarding his costs and profits. It was, he insisted, "impossible for us to open the details of our private business to the eyes of our competitors, and to the world."

His only recourse was to abandon a line of production which, in his opinion, "is not, and cannot be made a permanently satisfactory investment of capital. . . . We make about 150,000 tons of finished steel per month, & the two or three hundred tons of armor we make demand greater attention, give more trouble than all the 150,000 tons of steel. We shall be delighted if the Government will let us out of this armor business." He was willing to sell the plant at cost, and he believed the government was morally obliged to buy it since he had originally agreed to produce armor at the request of the Secretary of the Navy.[32]

But despite Carnegie's offer to sell, neither the Navy nor the Congress could or would make a quick decision to buy his plant or to build its own; they believed that the alternatives required further investigation. Yet for Carnegie, the longer the outcome remained in doubt, the greater the likelihood that new problems would arise.

Six months later, in December 1896, trouble came. A shipment of armor plate which had been tested and accepted by the Navy was declared to be unsuitable. The armor had been produced in exact accordance with Navy specifications, but naval officers at the Newport News proving grounds decided that the plates were not suitable for their intended purpose, and they proposed that the Carnegie Company replace them without charge. Carnegie was adamant: "we will not stand being called upon to stand any loss, either in money or in reputation in this matter." [33]

Ten days earlier, when Carnegie first learned of the Navy's proposal, he recommended that the company's best troubleshooter be assigned to deal with it. "I do think Mr. Schwab should be requested to give all of his time to this. . . . Proper management now, may save a great deal of trouble hereafter. There is one thing I strongly advise—the policy of standing up straight, at first protesting, and taking a strong stand against unreasonable requirements."[34]

Apparently Schwab resolved the dispute amicably, for no further trace of it can be found in the records. But the settlement did not resolve the larger issue—whether the Navy would buy Carnegie's plant or build its own—and from then on Schwab was embroiled in that controversy.

The idea of the government building its own armor plant and thus competing with private producers was too radical for most Congressmen to accept. Even so, there was wide support for the idea that the Navy should set a maximum price that it would pay for armor: $300 per ton—a substantial slash from the $450 to $650 range of previous years.

Schwab, as Carnegie Steel's armor salesman, had two choices: he could bid within the $300 limit or he could refuse to bid and thus encourage the Secretary of the Navy to appoint a new board to investigate the cost of a complete plant. Schwab recommended, and the Board of Managers agreed, that no bids be entered at the $300 limit. He decided to advise the Secretary of the Navy that the company simply could not supply armor at that price. He knew that Bethlehem would make the same reply; the two companies regularly consulted each other to agree on a bid price for Navy orders. Schwab hoped that his refusal to bid might encourage the government to buy Carnegie's armor plant.[35]

But when Schwab refused to bid, the Secretary of the Navy immediately appointed a board to investigate the cost of building an armor plant. The five-man board was headed by the new chief of the Bureau of Ordnance, Captain Charles O'Neil. Schwab was informed that Captain O'Neil planned to address identical inquiries to Carnegie Steel and Bethlehem: at what price would each be willing to sell its armor plant to the government?

Schwab was not encouraged by this; he believed that the Navy would not pay a fair price for the plant. An earlier Navy report had concluded that an armor plant could be built for $1,500,000, but Carnegie's plant had cost in excess of $3,000,000, including improvements. Schwab thought that a satisfactory solution was impossible. He predicted that Congress would not authorize a government-built plant or the purchase of the Carnegie or Bethlehem plant at a price agreeable to either company, nor would Congress agree to increase the price the Navy could pay for armor.[36]

Aware of Schwab's mounting frustration, Carnegie advised him to write to the Secretary of War, saying that "We are so anxious to be relieved from making Armor for the Government, that we beg to offer our

Armor Plant complete, as it stands, today, for the sum of $2 million."
This price, said Carnegie, would be $1,000,000 less than the actual cost
of the Carnegie plant, and certainly less than the cost of building a new
plant. "My idea," wrote Carnegie, "is that we should force our Works
upon the Government at a price that will appear to the Country much
below what would be required if the Government were to build." Carnegie explained why he was willing to accept an even lower price: "Indeed, I should offer the plant at one-half what it cost us and if it were
taken, we should have the supply of steel and natural gas, by arbitration,
which, I am sure, would net us handsomely." [37]

The best compromise would be to find a price per ton acceptable both
to the producers and to the government. Following months of discussions between Schwab, Secretary of the Navy John D. Long, and the
president of Bethlehem, an agreement (subject to Congressional approval) was reached. The two companies would bid an average price
of $400 a ton, with the order to be divided equally between them. [38]
Schwab was exultant; the contracts had been awarded on his terms. Best
of all, he told the Board of Managers, "we will probably have enough
Armor Plate work for the next two years at least." [39]

Schwab next turned his attention to wresting a larger share of the
armor market from Bethlehem. He was in a strong bargaining position.
If Bethlehem would agree to increase Carnegie's share of armor sales
from 50 per cent to 60 per cent, then Carnegie Steel would agree not to
encroach on one of Bethlehem's most lucrative lines of business, the
production of gun forgings.

There was a lively discussion at the next meeting of the Board of Managers. Some of the Carnegie partners wanted to enter gun forging production, but that would require an outlay of $2,000,000, which Schwab
was reluctant to recommend. He would be content to divide armor
60–40 with Bethlehem. Frick urged him to seek a 70–30 split, but
Schwab said this was unrealistic: "we have not any very good reason to
give to the Bethlehem Iron Company for asking that much. Forgings are
not any more profitable than Armor." [40] Lauder and Phipps agreed with
Schwab; it would be best to avoid a costly investment in gun forgings
and to receive a larger share of armor sales instead. Francis Lovejoy accused them all of being shortsighted. The size of the Navy would be
greatly increased in the next few years, Lovejoy claimed, and more guns
would be needed. Producing gun forgings would be a shrewd, forward-
looking decision. Schwab replied, "But if the Navy is increased, that will
mean additional Armor as well as Guns." Lovejoy answered that in a

few years there might be no market at all for armor. "It is being seriously considered whether unarmored, heavily armed, high speed vessels are not to be preferred to the ponderous battleship." [41]

This new possibility, technological obsolescence, was a relatively long-range problem. But a more immediate and serious threat had developed. Schwab and Frick had overlooked the fact that there was another manufacturer of gun forgings who would not remain passive when challenged by Carnegie Steel. The Midvale Steel Company, a small, well-managed, highly profitable Philadelphia firm which specialized in gun forgings, served notice on Schwab that it would enter armor plate production and would deliberately underbid Carnegie Steel and Bethlehem. Midvale offered Schwab a compromise: it would not produce armor plate if Carnegie Steel would not produce gun forgings. This seemed a valid quid pro quo to Schwab, so he recommended to his partners that "we make an arrangement and avoid the outlay for a gun factory." He encountered no opposition.

In return for leaving the gun forging field to Midvale and Bethlehem, Schwab expected he still could demand that Bethlehem give Carnegie Steel a larger share of the armor market.[42] He arranged a meeting with Robert P. Linderman, the president of Bethlehem, and Charles J. Harrah, the president of Midvale. He tried to obtain 65 per cent of the armor business, rather than 60 per cent. Perhaps the extra 5 per cent was intended only as leverage in bargaining. In any case, as he expected, no immediate agreement was concluded; Linderman and Harrah both said they had to consult with their partners. Schwab left them with a warning: "if no reply or no arrangement was made before the next order for gun forgings was given out, it was our intention to bid on same."

In reporting this meeting to Carnegie, Schwab was optimistic: he was "inclined to think that this will bring the matter to a crisis and something will be definitely decided soon." [43] But he misjudged the time factor. His proposed entry into the field of gun forgings would remain an empty threat until a bid on a contract was due, but no bidding was scheduled for months to come, so Bethlehem and Midvale delayed reaching an agreement with Schwab. During the interim, however, Midvale and Bethlehem were locked in battle. Midvale now insisted upon a larger share of the gun forging market in return for staying out of armor production. The only possible source from which either Midvale or Carnegie could extract such a claim was Bethlehem, since it was a producer in both fields.

Bethlehem, naturally, was not going to acquiesce in its own extinction, and it found an ally in Schwab. He decided that Midvale did not have a just claim to a larger share of gun forgings, that it was wrong for Midvale to employ the very same tactic Schwab himself was using on behalf of Carnegie Steel. He tried to mediate between Midvale and Bethlehem, but was unsuccessful. Bethlehem, to defend itself against Midvale, decided to teach that company an expensive lesson. The Navy had asked for bids on a small order for twelve special guns, and Bethlehem decided to bid a price so low that it would yield no profit, with the implicit threat that it would take similar action on all future orders until Midvale withdrew its demands and agreed to a reasonable division of the market. Midvale quickly saw the danger in its course of action and concluded an agreement with Bethlehem which preserved the status quo in the gun forging market. Schwab, however, did obtain a larger share of the armor market for Carnegie Steel in return for staying out of gun forgings.[44]

As Schwab had predicted, Congress authorized the Secretary of the Navy to pay $400 a ton for armor for four new ships—$100 more per ton than the maximum price legislated a few years before. By this authorization Congress had recognized that no one would produce armor at $300 a ton; even more, it had acknowledged that a price which was unacceptable to the producers would severely restrict the developing American Navy.[45] But Schwab and his partners believed that even $400 a ton was insufficient, and they were determined to make their case known in Washington.

Schwab directed the auditor of the Carnegie Company, Andrew Moreland, to prepare an elaborate statistical analysis, demonstrating that in the preceding year "the profits on the remaining business large and varied as it was, exceeded that of Armor Plate 123%." [46] This line of argument was perilous; statistics which showed that armor plate was 123 per cent less profitable than the rest of Carnegie's business did not necessarily prove that armor plate was not profitable. The same figures could just as easily be used to argue the opposition conclusion—that the other lines of business were earning "excessive" profits.

Moreland's analysis did reveal that armor plate had been an item of steadily decreasing profitability. The armor plant had originally cost $3,300,000 to build; the rate of profit on it had declined from over 30 per cent in 1893–94 to a mere 15 per cent in 1898.[47] Here was a statistic which Schwab might cite effectively in his campaign for higher armor prices. But in fact the figure could not be used; as Schwab told Carnegie, there was no way "to present these figures to any of the government officials without giving away the cost of Armor per ton." [48]

There is no way now to judge the rival claims that prices were either "punitively low" or "exorbitantly high"; precise information about the overhead expenses and variable costs of armor production is no longer available, and partisans on both sides used phrases such as "just price" and "fair profit" without defining them.

New bids on armor were scheduled to be made in April 1899. Schwab was unwilling to bid below the $400-a-ton limit, yet he was unable to prove that that limit was unreasonably low. Suddenly an ally emerged, and from a most unlikely quarter. Captain Charles O'Neil, the chief of the Bureau of Ordnance, began to advise Schwab about how best to present his case to the Congress.

America's need for warships had risen sharply as a result of the conquest of Cuba, the annexation of Hawaii, and the military occupation of the Philippines, and that need would increase further with the Navy's growing role in preserving free access for United States trade with China. Captain O'Neil realized that only the existing producers could meet the Navy's increased demand for armor. He also knew that neither Bethlehem nor Carnegie Steel was willing to accept reduced profits on the Navy's growing orders, and that the companies might refuse to produce armor unless they were offered a price they thought was sufficiently profitable. His personal opinion was that $545 a ton was "not out of the question," but he told Schwab that he was certain that Congress would never agree to pay more than $475 a ton. [49]

O'Neil discouraged Schwab from offering figures comparing the profitability of armor plate to other lines of business. The most effective statement Schwab could make, he thought, would be a simple declaration that Carnegie Steel was "unable to manufacture high class Armor at the price authorized by Congress." This might finally force Congress to act: either it would have to agree to raise the price limit or it would have to bear the consequences of depriving the Navy of armor. O'Neil believed that Congress might be willing to authorize a higher price without danger of establishing a precedent for the future because there was a very strong likelihood that in its next session Congress would appropriate funds for a government-owned armor plant. [50]

After he heard O'Neil, Schwab was sure that he could procure a final order for armor at $475 per ton rather than at the earlier limits of $300 or $400 per ton, provided that no new firm entered the market shared by Bethlehem and Carnegie Steel. The only serious potential entrant had been Midvale, but it had withdrawn after reaching an agreement with Bethlehem about gun forgings. Then, without warning, Schwab received word that Midvale intended to enter a bid on armor. Charles Harrah, Midvale's president, informed him that the British firm of Vickers' Sons & Maxim was about to buy Midvale and, not feeling bound by Harrah's earlier pledges to Schwab, had decided to bid on armor.[51] Two weeks passed before Schwab received a telegram from Vickers disclaiming any intention of entering into armor production in America.[52]

With Vickers-Midvale out of the competition, Schwab's confidence was restored. He then conferred with the Secretary of the Navy, and, after he did so, he was certain that his strategy would work. Instead of bidding on armor at a price limit of $400, he would submit a letter stating that Carnegie Steel was unable to enter a bid at that price.[53]

Carnegie Steel had stationed an observer in Washington to report on any developments relating to the armor question. He was a retired naval officer, Lieutenant C. A. Stone, who was paid an annual retainer for gathering and reporting information which otherwise might not be easily available to persons unconnected with the Navy Department. On May 28, 1899, Lieutenant Stone phoned Schwab to report that a "bid had been handed in, endorsed, in accordance with the advertisement—'Proposal for Armor Plate,' " but he had no way of determining who had submitted it. There were several possibilities: Vickers was reneging on its pledge, Bethlehem was trying to win the entire contract for itself instead of allowing Carnegie Steel to bid and then sharing the resulting order, or Federal Steel was entering the field. Schwab told his partners, "I think myself it is a fake bid, or from some crank inventor."[54]

The anonymous bid was not scheduled to be opened until June 2, three days later. Meanwhile, Schwab, sure that the bid would not be acceptable, prepared the draft of a letter to the Secretary of the Navy explaining why Carnegie Steel could not bid. Inasmuch as the letter was an act of defiance to Congress, which had set the price limit on armor, perhaps Schwab's letter was deliberately cast in a convoluted style:

We regret to inform you that, owing to condition, with reference to ballistic requirements, stipulated in the Circular in order to fulfill which would necessitate the Armor being manufactured under the latest Krupp system, thereby entailing a much increased cost of manufacture, we are unable to tender under the limiting price per ton set forth in the proposal." [55]

Schwab was eager not to lose the order to some other firm, so he held the letter in reserve. If it turned out that "any responsible party has or will bid," Carnegie Steel would reverse its stand and agree to enter a late bid under the $400-a-ton price limit.[56] After all, reduced profits were better than none. Schwab treated the Navy like any other customer; he tried to maximize profits on every order and was willing to cut prices only when it was necessary to meet the market.

When the anonymous bid was opened, Schwab was relieved to discover that it was from "an irresponsible party by the name of Burnstine of San Francisco, who would not promise to commence delivery until 1904. This was, of course, not considered." [57]

Schwab then handed the Secretary of the Navy his letter explaining why Carnegie Steel could not produce armor by the Krupp process within the price limit set by Congress. Two weeks later he received an order for 12,000 tons of armor at $400 a ton, to be produced by the Harvey process, which was less expensive than Krupp's. This was sufficient to keep the armor plant running at full capacity for the next eighteen months.[58]

The 12,000-ton order, however, only represented 25 per cent of the Navy's requirements. The Secretary of the Navy then received Congressional authorization to pay in excess of $400 a ton to obtain the remaining 36,000 tons, which were to be produced by the Krupp process. The increased price proved to be too tempting to Midvale; it broke its earlier pledge and sent in a bid. Even worse from Schwab's point of view was that, while Carnegie Steel had bid $490 a ton, Midvale had underbid them by $50. Schwab tried to get Midvale to withdraw its bid and he succeeded, although no one now knows precisely how he did it. Carnegie Steel was once again the sole bidder.

The final decision rested with Secretary of the Navy Long, who, Schwab believed, could "do as he pleases. He can advertise [for additional bidders,] close the bids, build an Armor Plant, or anything he likes, under the Act of Congress." [59] Secretary Long and Schwab met to negotiate a settlement. Long offered a compromise price of $450, which Schwab refused, believing that, with Midvale out of the competition, he could hold out for a higher price. Since there was a huge order at stake—

in excess of $15,000,000—Schwab consulted his partners about the final price he should agree to accept. He was determined that the issue be resolved quickly. "To close it up," he said, "means five years' work, while otherwise we run the risk of the order going elsewhere, or the establishing of an Armor Plant, or, what would be still worse, ordering a small quantity now and throw[ing] the whole subject back into Congress again." [60]

After a heated discussion the Board of Managers authorized Schwab to settle the contract at the Secretary of the Navy's price "if nothing better can be obtained." Schwab returned to Washington and bargained Secretary Long up $10 a ton—$360,000 more in profits. Now, finally, the matter seemed to be closed. Long would report the compromise to the President and then testify in support of it before the House Committee on Naval Affairs; if everything went as anticipated, a forthcoming appropriations bill would then authorize the contract.

But a new problem, a non-economic obstacle, arose. When the United States proposed to annex the Philippines, Andrew Carnegie was outspoken in his criticism. The leading Republicans in Congress were so outraged that they threatened to retaliate by blocking passage of the appropriations bill which authorized armor plate at $460 a ton. W. R. Balsinger, a Carnegie engineer of ordnance who was stationed in Washington as an observer, urged Schwab to come to Washington at once to "neutralize" the bad effect of Carnegie's criticism of President McKinley's foreign policy. "Our friends all think that you should come on here as soon as you can to see Mr. [Mark] Hanna, who can, if he will, again secure the earnest support of the Administration and its friends." [61] Schwab reported this development to Carnegie, but Carnegie refused to retract a word he had said. "Believe me," he told Schwab, "all men appreciate adherence to honest convictions. We sell our Steel, not our principles." [62]

Carnegie was right. Schwab was able to pacify the Republicans, and the armor contract was approved by the Committee on Naval Affairs without incident. Since it was an election year, the Republicans felt that it would be prudent to delay passage of the bill which authorized the final contract until after Election Day, so that the Democrats would find it difficult to make a political issue out of the armor settlement. [63]

Secretary Long had agreed originally that the government should pay the royalty to Krupp for the use of its patented process, but after the election he changed his mind. Schwab was not inclined to argue. Carnegie Steel would absorb the cost. In Schwab's view it was urgent that the contract be signed without further delay. "Congress meets next Monday," Schwab informed his partners, "and the matter must be definitely settled before that time." [64] It was. A contract was authorized for five years' work, and, as Schwab said, "The matter is now definitely closed." [65]

While Schwab was handling the thorny problems of armor negotiations and rail and structural steel sales, conflict had erupted within the hierarchy of Carnegie Steel. In late 1899 and early 1900 Frick and Carnegie had three bitter disagreements, the last of which led to a complete break between them.

For several years Carnegie had been hinting that he might be willing to retire from business if a suitable buyer for the Company could be found. In 1899 Frick was approached by a syndicate headed by William H. Moore of Chicago; Moore and his associates offered to purchase Carnegie's shares in the steel company and in the Frick coke company for a total of $157,000,000. Frick and Phipps were to act as intermediaries and to receive a joint bonus of $5,000,000 if the sale were completed. When Frick found that Carnegie was amenable to the syndicate's offer, he and Phipps personally invested $170,000 of the $1,170,000 which Carnegie had insisted upon as an option price. But the syndicate was unable to exercise its option before the expiration date because of a collapse of the money market. (The collapse had been triggered by the sudden death of the president of the largest brokerage firm on Wall Street.) Frick and Phipps regarded this monetary crisis as temporary, so they traveled to Scotland to ask Carnegie to extend the option deadline. But Carnegie refused. [66]

Frick was furious—not merely at the loss of his money, but also at Carnegie's stubbornness. Carnegie, in turn, was enraged when he learned that the two men had not told him and the Board of Managers that they personally stood to reap a $5,000,000 profit—the bonus from Judge Moore for negotiating the deal. Carnegie was convinced that Frick and Phipps had had a legal and a moral obligation to share any profits or bonuses with their other partners, and no right to conceal anything from them. He believed in the principle which is central to the concept of partnership: that one partner acts as the agent for all of the partners and cannot transact any business exclusively for his own profit. Hence, Carnegie considered both Frick and Phipps to be guilty of an unconscionable breach of trust. After their meeting, Carnegie wrote a note to himself: "Frick and Phipps. Secret bargain with Moores [the brothers, William and James] to get large sum for obtaining option. Never revealed to their partners." [67]

Frick and Phipps did something else which contributed to Carnegie's refusal to return their part of the option money, as he had originally promised he would do. Until May 10, after Carnegie had left America for Scotland, they had deliberately concealed the identity of the head of the purchasing syndicate. They well knew that Carnegie despised William H. Moore and his chief associate, John W. Gates; he regarded them as unscrupulous speculators and stock manipulators. Carnegie viewed his partners' failure to be frank with him as an act of deceit. His anger did not subside over the next five months. It increased. In writing to Schwab, he bitterly condemned "Mr. Frick's partnership with Moore by which he was to make millions—it was a betrayal of trust. He was [i.e. should have been] chairman negotiating for all his partners." And he elaborated several months later: Frick "was bound to take his partners in[to] the transaction. They had to be offered an interest, otherwise he had acted dishonestly." [68]

While Carnegie and Frick were still furious at what each considered the other's treachery, another battle erupted between them. Frick believed that he always had been fair in his coke dealings with Carnegie. Carnegie, however, did not agree, and in late October 1899 he made a remark which began this battle and, ultimately, precipitated their final break. Frick had offered to sell a tract of coke land to Carnegie Steel at a price below its estimated current market value but one which, nevertheless, would yield him a profit on his original investment. Carnegie said that Frick was making an excessive profit on the transaction. When Frick heard of the comment he became enraged, and at the next meeting of the Board of Managers he inserted an item into the minutes demanding that Carnegie offer him a personal apology as a precondition for the sale of the coke land. [69] As always, the minutes were sent to Carnegie, who was then in New York.

No apology was forthcoming. Nor would one ever be. Carnegie, as he informed George Lauder, planned "to tell Mr. Frick in kindest manner that I mean divorce under 'Incompatibility of Temper.' . . . It is divorce between us as far as management of our business is concerned. No feeling—only I believe our best business interests demand an end of quarreling." [70]

Carnegie did not demand that Frick be ousted from the company, but he was adamant that he be removed from the chairmanship. He now regretted his 1894 decision to create the position of chairman; it had led to a situation of divided authority and competing loyalties. In a letter to the Board Carnegie explained why Frick's removal was desirable:

". . . there is not the exclusive looking to the President as the agent of the Board which, in my opinion, is essential for obtaining the best results. In the recent unfortunate failure to float our company financially [i.e. the Moore syndicate venture] and indeed in most matters, the Chairman and not the President is most in evidence, giving to the public and more important to all, to our vast number of employes and officers, the injurious and false impression that not Mr. Schwab, our President, but the Chairman, whose duties are merely formal by [sic] our organization, is the real head of the company and the source of power and advancement." [71]

Carnegie had not protested Frick's increasing exercise of authority from 1894 on. He had recognized, albeit grudgingly, that Frick was a man of immense ability and shrewd judgment, one who could skillfully supervise the day-to-day activities of the company's president during the frequent intervals when he, Phipps, and Lauder were vacationing abroad. But Carnegie firmly resisted Frick's attempt to formalize the power he wielded. In 1897, soon after Schwab became president, Phipps approached Carnegie on Frick's behalf. He proposed that the chairman be made the chief executive officer and that the president be reduced to a subordinate role. Three years earlier Phipps had been appeased by Frick's appointment as chairman and by the election of Frick's protégé, Leishman, as president. But he had opposed Schwab's election. He believed that Schwab shared the same flaws he saw in Carnegie—impulsiveness and financial extravagance. Carnegie refused to raise the chairman above the president, explaining "that this would be treachery to Mr. Schwab—that it was against my judgment and never could be done as long as I held a controlling interest and had the support of my partners." [72] Schwab apparently never protested to Carnegie about Frick's increasing authority. On the contrary, as Carnegie told Lauder, "Schwab has behaved far too kindly to Frick," and he told the Board that "Mr. Schwab has shown the greatest desire throughout to sink [i.e. overlook] any personal objection—" [73]

Once Carnegie decided to remove Frick as chairman, he tried to win the support of Phipps, whose 11 per cent interest was second only to his own. He assured Phipps that he did not intend to humiliate Frick: "I shall never lay a stone in his path through life, never do him an injury—I wish to make all as easy for him as possible. Nevertheless my decision is unshakeable—There have been two outbursts. Never again a third." [74] Simultaneously, he attempted to persuade Schwab to apply pressure on Phipps. Carnegie advised Schwab:

> He [Phipps] will conform to anything but naturally wishes peaceful settlement but he said to me—if Mr. Schwab really prefers no chairman and we run danger of losing him, I'll not oppose [the ouster of Frick]—His attitude depends upon your reply *which you need not fear HCF* [Frick] *will ever know.* H.P. [Phipps] is not well and very weak [psychologically], but you have him with whatever you say *positively.*" [75]

Several courses of action were open to Schwab, all of them dangerous. If he had been an unscrupulous opportunist, he could have tried to pressure Phipps and to widen the rift between Carnegie and Frick. Or he could have remained silent and neutral, passively awaiting the final outcome—by taking no risk, seeming disloyal to no one. Or he might have pledged total obedience to Carnegie. But even this last alternative was perilous—Carnegie was a man of mercurial temperament; it was possible—though barely—that he might agree to another reconciliation with Frick. Schwab took none of these courses. He advised Carnegie to make an irrevocable decision before asking the members of the Board to seek Frick's resignation as chairman. Schwab wrote:

> Naturally the members of the Board would hesitate about taking any initial steps in this matter and if I were you, [I] would not ask them to do so until you have definitely instructed them as to your wishes. The boys are, I am sure, most loyal to you, but knowing Mr. Frick's power in the past, will hesitate to do anything against him, fearing the matter might ultimately be fixed up, and if it was would injure or end their career [sic]. [76]

But he left no doubt in Carnegie's mind where his own loyalties lay:

> Regarding myself, permit me to say, first, *I am always with you.* Aside from deep personal regard and feeling for you, you have heaped honors and riches upon me and I would indeed be an ingrate to do otherwise. My interests and best wishes will always be for you and the old firm, and when they don't want me any more I shall even then never give a thought to any other. Believe me, dear Mr. Carnegie, I am always with you and yours to command. [77]

Although Schwab and Carnegie sometimes had sharp disagreements about business matters, Schwab always felt respect and affection for Carnegie. Even years after, when he was no longer in Carnegie's employ, he still sought and cherished Carnegie's good opinion and personal friendship. He loved Carnegie and was loyal to him. Carnegie, in turn, considered Schwab to be the most brilliant of his young partners, and he took a father's pride in the achievements of his protégé, whose "subtle brain and sweet smile" he so admired and valued.[78]

But Schwab's affection for Carnegie did not render him indifferent to Frick's fate. This was perhaps the most painful episode to date in Schwab's career, for he was caught in the middle of a conflict between the two men whom he most respected and admired. Throughout his life, Schwab freely acknowledged that Frick had been consistently helpful and encouraging to him during his rise in the company hierarchy.

Seldom have two men of such diametrically opposite temperaments been able to work together so amicably. Although Schwab admired Frick and thought the feeling was reciprocated, he conceded that it was difficult to be close to him, because he was so aloof, detached, reserved, taciturn. Frick, said Schwab, was unlike anyone he ever had met: business seemed to be his only interest. He showed feeling only for his coke company and for his daughter, Helen. As Schwab recalled:

> He was to me a curious and puzzling man. No man on earth could get close to him or fathom him. He seemed more like a machine, without emotions or impulses. Absolutely cold-blooded. He had good foresight and was an excellent bargainer. He knew nothing about the technical side of steel, but he knew that with his coke supply tied up to Carnegie he was indispensable—or thought he was.[79]

Schwab made one final effort to effect a reconciliation: he knew that Carnegie was planning to demand Frick's resignation on Monday, December 4. Schwab suggested that he and Phipps meet with Carnegie in New York on Saturday, December 2. The two hoped to persuade Carnegie not to force Frick out of the chairmanship. But neither Schwab's eloquence nor Phipps's nervous anxiety could alter Carnegie's decision. They failed to negotiate a truce.

Schwab then attempted to reduce the intensity of the forthcoming battle. He sent Frick an anguished letter, urging him not to resist:

> I write you confidentially. I just returned from New York this morning. Mr. Carnegie is en route to Pittsburgh today—and will be at the offices in

106

the morning. Nothing could be done with him looking towards a reconciliation. He seems most determined. I did my best. So did Mr. Phipps. I feel certain he will give positive instructions to the Board and Stockholders as to his wishes in this matter. I have gone into the matter carefully and am advised by disinterested and good authority that, by reason of his interest [i.e. his majority ownership], he can regulate this matter to suit himself—with much trouble no doubt, but he can ultimately do so.

I believe all the Junior members of the boards and all the Junior Partners will do as he directs. Any concerted action would be ultimately useless and result in their downfall. Am satisfied that no action on my part would have any effect in the end. We must declare ourselves. Under these circumstances there is nothing left for us to do than to obey, although the situation the Board is thus placed in is most embarrassing.

Mr. Carnegie will no doubt see you in the morning and I appeal to you to sacrifice considerable [sic] if necessary to avert this crisis. I could say much more on this subject but you understand and it is unnecessary. Personally my position is most embarrassing as you well know. My long association with you and your kindly and generous treatment of me makes it very hard to act as I shall be obliged to do. But I cannot possibly see any good to you or anyone else by doing otherwise. It would probably ruin me and not help you. Of this as above stated I am well advised by one most friendly to you. I beg of you for myself and for all the Junior Partners, to avoid putting me in this awkward position, if possible and consistent.

I write you this instead of telling you because I cannot under the circumstances well discuss this subject with you at this time, and I wanted you to know before tomorrow. Please consider confidential for the present, and believe me
> As Ever
> C.M.S.[80]

To the infinite relief of all concerned, Frick tendered his resignation on December 5, 1899.

After Frick was removed from the chairmanship, Carnegie exclaimed to Schwab, "I hope you are as unclouded and delighted as I about the future." [81] But the future was not to remain unclouded. Frick still owned 6 per cent of the Carnegie Steel Company and 23 per cent of the Frick Coke Company. And Frick Coke was the major supplier of coke to Carnegie Steel.[82]

In December 1898 Frick and Carnegie had made a verbal agreement, setting the price of coke for Carnegie Steel at $1.35 a ton for the next five years. But by January 8, 1900, the price of coke on the open market had risen far above the $1.35 per ton price Carnegie had agreed to pay in 1898. The two men met that day to discuss price. Frick did not consider the 1898 agreement to be a binding contract; he wanted to raise the price of the coke his company supplied to Carnegie. Carnegie refused. To his mind, the contract, verbal though it had been, *was* binding. Even so, he proposed a compromise. Rather than following the terms of the five-year contract, Frick should continue to supply coke to Carnegie Steel at $1.35 per ton, but he should do so only for the next two years. Frick countered by offering either to buy out Carnegie's interest in the coke company or to sell his own interest in the steel company to Carnegie—and he suggested that an outside expert should be called in to determine the value of the holdings in question.

Carnegie firmly refused—he would have nothing to do with outside experts who would appraise the value of the Carnegie Steel Company. It was then that he revealed his ultimate weapon, one which would force Frick to surrender his holding in Carnegie Steel at the book value, which was several million dollars less than the actual market value—or it would have been if any Carnegie stock had been on the market. Carnegie's weapon was the Iron Clad Agreement, which had been signed by every one of the partners, including Frick. It required any partner to surrender his interest, at book value, upon the request of three-quarters of the voting partners and three-quarters of the total voting shares. This agreement had first been initiated at Phipps's suggestion in 1887 and was reaffirmed in 1892 and again in 1897 (though in that year Phipps refused to sign it, so Carnegie reverted to the agreement of 1892). The Iron Clad Agreement could be used to expel anyone—except, of course, Andrew Carnegie, since he himself controlled more than half of all the voting shares.

The meeting between Carnegie and Frick on January 8 ended in a violent argument, and they never again exchanged a word. (Both died nineteen years later, in 1919.) Schwab's office was adjacent to Frick's, and he heard the fight. Frick shouted at Carnegie, "For years I have been convinced that there is not an honest bone in your body, Now I know that you are a god damned thief. We will have a judge and jury of Allegheny County decide what you are to pay me." [83] As Schwab said later, "There were heated words. I could hear their voices raised and the air was thick. It was not a harmonious meeting." [84] This was a classic understatement: Frick was so angry that he came close to assaulting Carnegie.

The man who was responsible for obtaining the signatures of the voting partners was Francis T. F. Lovejoy, the secretary of Carnegie Steel. But Lovejoy refused; he believed that Frick was the victim of an injustice, and he would have no part in forcing Frick to give up his 6 per cent interest at less than its actual market value. He resigned instead. So, by default, the task fell to Schwab, the company's president. Schwab felt no qualms about it; he did not share Lovejoy's feelings on the matter. Beyond that, Frick himself had never protested when the Iron Clad had been used against other partners—among them, William Abbott and J. G. A. Leishman. So Schwab agreed to collect the signatures.

Frick filed suit in the Court of Equity to prevent his expulsion from the company, and he hired the same lawyer who had originally drawn up the Iron Clad Agreement to find a technicality which would now invalidate it. He was prepared to demand the dissolution of the company if he had to.

Schwab, of course, would have viewed a court-ordered dissolution as a calamity. While he did not believe it would happen, neither did he share Carnegie's confidence that Frick's suit would be dismissed. He feared that it would linger on for many months, and that because of it Carnegie would be distracted from making the major investment decisions that had to be made if the company were to remain preeminent in the industry. But when he told Carnegie of his worries, he received an almost flippant reply: "Don't overwork or what's worse *over think* yourself. Look at me. Happy as a lark and rejoicing that once again I am one of a band of trusting friends." Five days later Carnegie again reassured Schwab: "If we can't use the Iron Clad so be it. Then we have the result he [Frick] owns 6 per cent—, well, that needn't hurt anybody. Don't lose a moment's sleep in the matter." [85]

Despite Carnegie's euphoria, there was much to cause him—and Schwab—considerable anxiety. Because of the suit, the internal affairs of Carnegie Steel, especially its hitherto secret level of profits, were being discussed openly in the press. The American public learned, for example, that "the salary of President McKinley for an entire year is less by several thousands of dollars than the income which comes to Carnegie in a single day." [86] That figure and others made a mockery of any claim that the company would be financially crippled if Congress removed or reduced protective tariffs. Schwab was particularly upset by Carnegie's insistence that no dividends be declared until Frick's suit had been settled. He knew that passing dividends would demoralize the junior partners; even more, he realized that by passing them Carnegie was giving a signal: he would not authorize any new capital outlays until the suit was over. Carnegie wrote him, "I am just a little anxious about keeping strong financially, *very strong*—every dollar we can get, my boy—legal proceedings calculated to shake credit always." [87]

Because Schwab was so deeply concerned over what any prolonged moratorium on dividends and spending would do to the company, he worked hard at negotiating a settlement of the suit. Finally, he, acting as Carnegie's personal representative, and Phipps and Lovejoy, acting for Frick, held a secret meeting. There they reached an agreement: Frick was to drop his legal action, and Carnegie was to agree to the consolidation of Carnegie Steel and Frick Coke into a single new entity, the Carnegie Company, which would be capitalized at $320,000,000. Unfortunately, we know little about Schwab's role in this negotiation—no written minutes were taken at the meeting, and when Schwab reported to Carnegie he did so in person, not by letter. [88] Yet the agreement Schwab made was fulfilled.

In one sense, Carnegie Steel was dissolved, as Frick had threatened. But in its place stood the Carnegie Company, immensely stronger by the unification with Frick Coke. And Frick was out as a partner, though he was far more wealthy than he might have been. He obtained stocks and bonds in the Carnegie Company which were worth $31,000,000—$25,000,000 more than he would have received under the Iron Clad Agreement.

Schwab was also a major beneficiary of the settlement. Carnegie awarded him a 2 per cent increase in his partnership interest. Schwab, who had just passed his thirty-eighth birthday, now owned a 6 per cent interest in the Carnegie Company. It was a handsome return for twenty years of loyalty, ingenuity, and productive effort.

Schwab also gained new powers within the company and new prestige in the business community, both of which he found exhilarating. In later years he fondly recalled this period: it was the time when he was "number one," "an autocrat." Even so, within a few months after the internal conflict had ended, the Carnegie Company faced a new challenge, a powerful, external assault which threatened to topple Carnegie's empire.

6

Out from Carnegie's Shadow

The depression ended in 1897, and between 1898 and 1902 a wave of mergers and consolidations swept over the American economy.[1] In 1898 the Moore Brothers of Chicago consolidated a number of small steel companies into four larger firms and the banking house of J. P. Morgan organized three similar mergers of steel companies. A year later John W. Gates followed the trend by forming the American Steel and Wire Company, a consolidation which dominated barbed wire production.

Most of the new mergers in the steel industry involved firms which fabricated steel into finished products—tubes, hoops, wires, structural beams. There were three primary producers of steel upon whom the fabricators depended for their supply. The largest was Carnegie Steel, but the Morgan interests owned the next largest, the Federal Steel Company, and the Moore interests owned the third, the National Steel Company.

The three primary producers did not compete with one another. From 1898 to 1900, while the economy was recovering from the depression, there were enough customers for all of them. They had a Gentlemen's Agreement: they would not invade each other's territories or try to win away customers by price cutting. The market for steel was growing, and the Morgan and Moore interests did not have sufficient capacity to produce all the steel required by their fabricating companies; thus they were major purchasers of bars and billets from Carnegie Steel. Carnegie, in turn, was content to leave the production of most finished goods to others—as long as they were steady, major customers for the steel he produced.[2]

Early in 1900 the demand for fabricated steel products began to decline, and both the Moore and the Morgan consolidations had to cut prices to meet the market. As their profit margins shrank, they sought a way of reducing costs. They both reached the same conclusions—that it was necessary to reduce their dependence on Carnegie, that they must expand their own production facilities to meet their future requirements for steel.

In mid-1900, both groups advised the Carnegie Steel Company that they were expanding their own production facilities and that they would soon be reducing their future orders for steel. The survival of Carnegie's industrial empire was now at stake. With a few exceptions (mainly rails and structural shapes), Carnegie did not produce finished products; if the major fabricators did not buy most of their primary steel from him, the market for his output would be greatly reduced. Carnegie could either resign himself to losing his dominant position in the steel industry or he could seize the initiative and expand his own production to include a fabricating plant for each of the major finished products.

Carnegie was then vacationing at Skibo Castle in Scotland. Schwab sent him a series of cables, telling him that companies affiliated with Gates, Moore, and Morgan had begun to reduce their orders or had notified Schwab of their intention to do so in the near future.[3] Carnegie prepared for battle. "The situation is grave and interesting," he told Schwab. "A struggle is inevitable and it is a question of the survival of the fittest." Carnegie had no doubt that he would survive, but he preferred to avoid such a costly confrontation. Nevertheless, he knew that if he hesitated it might prove fatal. On July 11, 1900, he outlined his strategy to his Board of Managers: "I would make no dividends upon the common stock; save all surplus, and spend it for a hoop and cotton-tie mill, for wire and nail mills; for tube mills, for lines of boats upon the Lakes. . . . If you are not going to cross the stream do not enter it at all and be content to dwindle into second place." [4]

He ordered Schwab to prepare designs and cost estimates for a full line of fabricating plants. The first plan Schwab completed was that of a steel tube plant to be built at Conneaut Harbor, Ohio. The new plant was designed to eliminate the need for labor as much as possible. Schwab had recently negotiated for exclusive rights to a new process for making tubes, one which greatly reduced labor costs. Up to that time manufacturers had made a tube by bending a sheet of metal into a tubular shape and then welding it at the seam; the new method produced a tube without seams, by pushing hot metal directly through rolls. This method would save $10 a ton over the existing method—a sizable amount. Conneaut was to be the most technologically advanced fabricating plant in America; it would easily have eclipsed the older, less efficient plants of the Morgan-controlled National Tube Company.[5]

Morgan and the Moores knew that Carnegie intended to proceed with construction of the tube mill, and that this was the first battle in what would be a long and costly industrial war. They also knew that he had begun to acquire control of several railroad lines, lines which would carry his finished products from Pittsburgh to the Chicago area, where steel sales were dominated by the Morgan-controlled Federal Steel Company, and also eastward to the Atlantic seaboard.

The Moore Brothers and the House of Morgan were vulnerable to Carnegie's expansion strategy; they were overextended financially and could not bring their full capital resources to the aid of their threatened steel interests. Carnegie Steel had a well-deserved reputation for being the best organized company in the industry. It had the latest equipment, the lowest costs for raw materials, the most competent and most highly motivated managers and superintendents, and the lowest cost of production per ton in the industry. If Carnegie began producing finished goods and a price war ensued, he would be able to make a profit by selling at prices which his competitors could not match without taking losses, and they knew it.[6]

Carnegie recalled his past triumphs:

> In the former depressions we announced our policy, viz., take all the orders going and run full. Our competitors believed that we meant what we said and this no doubt operated to clear the field. One after the other dropped out; finally Pennsylvania Steel dropped out & only a few remained who could meet the lowest prices. We averaged $4 a ton profit on all our product in the worst of times.[7]

The statistics Schwab compiled for Carnegie gave clear evidence of Carnegie Steel's superiority over its competitors. In 1899 Carnegie's company had produced 75 per cent of all the steel products exported. While it sold more than 50 per cent of all the structural plate steel produced in America, it had more than 90 per cent of the export market for structural steel. The total iron reserves of the seven next largest firms in the steel industry were 99,000,000 tons; Carnegie held 162,000,000 tons. In 1900, Carnegie's five major rivals had a total steel output of

3,500,000 tons; Carnegie alone produced just under 3,000,000. And in that same year the seven major rivals' total estimated earnings were $48,000,000; Carnegie made $40,000,000. In Schwab's view, the company's dominant position in the industry was merely the base for limitless growth and profits.[8]

Once J. P. Morgan was convinced that Carnegie was serious about expanding, he knew that his own position in steel was endangered. He began to look for ways to stop Carnegie. According to John W. Gates, Morgan consulted a number of leading industrialists about devising a means to "stop Carnegie from building this railroad [network] and building this tube works."[9] In 1898 Judge Elbert H. Gary, president of Morgan's Federal Steel Company, had suggested to Morgan that he buy out Carnegie, but, as Gary later testified, he "did not receive any encouragement."[10] By 1900, however, Morgan was more amenable to the idea.

It was widely known in the American business community and in international financial circles that Carnegie might be willing to sell out if a suitable buyer could be found.[11] In his writings, Carnegie had often stated his philosophy: a man should spend the early years of his life amassing a fortune and the later years distributing it in charitable donations. In 1900, at age sixty-five, Carnegie's later years were at hand. He was willing for someone to buy Carnegie Steel. A suitable buyer ultimately was found—J. P. Morgan. Schwab played a central role in consummating the sale.

On the night of December 12, 1900, a dinner was held in New York, at the University Club. It was given in Schwab's honor by J. Edward Simmons, president of the Fourth National Bank of New York, who was returning the hospitality shown him when he had last visited Pittsburgh. Although Schwab was president of Carnegie Steel, he was not well known outside of the steel industry, so Simmons invited a group of his own close friends to meet Schwab. Eighty men attended the dinner. The guest list included many of the leading New York bankers and businessmen, of whom J. P. Morgan was the most prominent and influential, as well as four top officials of Carnegie Steel, including Carnegie himself. Also among the guests were E. H. Harriman, the railroad magnate; investment bankers August Belmont and Jacob H. Schiff; and the president of Standard Oil, H. H. Rogers.

It has often been said that Carnegie deliberately engineered this dinner to bring together the leading potential buyers for his business, and that the sales presentation was made by Schwab in his after-dinner remarks as guest of honor. There is no proof whatever that this was Carnegie's purpose. If it had been, he surely would not have left the dinner after being there for just a few minutes—he went to speak before a meeting of the Pennsylvania Society. Nor would he have included among the guests the Right Reverend Henry C. Potter, Bishop of New York, and the Right Reverend George Worthington, Bishop of Nebraska. And he never would have permitted one of his executives to telegraph details of the dinner and its full guest list to the *Pittsburgh Dispatch* for publication.[12]

After the dinner, Schwab, as the guest of honor, was called upon to speak for a few minutes. He had no prepared text; he began with the same opening he used in all his speeches. He said that he would talk about steel, because he could not talk about anything else.[13]

Schwab told his audience of his hopes for a greater and more profitable steel industry. The Carnegie Company, he said, had managed to reduce its costs of production to the lowest point conceivable; no more major economies of production could be attained. But substantial savings could be achieved elsewhere—in distribution.

Imagine, said Schwab, a huge firm with many plants, each specializing in a single product. If these plants were located in the same areas in which their products were sold, delivery charges to the customers could be reduced. Such a firm also could unify the separate sales forces which rival firms now maintained. And it could eliminate "crosshauling"—transporting products away from an area while similar or identical goods were being transported into it. Then too, the various plants of such a giant firm could compare their costs of production and devise ways of bringing performance of the laggards up to the standard of the pacesetters. Such comparisons would also allow the firm to identify its ablest managers, superintendents, and foremen; then the best of them could be advanced to positions of wider responsibility.

No such firm now existed, he said, but, conceivably, one could. And here Schwab was not offering a concrete proposal for consolidation; he was presenting a vision of what might someday exist. Manufacturers, he said, now thought in terms of markets in which they sometimes limited production, increased their prices, and thus made profits. But Schwab saw a new possibility.

If the steel industry were made as efficient as possible, if its plants were specialized, integrated, and centrally managed, and if its leaders were willing to cooperate for long-range mutual growth, then an ever-widening market for steel could be created. New uses for steel could be devised, new and improved methods of production could be discovered, and record profits could be attained. Both the producers and the consumers would benefit: producers would make greater profits on their products, while consumers would pay less for them, all because of increased efficiency and specialization.

Schwab's audience listened with rapt attention. His voice was mellifluous, his gestures were dramatic, and his message had immediate appeal. The businessmen and bankers responded favorably to the vision of soaring profits as well as to the possibility of avoiding the ruinous industrial war which seemed to be on the horizon. Schwab had intended to speak for only fifteen or twenty minutes, because, as he later explained, his listeners were old men who would be eager to get home. In fact, he spoke for forty-five minutes. Afterward, Morgan called Schwab aside for a half-hour conversation.[14]

The dinner had been a complete success, and Schwab had made an excellent impression upon the leaders of the New York business community. He took the next train back to Pittsburgh to carry on business as usual. To him, the dinner had been a pleasant interlude, a brief respite from the challenging job of planning Carnegie's new fabricating plants.

Three weeks later Morgan told John W. Gates that he was as eager as ever to dissuade or prevent Carnegie from building the new tube works at Conneaut Harbor. He told Gates that he planned to consult Henry Frick about the problem, but Gates said that this would be a calamitous mistake in view of Carnegie's recent feud with Frick. Instead, Gates advised Morgan to speak to Schwab. Gates later testified that he had said to Morgan, "there was only one man to talk to that had any influence with Carnegie, and that was Charlie Schwab."[15]

Gates himself spoke to Schwab; he suggested that Morgan and Schwab should meet. But Schwab was reluctant to agree to any prearrangement. It might in some way seem disloyal to Carnegie. Gates then said that Schwab might arrange to be at the Bellevue Hotel in Philadelphia on a certain night, and that perhaps by some happy accident he might run into Mr. Morgan there. Schwab answered that his schedule might very well permit him to do so.

Schwab went to Philadelphia, but when he arrived Morgan was not there—he was at home in bed in New York with a headcold. While Schwab was waiting for Morgan at the Bellevue, he received a telephone message from Gates, advising him that Morgan would not be able to show up and asking him to come to New York for dinner with the banking titan a few nights later. Schwab agreed; somehow his worries about appearing disloyal to Carnegie had receded. Morgan and Schwab had a private dinner, after which they returned to Morgan's mansion, where they were met by two of Morgan's partners, Robert Bacon and George W. Perkins.

Morgan told Schwab that his goal was to prevent Carnegie from expanding into fabricated products, not to buy him out. Schwab advised Morgan that unless Carnegie retired he *would* expand; Carnegie was a fighter, determined to meet and beat down all threats to his industrial supremacy. Morgan quickly became convinced that the impending industrial war could not be averted by a temporary truce; he had to make a permanent alliance. He and his partners explored with Schwab the possibility of forming a giant steel consolidation, one such as Schwab had projected in his University Club speech. Schwab once again enumerated the many advantages which would result from such a union. When the meeting ended at 3 a.m., Morgan asked Schwab to compile a list of the companies he thought should be included in a proposed merger.[16]

In making this request, Morgan did not commit himself to the venture. He knew very well that Carnegie was deeply suspicious of "Wall Street operators," of whom Morgan was certainly the most famous and in Carnegie's eyes the most notorious. Furthermore, even if Carnegie were amenable to selling out and retiring, there was no certainty that his asking price would be reasonable. And even if it were, Morgan, either alone or in syndicate, might not be able to raise the required sum. Finally, if the venture were to have the ideal characteristics described by Schwab in his speech, other companies would have to be included, and there was no certainty that their owners' terms would be reasonable.

Schwab spent the next few days preparing a memorandum for Morgan, listing the companies which should be included and the prices which ought to be paid for them. Years later, Schwab described how he had compiled the list:

> I knew exactly what each one was worth. Nobody in the world helped me with that list. I didn't use the ordinary book value, but based my estimates on earning capacity, good will, the physical state of the properties, and their potentialities as an investment. I left out many companies including Bethlehem, because they would have provided nothing but duplication, and the ideal corporation would have no duplication of any sort in it.[17]

When Schwab returned to New York with the memo, he conferred with Morgan and Gates, again until three in the morning. After carefully reviewing Schwab's list, Morgan said, "Well, if you can get a price from Carnegie, I don't know but what I'll undertake it." [18]

Schwab then faced an even more formidable problem: having found a buyer for Carnegie Steel, he had to convince Carnegie to sell. Louise Carnegie was eager for her husband to retire, so Schwab immediately enlisted her support. She suggested to Schwab that he should join Carnegie for eighteen holes of golf at St. Andrew's in Yonkers, New York.

Schwab and Carnegie spent several pleasant hours at golf, and by the time the game was over Carnegie was receptive to the idea of selling his business. Over a quiet dinner in Carnegie's cottage, the two men discussed Morgan's proposal. Schwab reminded Carnegie that if he did sell he would be free to undertake his lifelong ambition—organizing philanthropic activities for the advancement of knowledge and the promotion of world peace. By the end of the evening Carnegie had agreed to sell if Morgan could meet his price: $400,000,000.[19] Morgan accepted Carnegie's terms.

But Carnegie soon began to question his decision, and he expressed his misgivings to Schwab. Schwab told Carnegie that, although no formal contract had yet been signed, he could not break his promise to Morgan. Carnegie yielded, but he never was fully happy about it. He believed that giving away his money was his duty, but *making* money was his passion. However, once the contract was signed, Carnegie finally accepted the situation. It was irreversible. He told Schwab, "Of course there are regrets—must be—changes are all devilish—but must come. We ought to have perpetual youth—an option not to leave this Heaven below until we wish to." [20]

So the Carnegie empire ended—and the United States Steel Corporation was born.

After Carnegie retired, Schwab no longer felt it necessary to conceal his personal sources of pleasure. He had been doing so for years because of Carnegie's puritanical code of morality. But now, for the first time, he was free to live his private life in any manner he chose—or so he thought.

All of Carnegie's young partners had paid a high price for their success. Carnegie, quite deliberately, kept tight reins on them in an effort to insulate them from the triple perils of "worldly sin"—wine, wagers, and women. The young men knew that their status as partners depended not only upon their work performance, but also upon their avoidance of personal scandal. They could neither afford nor dare to be errant. Although Schwab had *not* conformed to Carnegie's moral standards in every respect, he had succeeded in concealing his wayward behavior.

Schwab had had to hide such aspects of his private life from Carnegie for the sake of his job, which explains why it is so difficult to find any contemporary sources dealing with his life-style from 1886, when he became superintendent of Homestead, to 1901, when Carnegie retired. And even when Schwab reminisced about his career in 1935, he barely touched upon the personal aspects of his life during that period. What little that is known comes chiefly from the recollections of his surviving relatives.[21]

Most of Schwab's "vices" were perfectly innocent. He never smoked, although he always carried cigars. Occasionally he would find a workman smoking in the steel mills, in violation of the company's safety rules; rather than firing him, or threatening to, he would give him an expensive cigar "to be smoked later, outside, when you are off duty." Nor was he a heavy drinker; Carnegie sometimes consumed large amounts of "old scotch" to aid his circulation and digestion, but Schwab drank only an occasional glass of brandy or scotch. Insofar as he did have a favorite drink, it was champagne—but only after it had effervesced. He was neither a gourmet nor a glutton; he was merely a hearty eater, with a preference for starchy and rich foods—the kind his mother cooked for him when he made his annual visit to Loretto each year on her birthday. Rana prepared the same kinds of foods, which accounts for his increasing portliness. He was not a flashy or foppish dresser; on the contrary, he was teased about how meager and ordinary his wardrobe was. After Schwab became president of Carnegie Steel, William E. Corey came to visit him. There was one man on the household staff whom Corey did not recognize, and Corey asked who he was. When he was told that the man was Schwab's valet, he exploded in laughter. "What for? He only has one suit."

During his years as superintendent at Braddock and Homestead, Schwab's favorite pastime was a Saturday night poker game. It was strictly private; Carnegie frowned on all forms of gambling, even for small stakes. Corey was one of the regular players; another was Johnny Harvey, a chemist at Homestead, whose love for poker was matched by his chronic inability to win. One evening Harvey had lost even more than usual, far more than he could afford. Christmas was approaching, so Schwab offered him a far-fetched wager in an effort to help him recoup his losses. He bet $100 against Harvey's trousers that he could correctly guess the top card which Harvey drew from a deck. The deck was neither marked nor prearranged, and Schwab took a wild guess: "It will be the nine of diamonds—the curse of Scotland." By an incredible coincidence, when Harvey turned over the card, it *was* the nine of diamonds. Schwab lent him a long overcoat to conceal his missing trousers.[22] It was a playful episode, one which even Carnegie might have found humorous, but he would have shown no mercy if he had discovered that Schwab, while working for Carnegie Steel, had played roulette in Monte Carlo every year since his first trip to Europe in 1886.

When Schwab accepted the presidency of Carnegie Steel in 1897, he moved from a modest house in Homestead to a small mansion in Pittsburgh. Its spaciousness reflected his new social status and enabled him to entertain in a style befitting the president of America's most prosperous steel firm.

Rana Schwab would have been perfectly content to be the wife of an ambitious workman, to keep his modest house and to bear his children. Instead she had married a man whose inexhaustible energy and ability had catapulted him to the presidency of the company at the age of thirty-five. His position put an end to the simple life she preferred; ahead lay a succession of mansions and private palaces, each staffed with a corps of servants ready to cater to every extravagant whim. But she had no such whims—only very modest desires. Tragically, her greatest ambition was unattainable: she was unable to have children. Schwab's most grandiose dreams were fulfilled, while his wife's, so much more conventional and seemingly achievable, were not.

Schwab was an unabashed hero-worshipper, and in his new position he met many celebrities. He came to include among his friends Commodore George Dewey, Mark Twain, O. Henry, and Thomas Alva Edison. Rana felt she had little in common with his new friends, and nothing at all in common with some of his business associates, whose notorious exploits he found so fascinating. Among those associates were James Buchanan Brady and John W. Gates. Brady, better known as "Diamond Jim," had seduced the famous Dolly sisters and had kept them as his mistresses simultaneously, after paying off their enraged husbands. Gates, whose nickname was "Bet-a-Million," would bet for high stakes on anything, even on which raindrop would descend first on a window pane.[23]

Although Rana had originally shared her husband's love of music and travel during the early years of their marriage, her interest grew sated, as his never did. She was increasingly incapable of keeping pace with him, physically or socially. Her love of starchy foods, coupled with a debilitating case of gout, caused her weight to climb to well over two hundred pounds. She became so self-conscious about her appearance that instead of accompanying Schwab when he traveled on business, she would stay at home. There she endured periods of intense loneliness.

Schwab was concerned about Rana's loneliness, and when she suggested that they raise her nephew, Carlton Wagner, whose parents had died at an early age, he agreed. It worked out well: Rana had a child to care for; Carlton was well taken care of; and Schwab developed a strong affection for the boy.

Charles's and Rana's relatives often came to the spacious Schwab house in Pittsburgh, sometimes to visit, other times to convalesce. One such visit affected many lives. Rana's youngest sister, Minnie, had contracted typhoid, and she was brought to the house to recover. Schwab summoned a friend of his, Dr. Marshall Ward, to attend his ailing sister-in-law. The Schwabs worked at match-making, and when Minnie was well again she married Dr. Ward, even though he was twenty-five years older than she.

But earlier, while Minnie was recuperating and Dr. Ward was attending his other patients, he sent a beautiful red-headed nurse to take care of her. Schwab had a clandestine affair with the nurse. Such liaisons usually leave no trace, but this one did—a baby girl. Schwab was in an agonizing position: if he legally recognized the girl, the resulting scandal might cost him his marriage; certainly it would cost him his job.

Schwab knew that Carnegie would not tolerate sexual promiscuity or marital infidelity in any of his partners. He knew that Carnegie's low opinion of J. P. Morgan was due far more to Morgan's reputed libertinism than to his activities as a stock market speculator. Carnegie could not fault Morgan for no longer being sexually attracted to his wife, but he was appalled by the rumors that Morgan kept a steady succession of mistresses, as many as seven at one time, and he was revolted by the rumor that Morgan had made a gift of land, buildings, and funds for the New York Lying-In Hospital in order to have some place to accommodate the women whom he was alleged to have made pregnant. To Carnegie's mind, these rumors far outweighed the well-known facts that Morgan was an active layman in the Episcopal Church and a patron of the arts.

In later years Schwab talked about the two men:

> The differences [between them] were marked in many ways, but in no way so strongly as in the matter of personal morals. Morgan had his "moments." He had not consorted with his wife for a long time, and he had his women. Carnegie was always faithful and could not understand Morgan's personal "freedom." Carnegie frowned on anything savoring of the flesh and the devil. He was very narrow in some respects, and he had no forgiveness for human weaknesses—because he couldn't understand them.[24]

Schwab managed to conceal his "moment" from Carnegie. He provided for his daughter's future education, and he made sure she had good medical care and physical therapy—she had been born with a deformed leg which later produced a slight limp. Although the affair between Schwab and the nurse did not continue, he managed to see his daughter once or twice a year, often taking her on a cruise with him to Europe, where he hoped the spa baths would help her condition. When the girl's mother died, Schwab increased his financial support: his daughter moved into a fine hotel in Washington, D.C., where she had both a chauffeur and a companion.[25] What she never could obtain, however, was what she undoubtedly would have valued most: the right to emerge from the shadows, to live with and be openly loved by her father, and to bear his name with pride.

Rana learned of Schwab's romantic involvement with the nurse and of the subsequent birth of his daughter, but the discovery did not end their marriage. As president of Carnegie Steel, Schwab could not consider a divorce. Nor did Rana want one; she had built her life around him. Schwab assured her that he had never been unfaithful before and promised that he would never be again. Their reconciliation enabled him to turn his full talents and energies to the new challenges facing him.

7

U.S. Steel: Schwab's Rise and Fall

On April 16, 1901, Schwab resigned from the presidency of the Carnegie Company. He had been named president of the United States Steel Corporation. At thirty-nine, he headed the first billion-dollar enterprise, a corporation which controlled nearly 50 per cent of America's steel-making capacity.

U.S. Steel was not an operating company, but a holding company. It controlled 213 steel mills and transportation companies, including 78 blast furnaces; 41 iron ore mines and a fleet of 112 ore barges; as well as 57,000 acres of coal and coke properties in the Connellsville region of Pennsylvania, with nearly 1000 miles of railroad tracks to service the region.[1] The new corporation soon acquired the label of "The Steel Trust," a pejorative which clung to it so tenaciously that even its friends soon referred to it by that name.

When the U.S. Steel merger was announced, press reaction was almost uniformly unfavorable.[2] Most newspapers in America and Europe viewed U.S. Steel with alarm, predicting that it would first crush all competition at home and then seek to dominate the world market. It was seen as a monstrous conspiracy, injurious to everyone except the clique of promoters who would profit from the sale of its watered stock and the coterie of insiders who would fleece a defenseless nation by charging exorbitant prices.[3]

Though Schwab had been little known outside of the business world, as the president of U.S. Steel he became a national celebrity. His new post brought him invitations to write for popular magazines, to speak to students and to major business groups, and to testify before Congressional investigating committees. His name became a symbol; he was either the predator who headed the Steel Trust or the self-made man who had worked his way up from the blast furnaces. His photograph appeared on many magazine covers and in hundreds of newspaper and magazine stories. Young, vigorous, and photogenic, Schwab did nothing to discourage the publicity.

One would think that the presidency of U.S. Steel would have provided Schwab with a magnificent opportunity to implement the ideas he had expressed in the University Club speech. In fact, however, his next four years were largely barren and unproductive. He did not adjust well to his sudden rise to national prominence. He fought with his colleagues, made ill-considered speeches, allowed wild rumors to go uncorrected, and generally comported himself in a way unlikely to inspire confidence in his maturity and sense of responsibility. His public image suffered when the wide disparity between his words and his actions became evident.

Schwab visited Homestead and Braddock before leaving for his new offices in New York; and he was accompanied by a newspaperman who reported his numerous acts of generosity toward individual workmen. Later that same day, on the train trip to New York, a reporter asked him what advice he would give to a young man who wanted to be successful in business. He answered with a pious homily:

> Work hard; make yourself indispensable; don't be afraid of working over time; persist; don't mind the hard knocks, you're only getting your share; be loyal to your employer, to every one; marry early, as early as you can; I favor the married man always; everything else being equal, I give him the preference; avoid drink; lead a moral life. The young man who does these things will succeed.

He was also asked for the secret of how he managed to get along so well with other men. He said:

> I treat men as I want to be treated myself. I never quarrel. If any difference arises I go to a man frankly and reason it out with him. If we disagree, we disagree amicably. Then I encourage men constantly. Every man, no matter how high his position, is susceptible to encouragement. That's the only secret.[4]

During the first six months of his presidency, Schwab traveled extensively. One of his chief duties was to visit the older and smaller steel plants in the new corporation and to recommend which of them should be closed down. In May 1901, Schwab made such an inspection trip through the Mahoning, Shenango, and Ohio valleys; it was quickly labeled a "dooming tour." [5] In fact, the newspaper accounts grossly exaggerated his powers. Some of the newspapers charged that if Schwab decided that a particular steel mill was obsolete and ought to be shut down, then he, singlehandedly, would cause unemployment, destroy flourishing towns, and ruin the livelihood of farmers and middlemen who supplied those towns.[6] But that was not so. Schwab alone could not decide to close down any plant. All of his decisions and appointments were subject to review and revocation. In fact, his responsibilities as president were clearly defined in U.S. Steel's bylaws:

> Subject to the executive committee, he shall have general charge of the business of the company, including manufacturing, mining, and transportation. . . . He shall do and perform such other duties as from time to time may be assigned to him by the board of directors." [7] (Italics added.)

Schwab was not satisfied with his position in the corporate hierarchy. He expected that he would have undivided authority—that he could be "an autocrat." He believed that his experience as a "practical steelman" should have given him preeminence over men whose training or background had been in law or finance.[8] Yet the U.S. Steel Corporation's structure was deliberately arranged so that no one individual could have undisputed authority. It was to be a business whose policies were made by committees of experts in the areas of finance and operations. These committees in turn were subordinate to the Board of Directors.[9] Ultimate power rested with the board, whose twenty-four members, meeting monthly, were responsible for electing the officers of the corporation, the members of the Executive and Finance committees, and the officers of the subsidiary companies within the corporation. U.S. Steel owned the stock of its subsidiary companies and derived its income solely from the dividends paid on the earnings of the subsidiaries. The corporation's Board of Directors chose the subsidiary officers who set the policies regarding production and sales. Since these officers could be removed, the board actually, although indirectly, controlled the policies of the member firms. Schwab, even as president, did not control the corporation.

While final authority rested with the Board of Directors, the actual policy-making was conducted by the members of the Finance and Executive committees. Robert Bacon was the first head of the Finance Committee. It had six members, including Schwab and the Chairman of the Executive Committee, Elbert H. Gary. Its jurisdiction over all matters of finance included final authority over any new acquisitions.

The Executive Committee had eight members, among them Schwab and the Chairman of the Finance Committee, Bacon. It was responsible for managing and directing the manufacturing, mining, and transportation operations of the corporation. When the committee was not in session, its authority could be exercised by its chairman, subject to the committee's review.

From the very outset, Schwab came into conflict with the Board of Directors and both committees.

Few members of the Board of Directors had ever had any connection with steel; most of them were outsiders who lent the prestige of their names to the controversial new enterprise. But they were not passive or absentee directors, and Schwab resented suggestions from any of them about how the corporation should be run. Marshall Field, the Chicago merchant, came to Schwab's office one day to insist that a proposed new plant be built in Chicago rather than in Pittsburgh, even though the plant's product was to be sold primarily in the eastern states. Schwab rejected Field's suggestion. He even balked at a proposal made by an experienced steelman, Percival Roberts, Jr., that U.S. Steel buy a certain property of which Roberts was then part owner. When Schwab refused, Roberts exclaimed, "And what is your interest in the matter, Mr. Schwab?" Schwab shouted, "None whatever Mr. Roberts, and if you do not resign from this Board at once, I will." [10] Roberts resigned from the board, at least temporarily. Yet in other conflicts Schwab was not as successful.

Schwab considered it essential that U.S. Steel immediately increase its holdings of ore lands, but he encountered opposition within the Finance Committee. As he later testified:

> In the early days of the Steel Corporation our Finance Committee was very reluctant to spend money for things of this sort [acquiring ore properties], or improvement of plants even. The members of our committee had not been educated to know the manufacturing business as I knew it . . . and it was with the greatest difficulty that I could get them to buy what I considered most valuable and essential to the company. . . .[11]

Schwab also came into conflict with Judge Gary. Born in Wheaton, Illinois, in 1846, Gary was graduated from law school in 1868. He was in private practice until 1882, when he was appointed county judge, a post he held for eight years. His first contact with the steel industry came in 1892, when he assisted John W. Gates with the legal problems involved in forming the Consolidated Steel and Wire Company of Illinois. Gary won considerable renown within the industry because of his skill in handling the legal intricacies of merging many small companies. In 1898 he served in a similar capacity and with equal ability in the formation of the Federal Steel Company, a consolidation undertaken by the House of Morgan. As a reward, he was chosen to be the first president of Federal Steel.[12]

Morgan respected Gary's judgment; he considered him a man of unimpeachable honesty, great intellect, and sound, conservative leadership. If Morgan had been looking for a model of respectability for U.S. Steel, he could have made no better choice. Gary neither smoked, nor drank, nor kept late hours, nor had disreputable friends; his reputation was untainted by any accusation of malfeasance or impropriety. In dress, in tone of voice, in choice of words, even in physical appearance, Gary more closely resembled an Episcopal bishop than a Gilded Age businessman.

Gary and Schwab could scarcely have been more dissimilar. Schwab was dashing, ambitious and self-assertive, fond of raucous horseplay and off-color stories, with a passion for personal publicity which Gary found increasingly abrasive. In Gary's view, Schwab was also tainted by his former association with Carnegie, whom Gary heartily disliked. Gary deplored Carnegie's policy of aggressive price-cutting, and he considered Schwab a co-conspirator in this "offense" against business stability and harmony.

Gary strongly opposed some of the policies which Schwab believed had been responsible for Carnegie Steel's leadership in the industry. As Schwab later said, "Judge Gary, who had no real knowledge of the steel business, forever opposed me on some of the methods and principles that I had seen worked out with Carnegie—methods that had made the Carnegie Company the most successful in the world." Gary did not believe in the bonus or "reward" system for workmen, nor in the policy of granting partnerships to bright young managers and superintendents.[13] And he could not see the wisdom or necessity of the old Carnegie-Schwab policy of expanding plant capacity whenever funds were available. He thought that the industry's capacity ought to be stabilized before any new expansion took place. He was more interested in having

Schwab inspect and close down small, old, and inefficient plants than he was in having him build new ones.

Schwab was responsible to the Executive Committee—and its chairman was Judge Gary. However, because of his desire to be "number one" and because he felt Gary was not knowledgeable about practical matters, Schwab tried to act without the authorization of the committee. Apparently he did this on several occasions, for on July 1, 1901, at a meeting of the Executive Committee, Gary introduced a formal resolution, stating that the president shall "furnish this Committee with full reports of the operations of the Company, and submit for their information and decision all matters requiring their supervision as set forth in the by-laws." [14] The committee approved the resolution. It was a clear indication of their intent to stop Schwab's insubordination.

The success of Gary's motion left Schwab dejected and embittered. He told Perkins two days later:

> Don't ask me to drive uptown with you this evening. My nerves are not in shape to enjoy the drive. I am simply heartbroken. I was so anxious to show that this company could be made a glorious success. And I am more enthusiastic to-day and surer of its success than I ever was before. I have suffered every torture on Mr. Morgan's account to make matters move smoothly until his return. I have been hampered, criticized and goaded by incompetent critics, who do not understand the whole steel situation, and who could only criticize and offer no suggestions. . . . Human nature could not bear the great load and suffer the indignities I have had to suffer. My heart and my whole interest is with this company and always will be and I feel most deeply for that great man, Mr. Morgan, who must find this condition upon his return. I'll do anything he may ask me no matter what, except that I will not continue as President of this company under any condition.[15]

Even so, Schwab did not try to resign; perhaps he expected Morgan to side with him against Gary.

Schwab accomplished very little as president of U.S. Steel and he made most of his innovations during his first months of office. He did create an integrated foreign sales department. To head this division he named James A. Farrell, a man who once had worked with him in Braddock and who one day would succeed him as president of U.S. Steel. Schwab also formulated corporate policy on freight shipments by boat and rail, and he directed an extensive search for new sources of ore. Perhaps his most important new policy was to have each plant submit weekly cost sheets which could be used for comparative purposes. This stimulated a sharp rivalry between the various superintendents, and the ablest of them were recognized and rewarded with bonuses for any cost-saving improvements they had introduced.[16]

Gary's move to restrict Schwab's powers in July 1901 was made immeasurably easier by the harvest of bad publicity which Schwab had reaped for himself and for U.S. Steel during the preceding months.

Schwab had made his debut as a public speaker on May 8, 1901, in an address to the graduating class of St. George's Evening Trade School in New York City. His subject was one he knew well: success. The content was completely uncontroversial: "success is not money alone"; "start early"; "exceed your duties"; "effort, not backing, is the key to promotion"; "go in and win on your own merits." [17]

But he provoked a storm of opposition when he said that a college education was not essential for boys who planned a career in business, that they would profit more by the additional four years of work experience. That "philistinic" remark was especially poorly timed, coming as it did just before the June graduation ceremonies. It served as the text for countless commencement addresses. From coast to coast, academicians and clerics told graduating students that Schwab's career and philosophy were an affront to the higher sensibilities of man, that many of the blessings in life could be obtained without cost and others money could not buy. Schwab had said that, of forty business leaders he knew, only two had been graduated from college; his critics said that this fact proved more than ever the desirability of going to college.[18]

William Jennings Bryan claimed that "Mr. Schwab's advice will do infinite damage," that he was the spokesman for blatant commercialism and plutocracy. A Lutheran publication cited Schwab's speech as evidence of America's decadence: a century earlier the nation's heroes had been Washington, Hamilton, and Jefferson, and at mid-century Emerson, Lowell, and Longfellow had held the place of honor. "Today are we to point only to these masters of money? Shall we have no great men but Mr. Carnegie and Mr. Schwab and Mr. Rockefeller?" America was facing the danger of a "delirium of material drunkenness." The editor warned his readers that if young men took Schwab as a model for their own lives, it would lead them from the high road of virtue and unrewarded sacrifice to the base pursuit of wealth and power.[19]

A few days later, on May 11, 1901, Schwab testified before the Industrial Commission, a governmental body established by Congress in 1898 to investigate problems of business and labor. He vigorously defended the protective tariff on steel, claiming that it was not only the primary cause of American prosperity, but also the safeguard of the high wages received by workers in the American steel industry. Schwab did not believe a tariff was necessary for all steel products; it was crucial only for those items of a highly finished quality whose production required a great amount of labor.[20]

Schwab was asked for his evaluation of a bill, then pending in Congress, which had been introduced by Representative Joseph W. Babcock, a Republican from Wisconsin. The bill, designed to encourage foreign competition, proposed to eliminate tariff protection on all items produced by the Steel Trust. "I do not see that that would do anybody any good," said Schwab. "It would not hurt anybody in those lines where we did not need a tariff, and the only persons it would hurt in those lines where we do are the working people." The anti-tariff newspapers seized upon this statement. They cited it as proof of an "impudent despotism" and attacked it as a thinly veiled blackmail threat: if the tariff were either removed or lowered, labor alone would suffer loss, in the form of reduced wages.[21]

Schwab offered another argument which was even more inflammatory in its effect, though that was certainly not his intention. A previous witness had stated that American steel producers sold their products overseas for considerably less than they did in the American market. Schwab was asked to confirm that testimony. He admitted that it was "quite true," but he offered an explanation: ". . . export prices are made at a very much lower rate than those here; but there is no one who has been a manufacturer for any length of time who will not tell you that the reason he sold, even at a loss, was to run his works full and steady." In other words, American producers, by selling abroad at very low prices, guaranteed that they could operate regularly at full capacity, which was essential to defray the heavy fixed costs of plant and equipment. In times of great demand at home there would be little surplus for the foreign market, but when the American market was depressed the steel producers, rather than having to curtail production or shut down, could sell their goods abroad. Since foreign governments erected tariff barriers against American steel, the selling price had to be the lowest possible, so that when the European tariff was added to the American price, the product could still be competitive.[22]

One anti-tariff newspaper called Schwab's explanation sophistry and "illogical gall." The paper said that the tariff allowed American steel producers to overcharge domestic customers ruthlessly. It claimed, without evidence, that domestic steel prices would fall to the *same* level as export prices if the protective tariff were eliminated, and that if domestic prices fell, labor would not have to suffer: lower prices would increase the demand for steel and thereby drive up wages.[23]

Schwab's testimony on the tariff certainly lacked the virtue of consistency. If, as he stated, export sales were invariably set at so low a price that they often entailed a loss, then why was he so eager to see foreign sales increased? It sounded like the old joke about losing a little bit on each sale but making it up on volume. Beyond that, as *The New York Times* claimed, the fact that U.S. Steel "is compelled to reduce its prices in foreign markets is conclusive proof that they would be good markets for our people to buy in. Our people should have the chance." [24] Another inference drawn from Schwab's testimony was that American customers were being forced to subsidize foreign purchases of American steel. According to the anti-tariff newspapers, European customers could purchase American steel below cost only because the tariff allowed the companies to charge such high prices in America that the inflated domestic profits offset the overseas losses.[25]

Schwab was not a skillful exponent of his position. He neither anticipated nor attempted to refute the conclusions which the free traders were certain to draw from his testimony. He damaged his own cause. This was not an isolated instance; he did the same thing in his testimony on labor, wages, and unionism.

Schwab said of the unions:

> Under the labor-union system all members are reduced to a dead level of equality, and the wage scale largely is determined by the worth and capability of the cheapest workman, instead of the most capable and highest priced. This narrows opportunity, dulls ambition and gives no man a chance to rise.[26]

Samuel Gompers, president of the A. F. of L., interpreted that statement as a declaration of war, and he fired off a public denunciation in reply. Unions did not seek to reduce all workers to a common level, Gompers declared. Rather, their objective was to set a minimum or lower limit on wages below which no employee could be compelled to work. Erecting such a base did not prevent anyone from rising above it. Furthermore, Gompers claimed, unions were essential for workers; otherwise they could not get redress of their grievances.[27]

While many newspapers and magazines were critical of Schwab's attack on unions,[28] the Hearst papers were outraged. *The New York Journal* devoted a full page to condemning Schwab; it was illustrated with a cartoon portraying him dangling from puppet's strings, mouthing the catchphrases of his masters, the Steel Trust. A signed editorial by William Randolph Hearst called Schwab "only a competent clerk." Prior to this, Hearst wrote, Schwab's remarks could be ignored, being merely a "young man's vaporings" which were "confined to excusable bursts of juvenile egotism." But now Schwab had testified under oath, and that testimony should be viewed as the official viewpoint of the Steel Trust, which, Hearst claimed, planned to destroy labor unions as soon as it had finished unloading its watered stock on a gullible public.[29]

Hearst also attacked Schwab's "industrious circulation of fictitious stories about his salary." Hearst reported that Schwab's salary was $100,000, not the $1,000,000 which had been popularly rumored. (In addition to his salary, Schwab received an annual bonus of one-half of 1 per cent of the company's net profits in excess of $80,000,000; when U.S. Steel earned $101,000,000 for the ten months ending October 31, 1902, Schwab's bonus was $100,000).[30]

Schwab was gaining notoriety, giving U.S. Steel the kind of publicity it could not afford. Judge Gary spoke out to deflate Schwab and protect U.S. Steel's image. "It is not a fact," stated Gary, "that Mr. Schwab is paid a salary of $1 million, or anything approaching that sum. He is a very wealthy man, a large holder of the stock of the company, and does not need and would not accept an extravagant salary." [31] (As an owner of 6 per cent of the Carnegie Company at the time of its sale, Schwab had received $25,000,000 in bonds of the new corporation.)

Schwab himself gave some credibility to the story of his million-dollar salary; in 1901 he began construction on a mansion which would be a monument to his success. The model for it was Chenonceaux, a chateau on the Loire which was confiscated from its owners by Francis I in the early sixteenth century. The site Schwab chose was a full city block at 72nd Street and Riverside Drive. The cash outlay for the land alone was $800,000. It took four years to build the mansion; Schwab spent $3,000,000 on it. The furnishings cost several million more. It was a four-story structure capped by a 116-foot lookout tower which gave a panoramic view of the city. The basement was huge. In it there was a sixty-foot swimming pool, as well as a power plant which cost $100,000 to build and which consumed ten tons of coal every day. It also had a wine cellar, a laundry room which could serve the needs of one hundred people, a bowling alley, and a fifty-foot gymnasium. The mansion had six elevators; they carried guests to over ninety bedrooms. All of the bedrooms were connected by a network of telephones linked to a central switchboard within the house. The master bedroom was twenty feet square, and in its bathroom there was a five-foot-square stall shower.[32] As Schwab said in 1935,

> When I built it, it was the most modern house in the United States . . . this was thirty years ago, yet it had an air-cooling system in it. It . . . was built of steel and granite. We opened up a quarry at Peekskill [New York] for the stone—the same quarry that supplied much of the stone for the Cathedral of St. John the Divine. . . . The thing that pleased me about it was that it was self-contained—had its own electricity generating plant. . . .[33]

Since every detail of the mansion had been carefully planned and executed, it cannot have been an accident that the end product made both Carnegie's and Frick's homes seem almost spartan. Schwab had a passion for owning the biggest and the best—homes, or automobiles, or private railroad cars. And that passion grew. The mansion was just one of the early expressions of it. His was the last, the largest, and the most lavish full-block mansion ever to be constructed in New York City. "Riverside," as it was called, made Schwab the object of sustained public interest. Such a house, declared *Harper's Weekly*, "may strike the average observer as a burdensome possession, oppressive to maintain, and likely to be embarrassing to heirs, but if Mr. Schwab can stand it, we can."[34]

In 1901, on the day after Christmas, Schwab sailed for Europe. Rumor had it that he intended to aid in the formation of an English equivalent of U.S. Steel, but Schwab denied this, saying that his purpose was purely pleasure. "I am going away for a complete rest. I have been on the go constantly for the last two years, and think that I am entitled to a vacation. My work has been hard and I need new strength. I am to visit Paris, the Riviera, Berlin, Vienna and London before I return."[35]

It promised to be a relaxing vacation. He had rented eight suites on the promenade deck of a French luxury liner, and he was accompanied by his wife and several personal friends, including Charles Schoen, the steel car manufacturer. As Schwab told a reporter, "My trip, I can most emphatically say, has no business significance whatever." This was quite true, but it proved to be a misfortune, for in seeking rest and pleasure he set off a scandal which shook his already precarious hold on the presidency of U.S. Steel.

On January 8, he reached Monte Carlo, having driven there from Paris in a sleek roadster he had purchased for the trip. He arrived on the same day that the Earl of Rosslyn was leaving, penniless. Some time earlier Rosslyn had announced to the world that he had devised a foolproof system to win at roulette; on the strength of his reputation, many Britishers had invested sums to make up his gambling stake. An entourage of British reporters accompanied him to Monte Carlo, where he was wiped out. Then a sharp-eyed reporter, starved for a story, spotted Schwab at the tables.[36]

Soon the cable wires to England and America were humming. Incredible tales of Schwab's breathtaking victories at roulette appeared on the front pages of American dailies. "Schwab Breaks the Bank," was the headline in the New York *Sun*.[37] Another account read:

> [Schwab] started playing maximums at Monte Carlo Thursday [January 9], the crowd pursuing him from table to table. He won $7,500. When he resumed playing yesterday the excitement was indescribable. After losing $10,000 at one table, Mr. Schwab went to the next, staking the maximum on No. 8, namely $36. Eight turned up and he won thirty-five times, the stake being $1,260. He then pushed the same stake on No. 9, winning again the same amount. His unprecedented luck made the audience frenzied.[38]

The frenzy of his Monte Carlo audience was nothing compared to that of his American associates. The press was filled with lurid accounts of his alleged exploits at the tables, and his personal stock sank to a new low. Schwab knew that the Steel Trust was extremely controversial; he should have anticipated that unfriendly newspapers would not pass up an opportunity to discredit or embarrass him. His action was comparable to that of a bank president visiting a race track—it was thoughtless, inconsiderate of the reputation of his institution, not evil *per se*.

Andrew Carnegie was the most enraged reader of these stories about Schwab's escapades. He felt personally wounded and betrayed, for he had a long-standing affection for Schwab and an even more long-standing hostility to any form of gambling. Believing Schwab to be guilty of a monstrous immorality, Carnegie cabled him:

> Public sentiment shocked. Times demands statement [that] gambling charge [is] false. Probably have [to] resign. Serves you right.[39]

On the same day Carnegie wrote to J. P. Morgan, calling for Schwab's resignation from the presidency of U.S. Steel. His letter, accompanied by clippings from the New York *Sun* and an editorial from *The New York Times*, was impassioned:

> I feel in regard to the enclosed as if a son had disgraced the family.
>
> What the Times says is true. He is unfit to be the head of the United States Steel Co.—brilliant as his talents are. Of course he would never have so fallen when with us. His resignation would have been called for instanter had he done so.
>
> I recommended him unreservedly to you. Never did he show any tendency to gambling when under me, or I should not have recommended him you may be sure.
>
> He shows a sad lack of *solid* qualities—of good sense, and his influence upon the many thousands of young men who naturally look to him will prove pernicious in the extreme.
>
> I have had nothing wound me so deeply for many a long day, if ever.[40]

Carnegie followed up his cable to Schwab with a long, chastizing letter. And Schwab received more than eighty cables from friends and colleagues in America, warning him that he was getting bad publicity from his Monte Carlo exploits. Many advised him to resign the presidency of U.S. Steel in order to spare the corporation any further embarrassment. While he did not wish to resign, he felt obliged to make a token gesture in that direction. The decision lay with J. P. Morgan, so on January 14 Schwab cabled a message to George W. Perkins, Morgan's partner and public relations aide. It was intended for Morgan. "Sensational statements of great winnings and losses absolutely false. Friends advise by cable that should resign. Of course will do so if Morgan thinks I should. Sorry." [41]

In a second cable, which he sent the next day, Schwab told Perkins, "Realize now it was a mistake to go there [to the gambling casino] at all but never expected publication." [42] (Schwab failed to recall, or perhaps he never knew, that his own press scrapbooks contained two stories about earlier gambling visits to Monte Carlo. The second had been witnessed by Mayor Tom L. Johnson of Cleveland, a former steel producer.) [43]

In his reply to Schwab's cables, Perkins ignored the question of resignation. He suggested that Schwab neither write nor cable anyone on the subject, and he reassured him, "Every thing all right. A.C. [Carnegie] and several others were very much excited but they did not make the slightest impression on Mr. M. [Morgan]. Do not give the matter any further thought or consideration. Go ahead and have bully good time." In reply, Schwab cabled, "Appreciate Mr. M attitude more than possible to express—am his to command always." [44]

Perkins then advised Schwab to issue a vigorous denial which could be circulated to the American press. But Schwab's denial was inherently deficient; he could only have absolved himself by saying that he had not gambled at all, and that simply was not true. He could only say that he had wagered for low stakes, that he had played casually ("indifferent play and simply for amusement"), that he had never once sat down at any gambling table, that he had visited Monte Carlo regularly for the past fifteen years because he admired its orchestra, and that on this last occasion his companions had included such pillars of propriety as Prince Metternich of Vienna; Baron Henri Rothschild; and Dr. Griez Wittgenstein, the wealthiest steel manufacturer in Austria. [45]

While Morgan was quite satisfied with these explanations, Carnegie was not. His response to Schwab's disclaimers, which Perkins sent on to Schwab, was both concise and accurate, "His denial is no denial." [46]

But Schwab believed that his explanation was adequate to quiet the tumult and, not having received word from Perkins that Morgan wanted his resignation, he assumed that the furor would die down. Accompanied by Dr. Wittgenstein, Schwab visited Vienna, where, at the personal request of Emperor Franz Joseph, he was presented at Court. He then chartered a special train to take him to Budapest, and from there he sped to Berlin for a special audience with the Kaiser. [47]

European newspapers ran stories about America's "Crown Prince of Steel," but American newspapers were filled more than ever with tales of the Monte Carlo imbroglio. When Schwab arrived in Berlin he received another stinging letter from Carnegie, who enclosed copies of critical editorials from the leading New York papers. Schwab replied:

> I am heartbroken and leave for home at the earliest available boat. My vacation has been utterly spoiled. The newspaper reports are absurd. Absolutely untrue. Of course, in going to Monte Carlo I invited this attack and will pay the penalty. But it is all absurd and ridiculous. Mr. Morgan must accept my resignation, but he has refused to do so as yet. He finds no fault. [48]

Morgan was nonchalant in his reaction to Schwab's gambling escapade because he believed that lurid gossip about a businessman's private life—

his own, or Schwab's, or anyone else's—should be ignored. His view was that the sensational stories about Schwab would soon be eclipsed by some new triviality which headline-hungry newsmen would exploit. He did not believe that Schwab's actions had adversely effected U.S. Steel.

Morgan was in a minority. Critics by the legion were busy chronicling Schwab's faults. A Chicago newspaper asked, "What guarantee can the holders of the stock of the great corporation of which Mr. Schwab is president have that their holdings are safe? None at all. Here is a moral delinquency that is openly scandalous." [49] *The Nation* devoted a full-page editorial to Schwab's errant behavior, declaring that "this is not the sort of man to be in a position of great trust and of vast financial responsibility." Instead of being "a shining object of emulation," instead of behaving as befitted "a man who had begun to wield large influence as a leader of public opinion," Schwab had acted "like a weak man in the first wantonness of newly acquired power to spend." [50]

Although Schwab remained out of reach of the denunciatory editorials and sermons, scolding notes from Carnegie awaited him wherever he went. One reached him in London, where Sir Thomas Lipton had presented him at Court, and he received another when he arrived in New York on February 16, 1902.

In mid-March, Carnegie invited Schwab to come see him. Schwab declined, saying: "I have not as yet been able to muster up sufficient courage to come to see you. Your very severe letters to me and especially your letter to Mr. Morgan has [*sic*] depressed me more than anything that has ever occurred. I cannot see how you could have so fully condemned me without ever giving me a hearing,—no one else did." [51]

Morgan had made light of the whole episode. In March 1902, at their first meeting after Schwab's return from Europe, Morgan told him, "forget it, my boy, forget it." [52] Schwab returned at the time when elections were to be held for new corporate officers. If Morgan had wished to oust him or to ease him out, here was a convenient opportunity. But Morgan had no such intention, and when he asked for proxies to re-elect the present officers there were only three letters critical of Schwab. Two were from clergymen; the third was from Carnegie. Schwab wrote to Carnegie:

> Of course, Mr. Morgan, much against my wishes, insisted upon my continuance. I consented, but have no heart in it until I can forget. As soon as the company is fully organized and reconstructed as it should be, I shall retire.

. . . I shall come to see you the first day I find sufficient courage and spirit to do so. I have been so depressed since my return that I only forget your condemnation when plunged in work. It is a nightmare from which I never seem to wake. I don't care for the newspaper criticism—I only mind yours.[53]

Schwab was not exaggerating. He actually became ill because of Carnegie's severe personal criticism. He had spells of insomnia and nervous irritability. He lost weight. He had a persisting numbness in his legs and arms, and sieges of fainting. He tried to convalesce at a summer cottage in Atlantic City, but his condition had not improved by the end of July. His doctors called it a nervous attack of the heart.[54] Whatever it was, it was probably triggered by acute anxiety, by an overwhelming sense of hurt and helplessness.

Schwab then tried the mountain air of his home in Loretto, but he still did not get any better. His doctors insisted tht he should have a total change of environment and complete rest, away from reporters and the solicitous inquiries of his friends and colleagues. The place chosen for his seclusion was Aix-le-Bain, in southern France. He sailed on August 21, 1902, after urging Perkins to deny "the nasty story" in the New York press that he would soon announce his retirement.[55]

Accompanied by his wife and his youngest sister, Gertrude, Schwab sailed to Le Havre and then motored to Paris, en route to Aix-le-Bain. When he reached the Ritz Hotel in Paris, a letter was waiting for him. It was another scolding from the relentless Carnegie. Once again Schwab had to defend himself, but this time against an "eye witness" account which Carnegie had heard.

Schwab reproached Carnegie:

That you would be willing to listen to, and believe the stories some one has seen fit to tell you was indeed a surprise to me and would not in all probability have been told if the hearer had not in some manner signified his willingness to listen and to believe.

. . . I am no gambler and if the stories published were true, of course I should be condemned. But they are not true. But be what I may, there is no condition of affairs that would make me even listen to a tale of such a character concerning you. I'd defend you or any of my friends until I knew the truth. I admit I have made a serious mistake and one I shall probably never be able to rectify—and I will pay the penalty. I have cabled Mr. Morgan again to-day saying that he *must* accept my resignation. My chief pang is not my loss of position, or loss of public opinion, but the loss of your confidence and friendship. Do not send for me for I should not come.[56]

Carnegie's letter was not the only blight upon Schwab's intended rest in Paris. Despite repeated denials issued by his physician, Schwab was reported to be dying. One newspaper, without actually claiming Schwab had died, nevertheless ran his obituary. A Paris newspaper ran a bogus story claiming that Schwab had announced that he wanted to give away his wealth before he died, and Schwab's Paris hotel was literally surrounded by beggars and fortune-seekers. He was also inundated by hundreds of letters, pleas from those who could not come in person to carry off their share of his money.[57]

Schwab's condition had not improved by late September. Yet newspapers were publishing lurid stories that he was engaged in high-speed auto-racing—which he denied as absurd. Finding neither seclusion nor serenity in France, Schwab decided to try Italy. He could promise no date for his return to U.S. Steel, and he had to tell Perkins that "as far as I am concerned any arrangement that will be for the good of 'Steel' will be entirely agreeable to me." [58]

Morgan did not ask for Schwab's resignation. But months of convalescence had made Schwab restless, so in February 1903 he decided to return to America, even though he was not fully recovered. There was a possibility, he told Perkins, that he might have to give up work for a very long period. His doctors called his condition "neurasthenia"; Schwab called it "damn bad." [59]

Perkins knew from Schwab's secretary, Oliver Wren, that Schwab was keeping in close touch with developments at U.S. Steel, and that, if his health did improve, he could easily step back into his job. However, Perkins doubted whether Schwab could resume the strenuous duties required of U.S. Steel's president. He claimed it would be a danger both to Schwab's health and to the welfare of the corporation if Schwab made a premature return.[60]

Schwab agreed with Perkins. Although his doctors said he was fit to resume work, Schwab doubted it. Nonetheless, he was planning to return to America in March, for his wife was gravely ill. (She had a progressively disabling rheumatism.) Schwab heard rumors that Perkins wanted him to resign, but he did not believe them. He wanted to be reelected president, but he promised Perkins on March 2 that he would resign if he found that he was unable to resume work by June 1. Perkins, however, was not pleased to hear that Schwab planned to return. He urged Schwab to stay in Europe, to avoid the reporters who would assault him upon his arrival in New York.[61]

Perkins believed that Schwab should be asked to resign the presidency, in part, because of his ill health; he proposed that Schwab be elected to a newly created post, Chairman of the Board of Directors, which would enable U.S. Steel to draw upon his expertise without burdening him with specific duties which might tax his stamina.[62] But Perkins also feared that if Schwab returned he would jeopardize the harmony which the corporation had enjoyed during his absence. As Perkins told Morgan; "We seem to have gotten by the period of having much internal friction. . . . I think we made an excellent move in the direction of harmony when we put [James] Gayley into the Board of Directors at the last meeting. It seemed to satisfy all the steel men and especially all the subordinate men in the Steel Corporation."[63] Gayley, a long-time Carnegie veteran, spoke for the steelmen when they clashed with the finance or non-operating men. Schwab had predicted that his absence would create no problems. He had told Perkins in August 1902, "Steel Company's affairs will move without any hitch. Mr. Gayley is quite capable of attending to all my duties. All my organization [i.e. the Carnegie veterans] are loyal and energetic and will do their best—will in fact try harder than if I were here."[64]

Morgan did not accept Perkins's suggestion; he wanted to keep Schwab in office. So Schwab was still president of U.S. Steel when he returned to America. It appeared to many that a conspiracy to oust him had failed. A leading New York newspaper summarized the abuses which Schwab had endured:

Mr. Schwab's commercial existence was at stake. He had been reviled, misrepresented and lied about. His acts had been misconstrued, his ability attacked, his health placed in a false light. Nothing that envy, jealousy and the cupidity of false friends could suggest had been left undone to render Mr. Schwab's position at the head of the Steel Trust untenable. . . . Wall Street has never witnessed anything meaner, more vicious and contemptible and far-reaching than this plot to overthrow Mr. Schwab. . . . a constant stream of stories was poured out from the headquarters of the cabal intended to place the young financier in a false and ridiculous light. His health was said to be permanently affected, his mental balance was alleged to be disturbed, his habits were attacked. He was charged with the most idiotic acts abroad. Spies followed him everywhere. A press bureau was established by his enemies, and every act was misconstrued. . . . The record of the persecutions to which Mr. Schwab has been subjected . . . is unique in the annals of Wall Street.[65]

This "plot" was traced to various sources. The likeliest suspects were an unidentified "anti-Carnegie faction" within U.S. Steel, men who were presumed to be trying to destroy Schwab in order to make Carnegie suffer. According to this theory, the ringleader of the faction was Henry Frick. Runner-up suspect was George W. Perkins, Morgan's "right-hand man," who was accused of sending Schwab notes of solace while simultaneously circulating slanderous stories about him to the press. Perkins was portrayed as a man consumed with a lust for power, secretly thinking of himself as George Washington Perkins. The least plausible of the suspects was a cabal, a group of Wall Street speculators who were aiming to destroy the head of U.S. Steel in order to depress the price of the stock so that they could cover a large short-sale interest. All of these identifications were no more than guesses, unsupported by substantive evidence.[66]

When Schwab returned to America, he did not immediately attempt to resume full-time activity as president of U.S. Steel. Instead, he concentrated on various personal matters and private financial opportunities. He had to make decisions about the furnishings for "Riverside," which was nearly ready for occupancy. He was also absorbed in a search for art works to adorn the mansion's two-story art gallery. And he supervised the installation of his most valued possession, an organ purchased for $100,000, its pipes carefully concealed behind vast tapestries which had been woven by one hundred Flemish women who had been brought to America for the purpose.[67]

Schwab was also involved in a conspicuous act of charity; he gave the town of Homestead, Pennsylvania, an industrial training school, a gift which cost him in excess of $200,000.[68] And he supplied $100,000 of venture capital to a young Canadian machinist, Joseph Boyer. Boyer had come to Schwab with two of his inventions: a pneumatic hammer and an adding machine. He needed financial backing. Schwab thought the pneumatic hammer showed the greater commercial potential; he invested in it and became a founder and director of the Chicago Pneumatic Tool Company. His profits were solid, but not spectacular; he earned thousands on the hammer. But he might have reaped millions. After Schwab turned down Boyer's adding machine, the inventor went to William S. Burroughs, who decided it was worth an investment. The adding machine was an instanteous success for the Burroughs Company. Years later Schwab said that he still felt like a "damn fool" for having chosen the hammer.[69]

Schwab also applied his skills as a negotiator for companies outside the steel industry which were interested in merging. As Professor Hans B. Thorelli has said:

> An outsider of sufficiently imposing personality and pleasant disposition could generally bring about an accord comprising [i.e. uniting] a number of mutually suspicious and jealous competitors much easier than could a member of the industry. From a psychological point of view it is but natural that many industrialists would prefer having an outsider pass on, say, the touchy question of evaluation of participating enterprises [rather than] entrusting one, or even a committee, of their fellows with their problem.[70]

Schwab was just such an "outsider."

One of the successful mergers created the International Nickel Company, and Schwab obtained its presidency for one of his ablest young protégés, Ambrose Monell, a former Carnegie veteran.[71] Another successful merger, combining eight firms in the Illinois, Ohio, and Pennsylvania area, resulted in the creation of the American Steel Foundries.[72]

Schwab continued to include his brother Joe in nearly all of his business activities. During Schwab's presidency of Carnegie Steel, Joe was elected a director; three months after Schwab became president of U.S. Steel, Joe was chosen as assistant to the president; and when Schwab became the largest single shareholder in American Steel Foundries, Joe was named its first president. Within a year, however, when an opportunity arose to sell their stock at a substantial profit, the Schwabs did so.[73]

Thereafter, the careers of the two brothers diverged sharply. Joe's one-third of 1 per cent interest in Carnegie Steel had brought him a sudden windfall when U.S. Steel was formed; he received more than $1,000,000 worth of stock. That, coupled with his profits from American Steel Foundries, gave him the capital to launch a new career as a stock market speculator. He purchased a seat on the New York Cotton Exchange, and eventually he became one of New York's biggest brokers in puts and calls—one of the most speculative, high-risk activities in the stock market.[74] Joe found a respectable outlet for his love of gambling; unlike his brother, he did not have to seek the chancy anonymity of the casinos at Monte Carlo.

Once freed from the watchful eye of Andrew Carnegie, Joe, like many another of Carnegie's young partners, was swept up into a new life-style made possible by sudden wealth.[75] One of the veterans went on a spending spree in New York; then he rented a private railroad car to carry a group of friends cross-country to Los Angeles, where he gave lavish parties for his new acquaintances. Another veteran, sober and circumspect in behavior while under Carnegie's restraining influence, began to drink heavily and to carry on numerous affairs. In 1935 David Garrett Kerr, a Carnegie veteran who had spent his post-Carnegie years as a vice president of U.S. Steel until his retirement in 1932, offered an explanation of this whirlwind profligacy:

> Many of them didn't know what to do with their money after they got it. There was a reason. They had been making steel all their lives. They hadn't much time or inclination to go in for outside interests. They hadn't even been counting the money they were piling up for themselves. As a consequence, they were amazed when the [U.S. Steel] Corporation was founded. They found themselves in possessions of millions. There was a let-down. All they could do was to try to devise ways to spend the money that, after all, meant nothing to them. As a consequence, many of them died comparatively poor. The stock market got some of them for they were babes in Wall Street. Poor business ventures took away some of their fortunes. Reckless spending and jovial adventures accounted for many more millions.[76]

As he passed his fortieth birthday with no loss of his good looks, Joe Schwab launched a series of "jovial adventures" with a succession of women. In 1907 he left his wife, Esther, whom he had married in 1886 when he was twenty-two, and his two children; for the next fifteen years he lived with a vivacious Broadway actress.[77]

Joe Schwab could afford to defy Carnegie and conventional morality; Charles Schwab could not. As long as he valued his post as president of U.S. Steel, he had to maintain a scandal-free reputation. He had to avoid conspicuous gambling, as the fury of public and press censure at the time of the Monte Carlo incident had taught him, and he had to be extremely discreet in his extramarital affairs. In this last he was held in check by three forces: his fear of scandal and censure, which would permanently alienate Carnegie; his desire to avoid offending his mother's moral sensibilities; and above all, his intense and genuine desire to avoid subjecting his wife to torment or public embarrassment.

But while Schwab successfully avoided any scandals concerning his private life, he was unwittingly embroiled in a new scandal involving his business career. In 1903 a new crisis erupted, one which made all of his earlier troubles seem trivial by comparison.

8

The U.S. Shipbuilding Company Scandal

Among the mergers in which Schwab was involved at the turn of the century, U.S. Steel, International Nickel, and American Steel Foundries proved successful and lasting. However, one merger was not. That was the United States Shipbuilding Company, a 1902 merger of seven shipyards on the Atlantic and Pacific coasts. By 1903 the company was verging on bankruptcy and Schwab was being accused of having deliberately engineered its collapse in order to obtain its assets for himself.

Any estimate of Schwab's guilt or innocence depends on which one of two underlying premises one holds. In their analysis of the collapse of the U.S. Shipbuilding Company, Professors Henry R. Seager and Charles A. Gulick assigned a major share of the blame to "the refusal of one of the participants [i.e. Schwab] to sacrifice his own financial interest to save the company . . . from bankruptcy." By contrast, in an earlier analysis of the same episode, Professor Arthur S. Dewing concluded that "Schwab had no prophetic vision which enabled him to foresee the collapse of the project, yet he was certainly entitled to protect his interests in case such a collapse should occur." [1] In other words, one can either condemn Schwab or exonerate him, depending on whether one accepts or rejects the premise that he had an obligation to sacrifice his self-interest for the sake of other investors.

The prime mover in the shipbuilding merger was John W. Young, one of the fifty-six children of Brigham Young, the Mormon prophet. In 1898, Young obtained options to buy various shipyards throughout the country, with the announced purpose of consolidating them into one firm; each yard would specialize in the construction of the type of vessel for which its facilities were best suited.

Young's chief associates in the venture were Lewis Nixon and Colonel John J. McCook. Nixon, born in Leesburg, Virginia, in 1861, had been graduated from Annapolis first in his class, had earned a solid reputation as a naval architect, had served as head of Tammany Hall in New York City, and was the owner-operator of a relatively prosperous shipyard in New Jersey. Colonel McCook, a partner in a leading New York firm which specialized in corporation law, handled the preparation of the prospectus, which was released to the public on May 7, 1901—a few hours before the stock market collapsed.

The firms which made up the proposed consolidation were a curious mixture. The one thing they lacked, individually and collectively, was a realistic prospect of earning sustained profits. Perhaps that was why their owners were so keen to accept the offer to consolidate, each firm hoping that somehow its association with the others would make it profitable. It was a perfect illustration of a phenomenon which Andrew Carnegie had described and denounced in his 1900 essay, "Popular Illusions About Trusts":

> Enormous sums are offered for antiquated plants which may not have been able to do more than pay their way for years. These are tied together, and the new industrial makes its appearance as a trust, under the delusion that if a dozen or twenty invalids be tied together[,] vitality will be infused thereby into the mass.[2]

One of the shipbuilding firms, located in San Francisco, was facing the competition of newer and more efficient shipyards; its profits, once substantial, were now diminishing. Another was so loaded with debt that it was approaching bankruptcy. A third, although it was modern, efficient, and well-managed, lacked one essential ingredient for profitability; it had few customers for its specialty product—heavy battleships. Another was saddled with superannuated equipment which was not even kept in good repair. By far the weakest link in this precarious chain was a firm which was not in any way involved in shipbuilding; Nixon had included it in the proposed consolidation as a favor to its owner, Charles Canda. Canda, after a decade of trying to produce steel railroad cars, still had progressed no further than the mastery of wheelmaking.

In April 1902 Young and Nixon thought the market was once again ripe for their venture. They issued a second prospectus, explaining that $9,000,000 in bonds were to be sold to the public to raise working capital. The remaining $7,000,000 in bonds, plus $10,000,000 each in preferred and common stock, were to be parceled out to the owners of the subsidiary plants, to the underwriters, and to the promoters.

John W. Young tried to sell $4,250,000 of the bonds in Europe. He set up headquarters in Paris and industriously proceeded to canvass members of the French nobility, seeking their prestigious names as a means of attracting other investors. With an eagerness which probably should have made Young suspicious, several noblemen agreed to subscribe to the bond offering. Young relayed the good news to his associate in America, who was handling the underwriting of the remaining $4,750,000. This was Daniel Leroy Dresser, founder of the Trust Company of the Republic, who was a newcomer to the field of bond underwriting. Dresser's most solid credential was that he was a brother-in-law of William H. Vanderbilt.

Nixon tried to expedite the American underwriting; he got in touch with a number of leading industrialists and investors, including Schwab, in May 1902. Nixon thought that Schwab, either as president of U.S. Steel or as a private investor, might like to purchase some of the bonds of the U.S. Shipbuilding Company. He was right. Schwab saw a way to benefit both himself and U.S. Steel. After reading the financial prospectus, and after learning that the foreign underwriting had been completed, Schwab decided to purchase $500,000 of bonds as a personal investment, provided that the Shipbuilding Company would purchase its steel from U.S. Steel. Nixon and Dresser agreed, and the deal was consummated.

One month later, on June 12, 1902, Schwab met Nixon and Dresser while they were lunching at the Lawyer's Club in New York City. According to Dresser, the three men had a brief conversation during which Schwab asked them, "Why don't you buy the Bethlehem Steel Company?" Schwab, however, testified later that it was Nixon and Dresser who first raised the possibility of including Bethlehem Steel in the shipbuilding merger. In any event, Nixon and Dresser came to Schwab's office the next day to find out if Bethlehem was for sale. They came to Schwab because in June 1901, less than eight weeks after he became president of U.S. Steel, Schwab had bought the Bethlehem Steel Company.[3]

At the time of the purchase, there was speculation in the press about Schwab's motives. Bethlehem had a favorable reputation; it was a small, well-managed, highly profitable company. It produced heavy steel forgings for guns and marine engines, and it was the only firm other than U.S. Steel which produced armor plate. Some newspapers said that U.S. Steel intended to achieve a monopoly in armor plate; they assumed that Schwab was acting for U.S. Steel in making the purchase. The Hearst newspapers were concerned: the U.S. Navy would now be totally at the mercy of the Steel Trust.[4] Other newspapers suggested that Schwab, in buying Bethlehem, had merely acted as the purchasing agent for E. H. Harriman, the railroad magnate. A third explanation was that U.S. Steel was planning a total monopoly in steel; the purchase of Bethlehem was only the first step in an insidious campaign to buy up all of the surviving independent firms, including Cambria Steel, Pennsylvania Steel, and Jones & Laughlin.[5]

But no one guessed the truth. Schwab had purchased Bethlehem for himself; it was an independent investment.

When Schwab met with Dresser and Nixon, he reviewed Bethlehem's balance sheet with them, pointing out that the company had $8,000,000 in bonded debts, carrying an annual charge of over $500,000. On the other hand, Bethlehem's earnings were substantial for a company of its size and specialization. In the preceding year it had earned $1,400,000.

Schwab's asking price for Bethlehem was $9,000,000, in cash. Dresser and Nixon, however, felt that Schwab just as well might have said $900,000,000. The proposed shipbuilding consolidation was desperately short of cash. Yet they wanted Bethlehem in the merger.

After negotiation, Schwab accepted a counter-offer: he would receive $10,000,000 in bonds, for which Bethlehem's plant and properties would be the collateral in the event of default. And he also would receive $20,000,000 in stock, half preferred and half common. Thus, instead of $9,000,000 in cash, Schwab was to receive a total of $30,000,000 in securities, a huge profit if the shipbuilding merger proved successful. If it failed, Schwab would at least re-acquire Bethlehem.

Schwab capitalized on his bargaining position; he intended to safeguard his investment in Bethlehem against any unforeseen difficulties. He insisted upon, and received, a second mortgage on all the properties of the shipbuilding merger. His bonds were to be given full voting power. As a final protection, he required the U.S. Shipbuilding Company to give Bethlehem Steel enough orders so that a 6 per cent dividend on Bethlehem stock would be guaranteed. Bethlehem was also to be guaranteed a working capital of not less than $4,000,000.[6]

Schwab received extremely favorable terms. Whether the shipbuilding merger succeeded or failed, he would lose nothing. For that reason, when the company showed signs of imminent collapse during the next year, everyone viewed Schwab as the villain.

Before Schwab concluded his negotiations with Nixon and Dresser, however, it was necessary for him to reacquire Bethlehem Steel from the J. P. Morgan syndicate. Although he had originally purchased Bethlehem as a personal investment, as president of U.S. Steel he had felt obligated to offer it to the Morgan syndicate, the underwriters of the U.S. Steel merger. He had told Charles Steele, one of Morgan's partners, "Now I bring it down to you; of course, it is at the disposal of the Steel Company if they want it," but he had advised Steele that owning Bethlehem would not be advantageous for U.S. Steel since it already possessed an armor plant of its own and did not intend to go into gun forgings.[7] Nonetheless, the Morgan syndicate had bought Bethlehem Steel from Schwab, perhaps because Morgan believed that Schwab was undertaking too much and would not be able to handle his responsibilities as president of U.S. Steel while simultaneously running Bethlehem.

When Nixon and Dresser agreed to his terms, Schwab informed Charles Steele that he had found a buyer for Bethlehem. Schwab arranged to pay the Morgan syndicate $7,246,000, in cash, for Bethlehem. Then he sold Bethlehem to the shipbuilding trust. As Morgan's profit on the transaction, Schwab transferred to the Morgan syndicate $2,500,000 each of his preferred and common stock in the U.S. Shipbuilding Company.[8]

During the next year, from mid-1902 to mid-1903, the shipbuilding merger suffered financial anemia. It had made grandiose claims about profitability, but profits failed to materialize. Then it was discovered that one of the shipyards had incorrectly reported its past earnings, and it was on the basis of that report that the yard had been included in the merger.

Of all of the properties in the merger, only Bethlehem Steel lived up to—and, in fact, exceeded—its anticipated earnings. Naturally, the ailing parent company, U.S. Shipbuilding, turned to its thriving subsidiary for financial aid. Very little was forthcoming. Schwab not only controlled Bethlehem's Board of Directors, which included his attorney, his brother, and his brother-in-law; he had also exercised his right to name three of the seven directors of U.S. Shipbuilding.

Bethlehem's Board of Directors and Schwab's representatives on U.S. Shipbuilding's board refused to allow Bethlehem's earnings to be used to provide working capital for the crumbling subsidiaries of the merger. Instead, the profits were used to build new furnaces at Bethlehem and to buy new ore properties in Juragua, Cuba. Despite substantial earnings, Bethlehem's directors did not declare a dividend at the end of 1902; all available funds were reinvested.[9]

Lewis Nixon, executive head of the Shipbuilding Company, was frantic. Unless Bethlehem's earnings were made available, the parent company would collapse. He tried to persuade Bethlehem's directors to put up the funds, but to no avail; they were acting on Schwab's behalf. He then tried to summon a meeting of the Shipbuilding Company's directors, but again he had no success; he could not muster a quorum, for the directors appointed by Schwab simply failed to attend. Finally, Nixon in desperation turned to Schwab's personal attorney, Max Pam, a short, bearded, Bohemian-born financial wizard who handled Schwab's business interests during his illness and his absences from America. Pam gave Nixon a sympathetic hearing, but almost no financial relief. Instead of letting loose a torrent from Bethlehem's reservoir, Pam released only a trickle, $250,000—just enough to enable the Shipbuilding Company to pay Schwab the interest due him on his bonds.[10]

When John W. Young failed to raise the bond money which had been pledged in France, Daniel Dresser made up the deficit by borrowing on the credit of his Trust Company of the Republic. By doing so Dresser overextended his company, and he was forced to turn over control of it to a group of investors headed by George R. Sheldon. In order to salvage their investment in the Trust Company, the new directors, who became known as the Sheldon syndicate, proposed that the U.S. Shipbuilding Company be reorganized.[11]

U.S. Shipbuilding needed $2,000,000 to provide working capital for the shipyards and to meet the interest payments on the bonds, which were to come due on July 1, 1903. George Sheldon appealed to Schwab on behalf of a number of bondholders and stockholders. Schwab replied that he would put up the $2,000,000 on one condition: in return for the money his present second-mortgage bonds were to be replaced with new first-mortgage bonds which would give him the primary lien on all the properties of the Shipbuilding Company. Nixon could see no alternative to this proposition, so he reluctantly agreed. The Sheldon syndicate, aware that the alternative to accepting Schwab's terms was the certain collapse of the Shipbuilding Company, also accepted his offer; this acceptance later made them vulnerable to the accusation that they were Schwab's lackeys.[12]

But one group of first-mortgage bondholders of the Shipbuilding Company had no intention of seeing Schwab awarded bonds with priority claims over their own, and they decided to give battle. On June 11, 1903, they filed suit in the U.S. Circuit Court at Trenton, New Jersey. The complainants, headed by Roland R. Conklin, charged that the Shipbuilding Company was insolvent, that it had been notoriously mismanaged, and that its original organization and promotion had been fraudulent. They had two objectives: first, they wanted to prevent the execution of the Sheldon plan of reorganization, which they said improperly favored Schwab at their expense, and, second, they wanted the court to appoint a receiver who would take over the company's affairs until its finances were put in order.[13]

The Conklin group hired Samuel Untermyer, one of the most brilliant attorneys of the era, to present their case. Born in Lynchburg, Virginia, and trained at Columbia University's School of Law, Untermyer, then forty-five, enjoyed a nationwide reputation as a trial lawyer. He had another asset which made him especially attractive to the Conklin group: he himself had organized a number of corporate mergers and trusts similar to the Shipbuilding Company, so he was no stranger to the mysteries and machinations of corporate finance and stock promotions.[14]

The Conklin group chose Lewis Nixon as their primary target. They accused him of having conceived the shipbuilding merger in order to unload his own firm, the Crescent Shipbuilding Company, at an inflated price. They also charged him with issuing heavily watered stock in U.S. Shipbuilding. Finally, they accused him of being the originator of the suggestion that Schwab sell Bethlehem Steel to U.S. Shipbuilding in return for a windfall profit.[15]

The Conklin group also attacked Schwab; they charged that he had made unconscionable profits in transferring a company worth at most $9,000,000 in exchange for $30,000,000 in stocks and bonds. Moreover, they condemned the fact that, under the proposed plan of reorganization, Schwab not only would recover all of the property he originally had sold, but that he would also gain control, through first-mortgage bonds, of all of the properties of the Shipbuilding Company. One of the attorneys for the Conklin group said:

> We want, first, to get this property out of the hands of Mr. Schwab and his associates and put it into the hands of the Court. Our eventual purpose is to defeat the attempt of Mr. Schwab to get this property from the first mortgage bondholders, to whom it belongs.[16]

On June 14, at Schwab's request, Max Pam issued a brief statement to the press, explaining Schwab's offer to put new cash into U.S. Shipbuilding. In the statement Schwab said that when he had first been approached by the Sheldon syndicate, he had said that he wanted to completely sever his connection with the shipbuilding venture. He had demanded that he be allowed to surrender all of his stocks and bonds in the company and that, in exchange, Bethlehem should be returned to him. The Sheldon syndicate had rejected Schwab's proposal on the ground that Bethlehem was the most valuable property in the merger. It was only when Schwab had found that he was unable to extricate himself that he had made the proposal to put up new cash in return for first-mortgage bonds.[17]

Schwab did not consider himself to be responsible for the financial deficiencies of the shipyards; he had taken Nixon and Dresser at their word and had not independently verified the accuracy of the financial statements and the earnings' estimates of the shipyards. He had, however, insisted on safeguards so that his interests would not be vulnerable. He believed that he was under no moral or legal obligation to gamble his assets on the dubious prospect of salvaging the shipbuilding merger.

On June 30 Judge Andrew Kirkpatrick announced his decision. He ruled that the company was insolvent. He added:

> It has been shown to my satisfaction that the board of directors are not taking steps to place this company on a sound financial basis, or to raise funds to meet its indebtedness. In view of all these developments, I believe the better plan for the protection of all parties concerned is the placing of this company's affairs in the hands of a receiver.[18]

Many applicants sought the post of receiver. It was a lucrative political plum, one which Judge Kirkpatrick could use to create or repay a debt. He chose to repay one: he gave the receivership to former U.S. Senator James Smith, Jr., boss of the Democratic machine in New Jersey, who had been responsible for Kirkpatrick's appointment to the bench.

The court ruling was a great blow to both Nixon and Schwab, although for different reasons. Nixon would have to face a public disclosure of the formation of the Shipbuilding Company; he would have to expose his own ineptitude in bargaining for control of Bethlehem Steel. Schwab, on the other hand, would have to make a full disclosure of the terms he had exacted; he would have to tell the world that he, and he alone, would not lose financially even if the Shipbuilding Company collapsed. The burden would be on him to prove that he and Nixon had not been co-conspirators in a scheme to defraud the public. Or, if Nixon turned against him and accused him of sabotaging the Shipbuilding Company, Schwab would have to prove that he had not duped Dresser and Nixon into turning over the company's assets to him.

On June 30, 1903, the same day that Judge Kirkpatrick announced his decision to appoint a receiver, the Finance Committee of U.S. Steel elected William E. Corey to the post of assistant to the president. The committee said that they had done so because of Schwab's continuing ill health, but the timing of the move makes that seem at best only a partial explanation. For months Corey had been Schwab's unofficial assistant, and his promotion on the very day that the court had condemned Schwab for "ruinous extortion" invites skepticism. A spokesman for Schwab announced that Corey's promotion did not foreshadow Schwab's ouster. "This does not mean by any manner of means that Mr. Schwab has resigned or intends to. He has been working even harder than ever before he went abroad. The appointment of Mr. Corey was made at Mr. Schwab's own request and in order to save him from a breakdown. Mr. Schwab will still be at the helm." [19]

However, it appears that Corey's promotion was in fact deliberately intended to be the first step toward Schwab's resignation, and that the initiative came from Schwab himself. Schwab recognized that his involvement in the shipbuilding scandal had severely compromised his position as president. Rather than risk the humiliation of being fired or asked to resign, he chose a two-stage withdrawal, offering ill health as his sole explanation. An immediate resignation would have been construed as an admission of guilt.

Once he had made the decision, Schwab sought to ease his exit. He enlisted the aid of Perkins, to whom he wrote:

I think the press treated me shamefully and I ask of you and Mr. Morgan to put me right in the minds of the public. Retiring from active business this

153

becomes important to me. I shall gladly do what I can in return—but I want you to have Mr. Morgan do this for me. A word from Frick would be much appreciated—but I have no right to ask him. I feel I have a right to ask Mr. Morgan to say the criticism is unfounded and that I acted entirely upon my own suggestion.[20]

Schwab felt he had no right to ask Frick for assistance because they had barely spoken since Frick's break with Carnegie. Although Frick was a major shareholder in U.S. Steel, and a director, Schwab had only minimal contact with him, lest he seem disloyal to Carnegie.

Perkins was able to report back that "I have spoken to both Mr. Morgan and Mr. Frick about the course you would like to have taken at the beginning of next month and they both will be very glad to put the matter as you have suggested." [21]

Newspapers were quick to point out what seemed to them to be the obvious moral: once Schwab abandoned reputable business for stock speculation, his career was ruined. And, of course, the origin of his undoing was the Monte Carlo episode, which had cost him the confidence of the public. One editorial writer asked, "How could they know [whether a] man who would ostentatiously stake thousands on the red, might not make ducks and drakes of their property by transferring the scenes of his gambling operations from Monte Carlo to Wall Street?" [22]

A month later, on August 4, Schwab formally submitted his resignation as president of U.S. Steel. Once again, the only reason he gave was ill health. His break with U.S. Steel, however, was not to be complete; he would still be a member of the Board of Directors and of the Finance Committee. Corey was elected to succeed Schwab as president and a new post was created for Judge Gary—he was made chairman of the Board of Directors.

Schwab chose to announce his resignation directly to the press instead of issuing a written statement. One reporter noted Schwab's changed appearance: "Although he was apparently in a cheerful mood, a slight quivering about the mouth and an indefinable lack of physical poise plainly showed that he was suffering from nervousness. His manner was restless and his eyes lacked their usual brilliancy." [23]

Schwab told the reporters:

The newspapers, I think, have treated me badly in reference to this matter. Every cause of my retirement that they have announced has been false. For instance, take the United States Shipbuilding matter. There never was any disagreement between me and Mr. Morgan, and Mr. Morgan never had anything to say in reference to my retirement. . . . Do not think that I am getting out of the corporation by any means.[24]

Both Frick and Morgan usually avoided contact with the press, but they joined in to help Schwab save face. Frick claimed that months before that time Schwab had asked him to obtain his release from the presidency. Morgan said that Schwab had "unequalled powers as an expert in the manufacture of steel," and he expressed regret that Schwab's health made his resignation necessary.[25]

Even after Schwab's retirement from U.S. Steel, the shipbuilding scandal continued to be front-page news, and Schwab, the best known participant, dominated the headlines. "Schwab Guilty of Big Fraud," read one headline; "Schwab a Wrecker," claimed another; "Deathly Grip of Schwab," a third declared.[26]

In October 1903, when U.S. Shipbuilding, already in receivership, failed to pay Schwab the $500,000 interest due to him on his bonds, he began a foreclosure suit. If successful, he would obtain not only the Bethlehem Steel Company, but also the properties of the U.S. Shipbuilding Company. To Untermyer and Senator Smith, Schwab's foreclosure suit seemed to be another link in the chain of evidence, his latest act of premeditated financial piracy. In their joint reply to the suit they asked that it be dismissed on the ground that Bethlehem Steel was worth far less than the bonds which had been issued to Schwab. Hence, Schwab had received the bonds "without consideration"—that is, without paying in a value equivalent to that which he had received in return. Untermyer and Smith also aimed a new series of accusations at Schwab. He was charged with masterminding the conspiracy to create an overcapitalized company and then systematically wrecking it in order to foreclose and seize its assets.[27] The worst of Schwab's transgressions, according to the two men, was his plan to turn Bethlehem over to the U.S. Shipbuilding Company in such a way that he not only would exercise control over the merger while it flourished, but also would be able "to absorb for his own benefit all of its property and assets to the injury of its creditors" when he brought about its collapse.

Everyone involved was accused of being Schwab's pawn. Nixon and Dresser were said to be Schwab's confederates in the original formation of U.S. Shipbuilding, as were each of the individual shipyard owners.

The promoters, including John W. Young, were charged with knowingly cooperating with Schwab in his plans to wreck the company and seize its assets. George Sheldon's reorganization committee was declared to be Schwab's puppet, subject to his manipulations. Young was called a "mere conduit," while Bethlehem's directors were labeled "mere dummies," and all were denounced for having "acted in bad faith and solely under instructions of said Schwab." [28]

Schwab's accusers offered no evidence to support these allegations, relying instead upon their *prima facie* plausibility.

One of the most serious charges was that Schwab had falsely represented that the 300,000 shares of Bethlehem stock were worth $30,000,000. This charge, however, was untenable. Schwab had originally offered to sell Bethlehem for $9,000,000 cash—this, then, was the value he set upon it. The fact that he subsequently accepted a total of $30,000,000 in stocks and bonds in lieu of $9,000,000 in cash in no way constituted either fraud or extortion. If Schwab had aimed to destroy U.S. Shipbuilding, he would have been committing an act of financial self-immolation, since the cash value of his common and preferred stocks would have been greatly reduced.

By October, the sparring and shadowboxing were over; the main bout was at hand. A court-appointed trial examiner, Henry Oliphant, held public hearings in New York City on the Conklin suit against Schwab.

The first witness was Daniel Dresser. He blamed Schwab for the failure of the U.S. Shipbuilding Company. He testified that he had attempted unsuccessfully to see Schwab and had finally sent a message urging him to transfer Bethlehem's earnings into U.S. Shipbuilding's treasury "where they rightfully belong." But any sympathy the public might have felt for Dresser was quickly dispelled. No one had deceived him or supplied him with false data or doctored statistics; instead, as he acknowledged under cross examination by Schwab's attorney, he had simply taken everyone and everything on faith. He made a series of remarkable admissions. He had never tried to ascertain the real value of the properties which were being merged in the Shipbuilding Company; he had never attempted to check on the accuracy of any accountant's statements which were sent to him; he had never felt it necessary to have first-hand knowledge, because another firm, which was underwriting twice as much Shipbuilding stock as his own Trust Company of the Republic was, seemed satisfied with the financial details of the venture.

Dresser's testimony amounted to a confession of self-induced blindness.[29]

Lewis Nixon testified on October 21. Untermyer pressed him for details of his conversation with Schwab about including Bethlehem in the merger. Nixon refused to testify that Schwab had said, "Why don't you buy Bethlehem?," but he did claim that it was Schwab who had first broached the subject. Nixon's testimony was lengthy, subdued, and dignified. He did not berate Schwab, nor did he try to make a scapegoat out of Dresser or Young or anyone else. He merely filled in details, neither strengthening or weakening Untermyer's attack on Schwab.[30]

Then, before Schwab had an opportunity to testify, Senator Smith, the receiver, issued a report denouncing him. Smith's statement read like a prosecutor's summation at the close of a trial—but in this case the defendant had not yet been heard. The report charged that everything which had occurred had been part of a premeditated conspiracy to cheat the public and "bag the profits"; that the "presiding genius" behind the scheme was Schwab; and, finally, that there was "no manner of doubt that all these details were carefully planned in the first place, that every step, from the issue of the shipbuilding prospectus in June, 1902, to the confession of bankruptcy twelve months later, was as familiar to the minds of the conspirators before it happened as afterward." [31]

Smith's report led to new attacks on Schwab in the press. One newspaper editorialized:

After wading through the mass of court evidence and saying patiently that judgment should be suspended until the accused as well as the accusers have had a chance to testify, most people will feel this morning that it is time to throw off all further restraint and speak out plainly about the abominable business of the Shipbuilding Trust. In the face of Receiver Smith's convincing report *it is no longer possible to believe that there can be another side* than the one which has already been held up in all its naked ugliness to the public gaze.[32] (Italics added.)

Schwab was painfully aware that Andrew Carnegie would read the grave accusations contained in the report, so he wrote to Carnegie:

Be assured, my Dear Friend, I would not have a dollar that was dishonestly made. I am sure you know that I am at least honest, however foolish. I never sold a share of stock in that concern [U.S. Shipbuilding] and when they failed to pay interest on my bonds I foreclosed, hence this difficulty. I am greatly worried as you can well imagine and most anxious to talk to you about it. I had far rather lose the money than your good opinion and esteem.[33]

On December 23, Schwab issued a denial of Smith's charges. He insisted he had not acted improperly in any respect, and he branded the charges as "false and malicious, and without justification and in disregard of well known facts." [34] He offered his own version of the genesis and collapse of the shipbuilding merger. He said that he had relied fully upon Nixon's and Dresser's appraisals of the properties: "Both Nixon and Dresser informed me that the properties were of great value, that they had been appraised by competent accountants and that in their opinion, after a careful investigation, the figures mentioned in the prospectus were safe and conservative." [35]

Schwab added that, shortly after the sale of Bethlehem, "I was taken seriously ill and compelled to abandon all business" and was unable either to run Bethlehem or to ruin the shipbuilding merger. Only upon his return, he claimed, did he learn that the estimates of assets and earnings were wildly distorted. (However, in making that statement he ignored the fact that all during his illness and absence his interests had been looked after by his authorized agent, attorney Max Pam.)

Schwab told of his meeting with George Sheldon and Charles Wetmore, the representatives of the reorganization syndicate: "I stated to them that as the shipbuilding properties did not have the assets and earnings which had been represented to me, I preferred to take back the Bethlehem stock and return to the Shipbuilding Company the bonds and stocks I had received." Sheldon and Wetmore refused to consider the withdrawal of Bethlehem on the ground that it was the most valuable property in the company, and they threatened to sue Schwab if he made any attempt to separate Bethlehem from U.S. Shipbuilding. With reluctance, Schwab then agreed

. . . to go into a plan of reorganization with them, and I suggested to them that an assessment be made upon the first mortgage bondholders, because the money required would have to be spent in and about the plants upon which their mortgage was a first lien, and because the Bethlehem Company would not require any new capital. They represented to me, however, that the first mortgage bondholders of the Shipbuilding Company would be unwilling or unable to pay any assessment whatever, even for the purpose of rehabilitating the shipbuilding plants. I finally agreed to supply $2,000,000 in cash for working capital, upon condition that I should be given a first mortgage bond [on the shipyards] for the amount of my advances and for the amount of the Bethlehem bonds.[36]

As an explanation, it was too little and it came too late. If Schwab had spoken out sooner perhaps he might have held onto the presidency of U.S. Steel. He could have pointed out that the accusations leveled by Untermyer and Smith were unsupported by any evidence, and that his efforts to protect his investment in Bethlehem Steel were not evidence of fraud and conspiracy, but quite the reverse; they showed proper business caution. Schwab had a narrow perception of his own self-interest, and it was probably that characteristic which most inhibited him from offering an outspoken defense. A self-styled "practical man," he wanted to resolve the controversy by making an out-of-court settlement with the bondholders in the Conklin group. He preferred negotiationg with the disgruntled bondholders behind the scenes rather than making any direct reply to the charges of criminal wrongdoing. His attorney, William D. Guthrie, counseled him that his approach would not provide a satisfactory solution:

> As I said to you at our last interview, if it were a mere question of settling with Mr. Untermyer's clients, who hold about $200,000 worth of bonds, it would have been cheaper to buy them out than it would be to meet their attacks; but such a proceeding would be but to invite innumerable suits and there would be no end to the attacks until we had bought up millions of bonds.[37]

Schwab replied to Receiver Smith's report two weeks before the day he testified at the Conklin suit hearings. When he did testify, on January 7, Untermyer goaded him relentlessly, causing him to lose his composure. Untermyer tried to get him to admit that he, personally, was responsible for the refusal of Bethlehem's directors to turn over Bethlehem's profits for the relief of the ailing parent company. Schwab claimed that he was not responsible, since he had given no such orders himself during his absence from the country. Untermyer then asked:

Q. When you went abroad, did you place anyone in charge of your affairs?
A. I did not.
Q. Who did you leave behind to take charge of your Shipbuilding affairs?
A. No one.
Q. You left it absolutely unrepresented?
A. Yes.

Untermyer next attempted to show that, following the election in April 1903 of Joe Schwab and Max Pam, the Board of Directors of the Ship-building Company was dominated by men representing Schwab's interests. He asked Schwab if the reconstituted board was selected in his interest. Schwab answered, "I don't know." Untermyer pursued:

Q. You don't know yet, do you?
A. I didn't select them; no sir, I didn't select those people.
Q. Now, let us see whether they were your nominees or not?
A. No man was my nominee in that directorate.
Q. Were they the nominees of your counsel and representative [i.e. Max Pam.]?
A. You must ask him that.
Q. You don't know, and you were never consulted?
A. My impression is that Messrs. Nixon, Dresser and Pam selected that Board, and if they didn't select people that represented me, they made a great mistake, I think.

Later, Untermyer asked Schwab if he controlled the Board of Directors of Bethlehem Steel.

A. I do not. They were my nominees and they were my people, and if that is control, then I did.
Q. What other kind of control can you conceive of?
A. I don't know.
Q. Then why do you say you do not control the Board?
A. Well, it would be pretty difficult to control some men.
Q. That is your best answer, is it?
A. Yes.

Schwab had been driven to doubletalk to evade the obvious fact that his interests in both companies were represented and protected by Max Pam and other directors chosen by Pam. It was a point which Untermyer hardly needed to establish, and one which Schwab could not deny, let alone so vehemently.[38]

Untermyer was able to make Schwab angry, but he could not trap him into any confessions of wrongdoing. Nor was he able to offer any evidence to sustain his charges of fraud, misrepresentation, and extortion. But these charges, after all, were mere hyperbole, designed to embarrass Schwab. As Professor Arthur S. Dewing wrote:

He [Untermyer,] as counsel for the Conklin protestants, did all that he could to exaggerate the fraudulent character of the enterprise and aggravate the already intense ill-feeling. He hoped, by turning the episode into a public scandal, to bring the pressure of public opinion to bear upon Schwab and the Sheldon Committee, so that they would feel themselves compelled to modify the original plan of reorganization.[39]

Following Schwab's interrogation, both sides to the dispute tried to compromise the outstanding differences between them. The Conklin group wanted the Sheldon plan of reorganization to be changed so that the first-mortgage bondholders' interest would not be subordinated to Schwab's. A "modified plan" was presented by Senator Smith and accepted by all parties concerned. In the final settlement, Schwab received $9,000,000 in preferred stock and $6,000,000 in common stock, while the first-mortgage bondholders (including the Conklin group) received $6,000,000 in preferred stock and $9,000,000 in common stock.[40]

If the original Sheldon plan had been approved, Schwab's $10,000,000 second-mortgage bonds would have been converted into first-mortgage bonds on the entire property of the Shipbuilding Company, including its subsidiary, Bethlehem Steel. Moreover, he would have been entitled to the $4,000,000 of Bethlehem's accumulated earnings. And if the mortgage on the Shipbuilding Company had subsequently been foreclosed, Schwab, as owner of the new first-mortgage bonds, would have received the $6,000,000 value of the seven shipbuilding companies which originally comprised the merger. The total value of his claims under the Sheldon plan would have been $20,000,000.

However, under the settlement that was finally approved, Schwab's $9,000,000 in preferred stock was worth only $6,750,000 at the current market price of $75 a share, and his $6,000,000 in common stock was worth only $1,500,000 at $25 a share. His holdings under this plan were worth $8,250,000.

On paper, at least, Schwab lost $11,750,000—the difference between the value of his assets under the two plans. This was a hard enough blow, but his monetary losses, in fact, were not merely "paper losses." Bethlehem had cost him $7,246,000. He had agreed to waive his claim to $500,000 which was owed him on the old second-mortgage bonds and $1,500,000 of the accumulated surplus of Bethlehem. The sum of his cash outlays and of the income he lost was thus $9,246,000—$996,000 more than the value of his assets under the final plan.[41] Even so, to Schwab, $996,000 was a small price to pay to quash this episode; if he had not paid it there would have been additional weeks or months of costly litigation and unsavory publicity. By this settlement, all litigation was stopped.

Schwab's losses, however, were not only monetary. The shipbuilding scandal had been front-page news for well over a year, and the publicity was almost unanimously unfavorable to him. The notoriety, far more than his illness, had certainly cost him the presidency of U.S. Steel. Schwab had been president of the world's two largest steel companies, and his new domain, Bethlehem Steel, was clearly a demotion. It was as if the King of England were deposed and sent off to become the ruler of Monaco or Liechtenstein.

But to Schwab, this was not an irreversible defeat. Like his hero, Napoleon, he had been banished. Now he was challenged to forge ahead, to restore himself to his former stature and power. And he succeeded. In a few years he transformed Bethlehem Steel into the foremost rival of the giant corporation from which he had been ousted.

9

The Transformation of Bethlehem Steel

Bethlehem Steel had been a small, specialty producer; within a decade after Schwab took control he had made it into the second largest and most diversified steel company in America. To magnify his own achievement, however, he repeatedly claimed that, before Bethlehem had come into his hands, it had been run-down and verging on bankruptcy. Across the years, writers who interviewed Schwab drew the conclusion that Bethlehem had been a "down-at-the-heels, decrepit, bankrupt, unimportant concern" before Schwab took charge, that it had consisted of "a few half-deserted buildings [which] were waiting to be stripped of machinery," and that it had been "a wreck financially and physically, a site of buried fortunes and financial hopes." [1] Schwab made similar claims in his reminiscences: "When I helped form the United States Steel Corporation, I took all the best. When I organized Bethlehem, I took all the leftovers, and made them into the best." Again, "Bear in mind that when I took Bethlehem it was a bankrupt concern, its plants were run down and obsolete, its buildings ramshackle." And, finally, "I found about eighty little structures scattered around the plant for this and that. They cluttered up the place. They were cheaply made, of lath and paper and what-have-you." [2]

But Schwab's claims were exaggerations, easily exposed. At a 1910 testimonial dinner for John Fritz, who had been Bethlehem's general superintendent, Schwab said that the buildings constructed at Bethlehem during Fritz's era were models of excellence which had withstood the years remarkably well. [3] Moreover, Bethlehem Steel had been the only profitable company in the shipbuilding merger, and it had been able to draw upon retained earnings to undertake major improvements and extensions.

The modern Bethlehem Steel Corporation traces its history back to the establishment of the Saucona Iron Company in 1857. That company's founder, Augustus Wolle, planned to build a blast furnace and to produce pig iron, drawing upon an iron ore deposit near the Saucon Creek in Pennsylvania. However, Charles Brodhead, a local attorney, persuaded him that the venture would be more profitable if the furnace was built near the Lehigh River and if the company concentrated on producing rails for the Lehigh Valley Railroad. Wolle accepted both suggestions, and for the next twenty-five years rails were the sole product of the Bethlehem Iron Company, a name change adopted in 1861.[4]

In 1873 Bethlehem adopted the Bessemer process and joined the Bessemer Rail Association, a group whose eleven member firms shared the use of patents and agreed not to infringe upon each other's sales territories.[5] Bethlehem's conversion to Bessemer production coincided with the onset of a severe depression. The demand for rails of both iron and Bessemer steel fell sharply, and Bethlehem, which had no regular customers for its new product, suffered from a steep decline in sales and profits until 1878, when the economy recovered.[6] Once the demand for Bessemer rails returned to the pre-depression level, Bethlehem's owners, always cautious and conservative, kept to their original policy—they made rails, and nothing else. They ignored the fact that the market for rails might collapse again, leaving the highly specialized firm in a precarious financial position.

But John Fritz, the renowned steelmaster who served as Bethlehem's general superintendent and chief engineer from 1860 to 1892, was convinced that Bethlehem should diversify its product line, and from 1880 on he made a series of suggestions to that effect. Invariably, his proposals met with stiff resistance from the company's owner-directors. First he proposed that Bethlehem produce structural steel shapes for bridges and buildings. In his *Autobiography* Fritz described the reaction:

> I urged the company with all the eloquent language I could command, but with no effect. Then I tried compulsion, by saying it was absolutely essential for the permanent success of the company to have some diversity in their business. . . . But it was no use, and some of the directors said I was never

satisfied, but must be at something new, and could not let well enough alone."

He then suggested that Bethlehem build a plate mill to supply a growing market for steel ships, but this idea also was rejected, as was his subsequent proposal that the company erect a forging plant to make steel shafts for ships and engines—although that idea was eventually accepted.[7]

Despite Fritz's repeated urgings, the company also failed to purchase sources of ore of the kind needed for making Bessemer steel,[8] so Bethlehem became increasingly dependent on foreign ore, from the Juragua mines in Cuba. The high cost of ocean freight from Cuba to the ports of Baltimore and Philadelphia and then inland freight by rail from the ports to the steel mill reduced Bethlehem's profit margins. But it was unable to pass along its higher costs in the form of higher prices to its customers; if it did, its customers might turn to other steel companies to supply them with Bessemer rails.[9]

Finally, in 1885, in order to offset the decline in profits from rail-making, the directors voted—again, at Fritz's urging—to undertake the production of heavy forgings. Fritz had convinced them that the profits on heavy forgings would more than compensate for the company's reduced profits on rail sales: a small amount of steel in the shape of guns or engine shafts would sell for hundreds of dollars per ton, whereas the large quantity of steel required for rails would only sell for tens of dollars per ton.[10] The directors, who had an aversion to any form of risk-taking, made this decision only after Bethlehem was virtually assured of a customer which would absorb its entire output—the United States Navy. Two types of heavy forgings were required by the Navy—hollow forgings for guns and solid forgings for the shafts of ships and engines. The Navy also sought a domestic supplier of armor plate.

In 1887 the Navy awarded the entire contract, valued at nearly $4,000,000, to Bethlehem, the sole firm bidding on both armor plate and gun forgings. The company established a reputation for quality and reliability in producing gun forgings; it even made deliveries ahead of schedule. Its success in this field earned it a contract from the United States Army in 1891—this for 100 large-caliber guns, also valued at nearly $4,000,000.[11] But while the company had had great success in forgings, it found that the construction of an armor plant involved numerous unforeseen complications and costly, embarrassing delays.[12].

After committing itself to produce for the Navy, the company built Siemens-Martin acid open hearth furnaces in order to manufacture the low-phosphorous steel required for armor and forgings. In August 1888 the first of these furnaces became operational, and then Bethlehem also began producing heavy forgings for ship shafts and pumps, open hearth ingots, and machine parts, as well as steel which was suitable for axles, springs, screws, and wire.[13] Bethlehem now had a growing number of military contracts from the U.S. Navy—as well as from foreign governments—and, after the initial production problems were solved, its profits from military work went up. In 1896 it decided to dismantle its Bessemer furnaces and totally discontinue rail production.[14] The sole surviving record of the company's sales and earnings for the early 1890's shows that the firm was solid and profitable.[15]

Nevertheless, Bethlehem was not as profitable as it might have been. Its operations were wasteful and inefficient, and the directors were unwilling to adopt cost-cutting methods. In 1898 Bethlehem hired Frederick W. Taylor, the controversial pioneer of scientific management, to introduce a piece-rate system to replace the company's existing day-rate wage system. Taylor first applied his time-and-motion study methods to the handling of raw materials in the yards; by the procedures he devised, only 140 men would be needed to do work which had previously required more than 400. However, Bethlehem's owners were not pleased. "They did not wish me, as they said, to depopulate South Bethlehem," Taylor later wrote. "They owned all the houses in South Bethlehem and the company stores, and when they saw we [Taylor and his assistants] were cutting the labor force down to about one-fourth, they did not want it." [16] Nor did Bethlehem's owners adopt Taylor's other suggestions which promised to cut costs and increase productive efficiency. Those included increased job specialization, standardization of work procedures, and salary increases for key personnel (in order to avoid the wasted time and unnecessary expense of training replacements if the key men left the company for higher paying jobs elsewhere).[17] The owners dismissed Taylor in April 1901, six weeks before Schwab bought the controlling interest in the company.

When Schwab took over Bethlehem, he discarded almost all of Taylor's policies. According to Taylor and his chief aide, Henry L. Gantt, Schwab replaced "scientific management" with the so-called "drive system." Taylor said that "the moment Schwab took charge of the Bethlehem works in 1901, he ordered our whole system thrown out. He saw no use whatever in paying premiums for fast work; much less in having time study men and slide rule men, 'supernumeraries,' as he called them, in the works at all." [18]

This was not at all out of character for Schwab. Even in his early days in steel he had never shown any interest in *how* results were achieved; his concern was that they *were* achieved. He himself had resented rigidly prescribed methods when they had been imposed upon him, and he did not insist that his superintendents and foremen follow uniform procedures. Schwab's method of management, as Taylor described it, consisted of placing

> . . . each department in command of a separate individual, who . . . is allowed to use practically whatever methods of managing the men he sees fit, the only check upon this man being that if he fails to make good in earning money, at the end of the year his head comes off. Under this plan you will necessarily have, in the same works, several kinds of management; and this accounts for the fact which has been asserted that a number of men in the Bethlehem Steel Co. are entirely satisfied with their treatment, while others feel that they are greatly abused. [19]

In August 1904, Schwab made his first official tour of the properties of Bethlehem Steel. He was accompanied by the company's president, Edward M. McIlvain, and its general superintendent, Archibald Johnston. He told a reporter who went with them, "I shall make the Bethlehem plant the greatest armor plate and gun factory in the world." Two months later he enlarged upon his promise:

> I intend to make Bethlehem the prize steel works of its class, not only in the United States, but in the entire world. In some respects the Bethlehem Steel Co. already holds first place. Its armor plate and ordnance shops are unsurpassed, its forging plant is nowhere excelled and its machine shop is equal to anything of its kind. Additions will be made to the plant rather than changes in the present process or method of its manufacture. [20]

In fact, however, Schwab did initiate a series of sweeping changes.

On December 10, 1904, Bethlehem Steel was incorporated in New Jersey, with a capitalization of $30,000,000. It, like U.S. Steel, was a holding company; it operated the Bethlehem Steel Company and the seven shipyards and one manufacturing company from the shipbuilding merger. Schwab became president of the corporation, while E. M. McIlvain remained president of the steel company. Schwab announced specific plans for enlargement so that the plant could produce all types of guns, gun forgings, and tools. His plan, he said, was to equal the output of Krupp in Germany and Vickers' Sons & Maxim in England.[21]

First, he strengthened and enlarged the corporation by selling off its unprofitable properties and investing the proceeds in those shipyards which had the greatest profit potential. The major operation was disposal. Schwab sold the Bath Iron Works Company and the Hyde Windlass Company, both in Bath, Maine. Then the facilities of the Crescent Shipyard Corporation in Elizabethport, New Jersey, were dismantled; later its site was sold. The Carteret Improvement Company, also in New Jersey, showed no prospect of living up to its name, so Schwab ordered it to remain idle until it could be sold. He made the same decision regarding the Eastern Shipbuilding Corporation at Groton, Connecticut, and the Canda Manufacturing Company of Elizabethport.

Then he turned to the properties worth saving. The corporation's three most valuable shipyards were the Harlan & Hollingsworth Corporation in Wilmington, Delaware, the Union Iron Works Company in San Francisco, and the Samuel L. Moore Corporation in Elizabethport. Each was equipped for shipbuilding and for marine repair work; the latter was a lucrative line of business which was not subject to cyclical fluctuations. At these three yards, Schwab replaced the top managers and allocated funds for expansion and modernization.[22]

The corporation's first annual report covered its activities during 1905. It struck a tone of buoyant optimism. Schwab announced the successful reorganization of Bethlehem's major source of iron ore, the Juragua mines in Cuba. He himself had spent many months examining potential ore sources in America which Bethlehem might buy or lease, and he had ordered a search made for coal and limestone deposits so that Bethlehem could become self-sufficient. Unlike the earlier owners of Bethlehem, Schwab understood the necessity of securing sources of raw materials long before those materials would be needed.

The report also announced the new direction in which Bethlehem Steel was moving. Schwab had authorized the first of a series of improvements and additions to the old plant. These included a crucible steel plant for making special steel alloys, a drop-forge shop for producing medium and light forgings (thus complementing the company's older heavy forging facilities), a machine shop for manufacturing large hydraulic presses and pumps, and a rolling mill for open hearth rails.[23] Out of the $2,900,000 which had been appropriated for improvements early in 1905, over $2,000,000 had already been spent by the end of the year.

The production of open hearth rails, which were more durable than Bessemer rails, was Schwab's first major departure from Bethlehem's traditional product line. Despite the fact that open hearth rails sold for $6 more per ton than the Bessemer rails, the new rolling mill at Bethlehem was soon operating at full capacity.[24] U.S. Steel, the nation's largest rail producer, did not follow Schwab's lead; it would have had to replace its Bessemer facilities with open hearth equipment. Being a late starter, Bethlehem enjoyed a clear advantage: with no heavy investment in obsolete equipment to protect, it could adopt the newest and most efficient technological processes.

At the close of 1905, Schwab said, "What I am going to do is to make it [Bethlehem] the greatest steel plant in the world. By that I mean the largest, most modern, the best equipped, the most highly specialized, and I fear also the most expensive steel works anywhere." [25]

A year later, Schwab published his second annual report. Despite its tone of optimism, it read like a chronicle of disaster. In 1905 the corporation's net income had been $2,600,000; it fell to $762,000 in 1906. The problem lay with Bethlehem's largest single customer, the federal government. In 1906 it had sharply reduced its purchases of armor plate and ordnance. Government orders had totaled $4,455,470 in 1905; they declined to $4,056,062 in 1906 and to only $2,635,054 in 1907.

An even more serious problem than the securing of new orders was the fulfillment of old ones. Government contracts for cruisers had been made by Bethlehem's predecessor, the United States Shipbuilding Company, and Bethlehem had to fulfill those contracts. Every one of those cruisers was built at a loss. But not all of the losses were legacies of the past. Shipbuilding contracts made in 1905, under Schwab's aegis, a year later turned out to have been made at prices too low to yield any profit.[26] Bethlehem's profits also were eroded by a natural disaster—the San Francisco earthquake. The Union Iron Works was in San Francisco, and in January 1906 it was partially destroyed by the fire which followed the earthquake.

Schwab realized that Bethlehem's economic future would be precarious if it depended almost exclusively on government contracts. The company needed a new commercial product, and that would require increased plant capacity and new land. Schwab began to buy property between the town of Freemansburg, in the Saucon Valley area, and the company's site in South Bethlehem.[27] He also recognized that he could not effectively run a business in South Bethlehem from an office on lower Broadway in New York City, so he decided to move to Bethlehem. There he, personally, could direct the company's growth. He closed down his mansion on Riverside Drive and moved into a modest fifteen-room house in Bethlehem.

When he began to take a more direct role in the company's management, Schwab discovered that he and McIlvain could not agree. Their views about the company's future fields of production were irreconcilable. The two men argued for months. Then, on July 1, 1906, McIlvain resigned. His successor as president of the Bethlehem Steel Company was Archibald Johnston, a Lehigh University graduate who had served the company for nearly a decade.[28]

When Schwab took over active direction of Bethlehem in 1906, the company was operating under two handicaps: it had no cost accounting records (the firm's cost sheets on which it based its prices had been destroyed in a fire early in 1906) and its organizational structure was ineffective. But Bethlehem suffered from another, more serious deficiency: there were no precisely defined areas of authority within the executive hierarchy. Even so, Schwab was reluctant to impose sweeping changes until he had mastered the operations of his new company, until he had learned the capabilities and limitations of his executives, managers, and foremen, and until he was able to inspect all of the company's properties.[29]

Early in 1906 Schwab asked his assistant, James H. Ward, which man Ward considered to be "the most outstanding fellow in this Bethlehem group."

"That's a foolish question, there is only one choice."
"All right, who is he?"
"E. G. Grace,"
"That's the one." [30]

Eugene Gifford Grace, born in Goshen, New Jersey in 1876, was graduated from Lehigh University in 1899. He was a fine baseball player, good enough to become a professional. He also had an engineering degree. Industry won out. Grace took his first job with Bethlehem Steel in 1899, working as an electric crane operator. After observing and suggesting ways to eliminate the waste and bottlenecks in the flow of raw materials and finished goods within the Bethlehem yards, he was promoted to supervise and overhaul yard traffic. It was in this capacity that he first met Schwab in 1904. He also handled the storage of Schwab's private railroad car. Schwab was immediately impressed by Grace's quick, inquisitive mind and by his obvious interest in and intimate knowledge of all aspects of Bethlehem Steel's facilities.[31] Grace, although he was somewhat shy and reserved, was also undisguisedly ambitious, and Schwab treated him as he himself had been treated by Captain Jones a quarter-century earlier.

In January 1906, Schwab chose Grace for an important assignment: to reorganize the Juragua Iron Mines in Cuba, which were Bethlehem's primary source of ore. Grace realized that it was a step up, but he feared that if he succeeded at it he might be given a permanent position in Cuba. He told Schwab,

> If you want me to do it, I'll do it. But I want to make one condition. I don't visualize my future as running an ore mine in Cuba. I will go down and stay just as long as it is necessary to get the Cuban mines operating successfully—until you are satisfied. If I can't do it, then you can fire me. But my ambitions lie back in Bethlehem.[32]

Schwab agreed, and Grace went south.

Grace found the Juragua mines to be a wasteful and inefficient operation, largely due to its lack of mechanization. He took charge of mechanizing the mines in every possible respect. He thereby reduced Bethlehem's cost per ton of ore substantially. Cuban ore was richer in iron and lower in phosphorus than was the Mesabi range ore used by U.S. Steel. It also had another advantage: it contained large amounts of nickel, so that Bethlehem could produce nickel steel at no extra cost. For a ton of iron Bethlehem's cost was $4.31; U.S. Steel's was $7.10.[33]

Then Grace returned to America. After he completed his report to Schwab, he said, "I just wanted to ask you if it is necessary for me to return to Cuba. The job is done there. I am sure that it will take care of itself properly with the men who are in charge, but Mr. Snyder and Mr. McIlvain [officials of Bethlehem] insist that it is my duty to go back there. I feel that it would be a waste of effort."

Without hesitation, Schwab told him, "You go back to Bethlehem right now, and stay there. I shall have work for you to do."

J. H. Ward, who witnessed the conversation, later recalled, "I shall never forget the expression of relief and elation on Mr. Grace's face. He looked as though he had seen a vision of things to come."[34]

Schwab liked Grace. He was enthusiastic about Schwab's plans to modernize and improve Bethlehem. In that respect, he was an exception, for many of the veteran Bethlehem executives preferred the old, pre-Taylor and pre-Schwab way of doing things. They resented Schwab; he was an intruder. One man went so far as to say that Schwab was crazy, and that his production goals were fantastic. Schwab singled out that man. He told him, as Ward recalled, "If we can get our blast furnace operations to such and such a state of efficiency, I'll pay off the mortgage on your house." The skeptic fell into line, pushed himself and his subordinates to reach the goal which he had said was unattainable, and, in a matter of months, found himself with a mortgage-free house. The lesson, which was well publicized, was not lost on the other old-timers.[35]

Expansion into the field of commercial steel was foremost among Schwab's plans for the growth of Bethlehem. But instead of choosing safe and familiar items, he decided to risk his fortune and the future of the company on a new and largely untried product.

It was a new form of structural steel, the invention of Henry Grey, who had been born in England in 1849 and had emigrated to America in 1870. Grey was serving as general superintendent of the Cleveland Rolling Mills in Newburgh, Ohio, when he learned of efforts to roll a structural beam directly from an ingot. In November 1896, the Ironton Structural Steel Company in Duluth, Minnesota, hired Grey and his associate, George W. Burrell, to try to carry to completion some earlier experiments which had been made by Levi and James York. The Yorks had been trying to produce a wide-flanged beam which could be rolled as a single section instead of being riveted together, as conventional beams then were.[36]

In 1897, Grey succeeded. The Ironton Mill had intended to perfect the method and then to sell the rights, so it closed down operations and Grey moved to New York. There he sought companies to which he could license the structural steel patent or which he could interest in purchasing the patent outright. In June 1897, Grey published an article in a leading steel trade journal. In it he described in detail the new process and its advantages. "In the construction of the iron work of buildings, bridges or other structures," he wrote, "it is highly desirable to secure the greatest possible strength with the least dead weight and at the lowest cost. This the Duluth girder insures." [37] At the end of the article Grey wrote, "This should be of interest to engineers and builders everywhere." But he misjudged his audience. The engineers and builders showed no interest; instead, they looked upon the beam with skepticism and indifference. As a later writer observed, "notwithstanding the fact that this new type of a mill showed distinct advantages over any other type which had been used for rolling wide flanged beams, manufacturers in the United States refused to acknowledge these advantages and even went so far as to refuse to believe in the possibility of rolling wide flanged beams satisfactorily in the mill in question." [38] The first plant which did utilize Grey's invention was built in a German-owned steel mill in Lorrain, in the Duchy of Luxembourg. Max Meier, the general manager of the Differdingen Works, had learned of Grey's work, and he hired him to supervise the construction of a Grey beam mill. It became fully operational in 1902. [39]

Schwab first became interested in the Grey beam while he was president of U.S. Steel. He thought the invention showed commercial possibilities, and he urged U.S. Steel to purchase the rights. But U.S. Steel's Finance Committee rejected his recommendations. In 1905, however, Schwab, as president of Bethlehem, decided to pursue the idea. He, along with Grey, inspected the Differdingen mill. A few months later he went through the mill again, this time with George Blakeley, Bethlehem's expert on structural steel. Then, in December of that year, Schwab announced that Bethlehem Steel would undertake the production of Grey beams. [40] He later chose Eugene Grace to direct the construction of the new plant.

Inasmuch as the invention had first come to his attention while he had been president of U.S. Steel, Schwab felt obliged to inform Judge Gary that Bethlehem planned to develop Grey's invention. He did so in 1906. After referring the question to a committee of experts who recommended against adopting the invention, Gary told Schwab that U.S. Steel was not interested in the new process. [41]

Schwab intended to construct the new mill on a site purchased in the Saucon Valley, about a mile from the main Bethlehem blast furnaces. Bethlehem needed $4,500,000 to build it, and, in addition to the royalty for the use of the patent rights, the company had to pay Henry Grey and his son, Charles, a fee of $85,000 for designing the plant. [42] Schwab decided that the most effective way to raise the money was to float a bond issue. He approached his friend Pliny Fisk, an investment banker.

After hearing Schwab's presentation, Fisk committed his firm, Harvey Fisk & Sons, to underwrite a $5,000,000 bond issue. At the end of their meeting, Fisk asked Schwab to put his proposal in writing.

Schwab sent him this description of the Grey beam mill:

> It is designed primarily for the construction of a special character of structural material, which is not now made by any other mill in the United States, and which will enable consumers to effect a saving on their buildings, as compared with present shapes and methods of construction. Columns which are now riveted together at an expense of $9 to $14 per ton will be rolled in one solid section, facilitating time of delivery as well as adding to the efficiency and strength of the material. . . . To one familiar with the trade the enormous advantages of such sections are quickly appreciated.[43]

When Fisk read this, he telephoned Schwab and said, "that isn't what you told me at all." Schwab later recalled that he went right over and showed Fisk point by point that that was exactly what he had said. Fisk replied: "It doesn't sound the same. I think I shall have to have you say it all over again in a phonograph and then I can sell those bonds."[44] Schwab was eloquent, but not as a writer. His salesmanship depended upon his personal enthusiasm, and that was transmitted through the spoken word.

Armed with Schwab's recording, Fisk's salesmen might easily have sold the bonds. But they never got the chance. A panic began in September 1907, and the American stock market collapsed. Stock prices fell sharply, and new stock and bond offerings were stillborn.[45]

In fact, the economic outlook was so bleak that many of the officers of Bethlehem Steel told Schwab that he should abandon his plans to build a Grey mill. Eugene Grace later recalled Schwab's reaction:

> This gave him pause. He believed in giving men leadership, not driving them. In most instances he was inclined to accept the advice of his colleagues, or if he could not sell them when he disagreed he was inclined to feel that his own decision needed to be reconsidered. In this instance he did at first accede to the opinion of his boys, as he called them, and gloomily took the train into New York to consider how a retreat could be worked out. The next morning he phoned his secretary very early and said, "Get up, Wardie, we are going back to Bethlehem and talk to the boys. I've thought the whole thing over, and if we are going bust, we will go bust big."[46]

From his experience at Carnegie Steel, Schwab knew that in years of depression, when the costs of labor and raw materials are lower than in normal times, a company is in a good position to expand. At Carnegie Steel such expansion had been financed by drawing upon undistributed profits earned during earlier periods of prosperity. Bethlehem Steel, however, had no such capital reserves to draw upon. For that reason Schwab approached various people who might give him loans based on his personal reputation and on the merits of his proposal. A bondholder might lose his investment if the Grey beam were a financial failure, but if someone gave Schwab the money as a personal loan and the beam failed, then Schwab would have a personal obligation to repay the debt.

He turned to some of his wealthy friends, but only Thomas Fortune Ryan, the banker, and Levi P. Morton, who had been Vice President of the United States under Benjamin Harrison, agreed to accept his personal note. Ryan and Morton each gave him $250,000. In return, as partial collateral, he gave each of them 2500 shares of the common stock of Bethlehem Steel (worth about $125,000) out of his personal holdings.[47]

But Schwab was impatient with this piecemeal way of raising the $5,000,000 he needed. It took too much time. So he suggested an unusual alternative to his contractors and suppliers: that they provide him with their services on credit until the Grey beam mill had begun to show profits—that is, that they work without immediate payment. In a sense, he wanted them to put up the money themselves.

He was successful. In one instance, he sent W. H. Tobias, Bethlehem's chief purchasing agent, to Philadelphia to negotiate with the contracting firm which Schwab had chosen for the excavation work. The owner, Captain F. H. Clement, was asked to accept Schwab's personally endorsed notes in payment for the work. Clement not only agreed, he told Tobias, "Yes, I'll take your notes, but if you fellows want any money, I'll lend you a million." [48]

Schwab himself talked to the directors of the Philadelphia and Reading Railroad and the Lehigh Valley Railroad, the two lines which served the South Bethlehem plant and which would serve the new Saucon plant as well. He reached an agreement with both lines: Bethlehem Steel would pay only 50 per cent of its total monthly freight bills. The other half, not to exceed $470,000, could be paid in drafts on Drexel & Company, a prominent Philadelphia banking firm. Drexel, in turn, would sell $1,000,000 worth of notes of the Bethlehem Steel Company at par value. If Drexel could not find a buyer, the railroads themselves would buy the notes. By this circuitous process, Schwab arranged for his suppliers to underwrite his new mill.[49]

Schwab's Grey beam mill became operational in January 1908. However, in July of that year, Bethlehem Steel faced a new financial crisis. It needed at least $1,500,000 in operating funds for the new mill and for meeting the interest payments on its bonds. For a growing company, Bethlehem was critically short of working capital. Schwab appealed to bankers and brokers of his acquaintance, but he was unable to raise the funds he needed. He then turned to a new source, the Chase National Bank, whose president, A. Barton Hepburn, gave his loan application a full hearing. Hepburn refused the request, and he explained to Schwab why he had done so:

> The large contracts you have and will have run into large amounts of money and it does not seem to me that $1,500,000 is at all adequate as a working capital for the conduct of your prospective business; twice that amount would seemingly be nearer a proper estimate, and yet I should think that would be rather small. It is very apparent, in fact your conversation made it explicit, that the object you have in extending your banking connections is to extend your borrowing facilities. A long banking experience has impressed me with the conviction that borrowing accounts should only be taken on by banks to a very limited extent. In periods of easy money they are desirable and they are correspondingly embarrassing when an opposite condition of the money market obtains. Therefore . . . the account seems to me to be undesirable from a banking standpoint.[50]

After having exhausted all other possibilities, Schwab turned to Andrew Carnegie. During the first four years following the Monte Carlo episode, relations between the two men had been strained, but in March 1906, Carnegie had learned that Schwab was ill. The old Scotsman then wrote to his former protégé, expressing his concern and offering fatherly counsel: "Have been on the eve of wiring several mornings to ask for the truth about you but desisted fearing reply might reach the Press. Don't like abscess on the thigh – in one so young—Blood out of order. Be a model like me and keep well." He then prescribed a diet which would surely restore Schwab's health:

> Baked apples – Oatmeal – chicken   Breakfast
> Fish or chicken – oysters *when in season*   Lunch
> 2 spoonfuls Old Scotch
>
> Soup, oysters – no meats – simple pudding   Dinner
> Sago, Tapioca or Rice
> 2 spoonfuls O.S.

Charles Schwab's birthplace, Williamsburg, Pa.

Charles Schwab at sixteen (1878).

Schwab's mother, Pauline, and his father, John.

Charles and Rana Schwab on their honeymoon in Atlantic City, 1883.

The Schwab's first home, Braddock, Pa.

Captain William R. Jones.

Henry Clay Frick.

Henry Phipps and Andrew Carnegie.

President Schwab of United States Steel.

Elbert H. Gary.

J. P. Morgan.

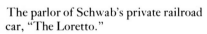

The parlor of Schwab's private railroad car, "The Loretto."

"Riverside," the Schwab mansion.

Eugene G. Grace (above, left).
Henry Grey (above, right).
Ernest R. Graham (left).

Cartoon of Schwab's appointment to head the Emergency Fleet Corporation.

Ivy Lee.

Edward N. Hurley.

Schwab and champion riveters at Hog Island
Shipyard, with President Woodrow Wilson looking
on, August 1918.

"Now, boys, I am a very rich man."
Director General Schwab addressing
members of the I.W.W. at a Seattle
shipyard in 1918.

Josephus Daniels.

Quadruple launching at Alameda Yard, July 4, 1918.

"Immergrun," Loretto, Pa.

Charles Schwab on his seventy-second birthday.

Charles and Rana Schwab at the celebration of their golden wedding anniversary, 1933.

Exercise every day, golf often –
Retire about 10 or 10:30
No wine – no liquor between meals.
Listen to Madam – she will approve this, I'm sure. . . . .[51]

Schwab may never have tried Carnegie's diet, but the letter itself was therapeutic. It was the icebreaker which restored their friendship. The next month Schwab placed his private railroad car, "The Loretto," at Carnegie's disposal for a visit to the midwest. It was a tour which Carnegie praised as the most enjoyable of his life.[52] Thereafter Carnegie and Schwab resumed their close relationship, mutually agreeing to obliterate the memory of past disappointments.

Then, in 1908, Schwab asked Carnegie for help. When Carnegie saw the Grey beam, he too was impressed by it, and he agreed to lend Schwab U.S. Steel bonds to use as collateral for bank loans.[53] A few months later, Carnegie paid public tribute to Schwab. While testifying before a Congressional committee, he said:

"I think Mr. Schwab deserves a vote of thanks by Congress. . . . He visited Germany, and he found in one mill the practice of rolling girders of scientific form. There is not a girder made in America that does not charge the customer for 15 per cent of steel in that beam which is useless. The form is not scientifically right. Mr. Schwab is a genius. I have never met his equal. . . . And that one fellow, not in the combination [U.S. Steel], and struggling against them, beset by financial difficulties that would have almost overwhelmed any other man, nevertheless resolved that his mill should have a beam mill equal to any in the world, and he has improved upon the German [mill] and is making beams to-day scientifically correct [sic]." [54]

Earlier, while Schwab was still trying to locate a new source of working capital for Bethlehem, he had turned his attention to the problem of executive reorganization. Because Grace had been so successful in building the Grey beam plant, Schwab had placed him in charge of the company. "I have every confidence," he told Grace, "in your ability to handle this whole situation, and leave it entirely to you." Schwab's role, by his own choice, was largely advisory.[55]

Schwab proposed to make Grace, then thirty-two years old, the president of the Bethlehem Steel Company, but Grace refused the offer. He later explained: "I refused to take the title of President believing that the company needed his [Schwab's] leadership indicating he was the head of it, and I was young and unknown. I said I would take the job and do the

177

work but I wouldn't take the title." [56] So Schwab made Grace the chief executive officer, with the title of General Manager.

The corporate structure of Bethlehem had been inherited from the old management, and it had a number of serious deficiencies. The worst problem was that the sales and operating departments functioned autonomously. Grace and his assistant, C. A. Buck, offered a solution: this was unacceptable to Schwab, but he could not come up with a better alternative.

> I invariably come back to my original objection that it [Grace's plan] throws practically all the responsibility and work on the general manager. In other words he is the Supt in charge of operations and yet responsible for all that selling as well. I am afraid that one will detract from the other. I realize the great importance of operations to sales—but is there any way that you and Buck with your fertile brains can devise of doing this without making yourselves the whole show. It may be the only way but . . . it makes me fearful of the awful load one man must carry. I don't want to seem vacillating but this is too important a matter to make a mistake in. . . . Could you not devise a plan whereby we could have a general sales agent who must work in close touch with you somehow but let the responsibility rest with him. [57]

On October 21, 1908, Schwab announced to the Board of Directors that he had "decided to make a permanent organization which shall go into effect at once, and in which the duties of each officer are so clearly defined as to permit of no misunderstanding in the future." His aim was to remedy what he regarded as a "serious defect" in the company's management structure—namely, "the lack of cooperation between the Works and Selling Department." He announced the appointment of Eugene Grace as "General Manager, who . . . [would] have general supervision of manufacturing as well as the sales and purchases"; of Archibald Johnston as first vice president, who would exercise complete authority over all sales involving government work, at home and abroad; and of Henry S. Snyder as second vice president, who would supervise all financial, legal, real estate, and transportation matters for both the corporation and the company. [58]

Production was now under Grace's direction; Schwab and Archibald Johnston shared the sales end of the business. Johnston was an expert in foreign sales, while Schwab concentrated on the domestic market. Negotiations with foreign governments for rails, munitions, or battleships often kept Johnston out of the country for six months at a time, but he got the contracts he wanted. He had both of the skills required of a first-rate salesman—patience and daring. He and Schwab were, in fact, very much alike in that respect.

Johnston's most daring exploit was his negotiation for Bethlehem's share of the European armor market. The European producers customarily allocated a small share of the market to Bethlehem in exchange for its promise not to undercut the pool price. In 1906, Schwab urged Johnston to try to obtain a nominal increase in Bethlehem's allotment. Instead, when Johnston met with the Europeans, he demanded a more than sixfold increase. He told the startled representatives, "We want $2,000,000. That is our lowest figure. We are not here to haggle and engage in any undignified bargaining. If we don't get it, we're not going to make any compromise. We shall merely withdraw." Basil Zaharoff, the leading spokesman for the Europeans, rose to answer Johnston's ultimatum. After a moment of hesitation, Zaharoff said, "Well, gentlemen, Mr. Johnston has made his proposal, and I believe there is no good in arguing against it." Johnston immediately sent a coded cable to Schwab, announcing his audacious coup. Schwab replied, "Message all garbled, incredible amount mentioned. Avoid code, send message straight." [59]

When Schwab finally realized that the code had not been "garbled," he was delighted, both with the deal and with Johnston. The two men remained close friends and business associates until Schwab's death in 1939. (Johnston died in 1947.) Johnston understood Schwab better than most men did. He realized that Bethlehem's Board of Directors was sometimes swayed more by the eloquence of Schwab's proposals than by their actual merits. On one occasion, in 1906, he told this to Schwab, who did not believe it. Johnston then suggested that when they returned to the meeting room after the recess Schwab should slowly and imperceptibly begin to argue *against* a proposal for which he had just won the board's approval. Schwab accepted the challenge, and, to his amazement, Johnston proved to be correct. The episode confirmed Schwab's high opinion of Johnston's shrewdness and independent judgment.

A year earlier, however, there had been an order of such a size that Schwab chose to handle it himself. In 1905, the Russian government was building the Trans-Siberian Railroad, and American rail manufacturers were eager to obtain the order for 65,000 tons of rail. Since no single firm had facilities which were large enough to produce the entire order, they agreed to establish a pool. They knew that the deciding factor in obtaining the order would not be the price per ton nor the ability to meet the delivery deadline; rather, it would be awarded by "the usual Russian method": a personal "gift" had to be given to the Czar's nephew, the Grand Duke Alexis Aleksandrovich. The members of the American rail pool selected Schwab as their emissary and jointly contributed $250,000 for him to use at his discretion.

Schwab had met Duke Alexis before, and he knew that it was best to approach him indirectly. On his way through New York he stopped at Marcus's, the famous New York jeweler, where he purchased "the most beautiful necklace I ever saw—of pearls, with a gorgeous diamond pendant." It cost $200,000. Schwab was not going to have a business conference with the duke, but a private supper with the duke's mistress. Throughout the evening Schwab never raised the subject of the order for rails. The two dined and chatted. Just as he was leaving, Schwab said; "Here is a small gift from your many admirers in America, among whom I count myself," and he handed her the package. She did not open it, but merely placed it casually on a near-by table. But his "gift" produced its intended effect; several days later he was invited to dine with the duke, and during that dinner he was handed a contract for the entire rail order.[60]

As soon as Grace had the management of Bethlehem firmly under control, Schwab was able to turn his attention to finding clients for the Grey beam. News of the beam's availability and cost advantage had brought in a few orders—8000 tons for the State Educational Building in Albany, New York, 3000 tons for a sugar refinery in Boston—but these were hardly adequate to sustain the plant whose actual cost of construction had been nearly triple the original estimate.[61] One of the first big buildings to specify Grey beams throughout was a taxicab garage in New York City. When the owners asked Schwab if he was certain that the beam would actually perform as he claimed, he said without hesitation: "There's no such thing as fail." Yet his confidence in the capability of the beam was not fully shared by everyone connected with its production; he later recalled that "even the most sanguine of the steel men in Bethlehem had their fingers crossed when they predicted success." [62]

In order to demonstrate the excellence of Bethlehem's newest product, Schwab needed to make a showcase sale to a leading architect for a major building project. Since his wide circle of friends and acquaintances included some of the nation's leading architects and construction engineers, he conferred with many of them, urging them to try the new beam. He nearly succeeded with Daniel R. Burnham, who was commissioned to build the Field Museum in Chicago. Schwab persuaded Burnham to build certain sections of it with Grey beams. But when the building of the Field Museum was postponed, Schwab lost his showcase.[63]

It was Pliny Fisk who came to Bethlehem's assistance. Fisk was handling a bond issue to finance construction of new headquarters for the Gimbel Brothers' department store in New York. The contractor, Louis Horowitz of the Thompson-Starret Construction Co., was afraid that the continued depression in the money market might cause Fisk to postpone or even abandon the Gimbel bond issue. When Horowitz pressed Fisk to close the deal, Fisk imposed one new condition: Charles Schwab must have the order for structural steel. Horowitz agreed to meet with Schwab. At their meeting Schwab said, "This is an important building. My sections are new and I want you to introduce them. I will name you a favorable price." Horowitz answered: "If the architect is satisfied, and if you make the price right, we will buy your steel." [64]

The architect, Ernest R. Graham, recognized the advantages of the Grey beam and agreed to use it. Horowitz and Schwab then negotiated a price—$32 a ton. After closing the deal, Schwab left for Europe. During the six weeks he was away, steel prices fell sharply, and the $32 a ton for Grey beams became higher than the price for conventional riveted beams. But Schwab was determined to meet the market. On his return from Europe, he went to see Horowitz. He said that he supposed "we will have to re-trade the Gimbel job." Horowitz refused. He said that if prices had risen during Schwab's absence, he would have held Schwab to the lower price, and that the original price they had agreed upon would stand. This contract, for 12,000 tons, valued at $384,000, was Bethlehem's first major sale of Grey beams.[65]

Even after the completion of the Gimbel building, Schwab faced the problem which confronts most innovators—the battle against inertia, the fight to persuade people to abandon a routine method for a new one.[66] Schwab worked out a new sales campaign; he persuaded architects to submit two cost estimates to their clients, one based on the use of conventional riveted beams, the other on Bethlehem beams—the name by which the Grey beam became more widely known. The new beam promised substantial savings in steel tonnage and in labor costs because it eliminated overlapping sections of steel and the need to rivet them together. Because of these significant advantages, and because of Schwab's relentless sales campaign, the Bethlehem beam began to win acceptance.[67]

Schwab had one client whose purchases were large enough to make the Bethlehem beam a commercial success. That man was Ernest R. Graham, the architect who had designed the Gimbel Brothers' store. Graham had bought substantial quantities of structural steel from Schwab when Schwab had headed Carnegie Steel and, later, when he had been president of U.S. Steel. Graham pooled the orders of his various clients; by doing so he was able to obtain a sizable discount, often as much as 20 per cent, from the standard price. But after Schwab left U.S. Steel Graham found it more and more difficult to obtain his usual discount from the corporation. Henry Bope, the head of U.S. Steel's sales department, told Graham that on future orders he would have to negotiate directly with Gary, and Gary told Graham, "We'll make no more contracts with you; we consider you a rather undesirable customer; you make us cut prices." This only served to drive Graham toward Bethlehem. On one large order, involving nearly 100,000 tons of structural steel, Graham advised Schwab that the order would go to the lowest competitive bidder. Bethlehem got the order; its bid was lower than U.S. Steel's. And Graham saved $35,000 on the deal.[68]

By August 1909, Bethlehem's sales of structural steel and open hearth rails had more than offset the continuing decline in armor and ordnance sales to the government. Schwab told a reporter for the *Wall Street Journal* that "all the departments of the Bethlehem Steel Corporation, with the exception of the ordnance plant, are running fully. We have been forced to refuse additional orders for structural steel, owing to the already congested condition of our mills." Three days later *The New York Times* carried a news item announcing Bethlehem's plans to enlarge its structural steel facilities. Schwab told the *Times*, "This year I will spend $5,000,000 improving the Bethlehem plant. There will be a new beam mill to roll smaller structural steel—six to twelve inch beams—so that I can run the original mill on larger sections." Commenting on the growing market for Bethlehem beams, Schwab said:

> We have found a steady demand for this new structural steel and the present necessity for rolling small sections handicaps the larger work for which the new mill was planned. No other concern can manufacture steel that way, as I have the exclusive patent rights for Bethlehem, but I have been agreeably surprised at the way this steel has found its place.[69]

The new mill was completed in 1911, and when it went into operation Bethlehem became the largest producer of structural steel in the eastern region of the United States. Commenting on the success of the Bethlehem beam, its inventor, Henry Grey, paid tribute to its promoter: "To Charles M. Schwab belongs the credit for introduction of this mill to the United States, as without his courage and backing the new structural mill and the new structural shapes would in all likelihood have been allowed to lie dormant for many years."[70]

Ernest R. Graham was so convinced of the superiority of the beam that he used it on all of his major commissions. In fact, the city of Chicago became an advertisement for the Bethlehem beam. Among the buildings Graham constructed with the new beam were the Merchandise Mart, the Field Office Building and Field Museum, the Chicago Opera House, the Insurance Exchange Building, and the Twenty Wacker Drive Building. Graham labored with missionary zeal to persuade other architects to try Schwab's newest product, but he met with considerable resistance. As he later observed,

> Architects are sometimes the most conservative and least progressive men in any profession. Their own formulae and mathematics could have told them the advantages of the Bethlehem beam, but they were too old-fashioned to see it until we proved by use exactly what an improvement it was.[71]

Once architects and builders saw how good the Bethlehem beam really was, Schwab obtained a number of large orders almost without effort. In New York City, the Metropolitan Life Insurance Company Building utilized the revolutionary beam, as did the new headquarters of the Chase Bank. Schwab never fully understood how he obtained large orders for the beam. When he tried to explain how he had secured the orders for Metropolitan Life and Chase, he said it had been "no trick at all," that they had been obtained through *friendship*. "I went in to see the president of the Metropolitan and never even mentioned steel. He brought the matter up and told me I was to have the contract. The same with my friend Wiggan [Albert G. Wiggan] of Chase National Bank."[72] In his explanation Schwab ignored the fact that he had earned a solid reputation for his company and its products; friends would never have awarded him huge contracts if they had doubted the value of his products or his ability to deliver on time, no matter how much they liked him.

Ernest Graham understood Schwab's abilities as a salesman. Schwab, he said, had "complete knowledge of his business. He never fooled himself. He knew what he could deliver and what he could not. There was never a moment of delay in deliveries of any steel we bought, and we always knew it was right." [73] Graham had begun his career as an architect in Chicago in 1888, and he first met Schwab around the time Schwab became president of Carnegie Steel. Their friendship and business dealings stretched over forty years. According to Graham, Schwab was so reliable that they could conduct their business on a casual, informal basis—he had such great integrity that signing a contract with him was a mere formality. In fact, Graham said, one contract between them simply read "50,000 tons," and it was written on a torn slip of newspaper. [74]

Although the Grey beam was Schwab's most successful innovation, it was not the only source of Bethlehem's rising sales and profits. Ironically, in 1909, after the Grey beam's commercial success was assured and the company no longer needed to depend on ordnance, there was a sharp upturn in military orders. Bethlehem received contracts from the U.S. Navy for 7731 tons of armor, valued at $3,200,000, as well as new contracts worth $2,300,000 for guns, gun mounts, target shells, and armor-piercing shells. [75] And early in 1910, Bethlehem received the largest single order it had ever obtained: a contract valued in excess of $10,000,000 to manufacture armor, guns, and shells for the navy of the Argentine Republic. [76]

While rising sales contributed to Bethlehem's growing profits, so did increased efficiency. Schwab adopted a major cost-reducing innovation in 1910. Before that time, Bethlehem had lacked its own coke-making facilities and had bought coke from the Lehigh Coke Company. Then Schwab signed a contract with a German-owned firm, the Didier-March Company, which agreed to build and operate a coke plant in South Bethlehem, with a daily capacity of 2000 tons. Didier-March was to be supplied with coal from Bethlehem's own mines, which would permit a considerable increase in the scale of Bethlehem's coal operations and cut costs substantially. Once the coke plant was operational, a further savings was achieved. American producers had always believed that the gases produced during the manufacture of coke were a useless by-product. But Schwab, while visiting a steel mill in Europe, saw something new: the Europeans were using those gases for fuel. Schwab arranged to purchase all the waste gas produced by Didier-March. This gas was piped into the main Bethlehem plant for only 8¢ per 1000 cubic feet, and it became the chief source of the plant's heat and power. [77]

Schwab had shrewd business sense; he was bold; he was willing to undertake high-risk innovations; he was determined to surmount all obstacles. Judge Gary of U.S. Steel was of a different breed. Gary was willing to tolerate the loss of a portion of U.S. Steel's sales to smaller rivals such as Bethlehem. By failing to match Schwab's aggressive policies, Gary contributed to Bethlehem's growing strength.

Schwab, unlike Gary, was an avid proponent of bonus plans; he believed that they encouraged extra effort and efficiency. In 1901, during his first brief ownership of Bethlehem, he had introduced an experimental bonus system for machinists, a system whose coverage he extended when he took over again.[78] Schwab's bonus plan was calculated on what the most and least capable machinists could produce during a given work period. The norm or base point was set at 80 per cent of the best man's output. Any man who was unable to reach the 80 per cent output figure most of the time was either dismissed or demoted to a less demanding job; any man who exceeded that figure received a bonus, and so did his foreman and his department head.[79]

In 1906 Schwab extended the plan, establishing a bonus arrangement for salesmen. This was designed to discourage them from cutting prices in an effort to increase the size of an order. The salesmen's bonuses were based not on the number of tons ordered, but on the amount of profit per ton.

In calculating the bonus of a superintendent of a blast furnace, the average monthly cost of converting a ton of ore into pig iron was taken as the base point. If he reduced the monthly average cost by 5 per cent, he received a bonus of one-fifth of the savings; if he reduced it by 10 per cent, he received one-fourth; and if he reduced it by 15 per cent, he received one-third.[80] Schwab also established a system of special bonuses to improve performance in five categories: breaking production records, saving time on repairs, achieving improved quality of steel made by the open hearth melters, reducing the amount of off-grade or second-class material produced, and recovering salable waste materials.[81]

Years later, Schwab described his system:

As the corporation has grown, the original plan has been extended as far as practicable throughout the organization. Bethlehem has made wide use of piece rates, tonnage rates, premiums and group incentive plans for wage earners, and various forms of incentive payments for supervisors and others in the ranks of management. Do so much and you get so much; do more and you get more—that is the essence of the system.[82]

Bethlehem Steel's bonus system directly and immediately rewarded any employee who increased his efficiency. The bonus system at U.S. Steel, which had been proposed by George W. Perkins and approved by Gary in 1903, was quite different. Under the U.S. Steel plan, the percentage of profits set aside for distribution to workers was based on the general progress of the company, not on the performance of the individual worker. Instead of paying bi-weekly or monthly bonuses directly to the worker, U.S. Steel reinvested the bonus money in shares of its own preferred stock. If a worker left U.S. Steel or was discharged for any reason, he forfeited all claim to the bonus.[83]

The major policy difference which separated Gary and Schwab, however, was not in their bonus systems. Gary had an aversion to any form of price competition; he considered it to be immoral and unprofitable, and he believed that Schwab had been a co-conspirator with Carnegie in this "offense" against business stability and harmony. From 1901 on, U.S. Steel not only established fixed prices for its products; it announced those prices in the steel trade journals. The giant firm generally resisted any price changes, whether times were prosperous or depressed. Gary believed that everyone in the industry would benefit from control of prices. Price and wage stabilization were major innovations within the industry; U.S. Steel led, and other firms followed.[84]

But not all firms followed, and not all of the time. Knowing U.S. Steel's prices, smaller firms secretly cut prices, while other firms entered new fields of production with confidence because they knew they would not face price competition from the giant of the industry. Those firms which sold below U.S. Steel's price sometimes received orders to the limit of their productive capacity—then the remainder would go to U.S. Steel. Although this "remainder" was as much as 50 or 60 per cent of the market, depending on the product, because of the competition from the smaller firms some of U.S. Steel's plants were forced to run at less than full capacity.[85]

Gary's price stabilization policy frustrated one of the Carnegie veterans, William B. Dickson, who was second vice president of U.S. Steel. In 1904 Dickson complained bitterly to President William E. Corey: the steel corporation was losing the respect of its competitors and the patronage of its largest customers, primarily because of its price policy on rails, structural material, plates, billets, and bars. Dickson explained:

> In economy of operations, reduction of cost, and improvement in the efficiency of our operating force, we have, I feel, accomplished even more

than was expected when the Corporation was organized. All this good work is, however, to a large extent neutralized by the irregularity of operations. For some time past we have been operating at not much over 50% of our capacity, including blast furnaces. As an inevitable result, costs have increased; works standing idle have deteriorated relatively more than if they had been in operation; the men are disheartened and a certain amount of apathy exists.[86]

Dickson believed that U.S. Steel's rigid adherence to stable prices in the face of price-cutting by its rivals simply built up the competition. The corporation, he said, had two options: it could buy out its rivals at high prices, or it could drive them out by resuming brisk competitive price-cutting policies. He counseled the latter course, saying, "we are even now looked upon by vigorous competitors as an 'easy mark.' . . . it is better by all odds to make . . . profit on a full output at competitive prices than by half output at artificial prices." [87]

Gary's blueprint for stability within the industry often worked against the dominant position of his own corporation. If, during the depression of 1907, U.S. Steel had cut prices as Carnegie Steel had done during earlier depressions, smaller firms such as Bethlehem would probably not have been able to survive. It was Gary's policies, at least in part, which enabled Bethlehem to withstand the depression and to introduce new commercial products. Fixed prices served as a magnet to draw aggressive smaller firms into price and product competition with the giant steel corporation. In 1906, for example, Schwab started to produce open hearth rails at Bethlehem. If U.S. Steel had cut prices on its Bessemer rails in the face of actual or even potential competition, he might have been deterred. But Schwab knew that Gary would ignore the challenge, that he would adhere to the prevailing price. U.S. Steel's high prices were a financial buffer: they made it possible for Bethlehem to absorb the heavy start-up costs of the new open hearth rail mill and to put up with higher unit costs until it could match U.S. Steel's economies of scale.

By his policy of price stabilization Gary created massive dissatisfaction among the managers, superintendents, and sales managers of U.S. Steel's subsidiaries. They repeatedly urged him to end fixed prices rather than continue to lose orders. Finally, in February 1909, at the third annual "Gary Dinner," which was attended by leaders of the steel industry, U.S. Steel announced that it was abandoning its price stabilization policy. Reluctantly, Gary had yielded to the men within the corporation who wanted to be able to cut prices in order to reduce competition from smaller firms.[88]

Although U.S. Steel dropped its fixed price policy, Gary continued to condemn "cut-throat competition" and to urge "friendly cooperation" within the industry. On October 15, 1909, he lashed out against the "savage" competition which had prevailed before the formation of U.S. Steel:

> A competitor was treated as a common enemy. Methods for his defeat and overthrow were used regardless of good morals or good policy. . . . Certainly, it was not permanently beneficial to the general public; and, from the standpoint of good morals, was a shame and a disgrace. . . . there was in some cases lack of confidence, a withholding of information, a piracy of business, an indiscriminate and reckless cutting of prices . . . an overbearing, unfair, destructive competition which drove many out of business, kept many others on the ragged edge of existence and brought demoralization to the industry.[89]

Gary's attack on old-style competitive tactics received warm praise from Schwab, John A. Topping of Republic Steel, Willis L. King of Jones & Laughlin, and other leading steelmen. They were undoubtedly aware of the fact that if they had been facing Andrew Carnegie the story would have ended quite differently. Understandably, they were grateful to Gary. His passive, tolerant strategy had provided their companies with a protective umbrella. But despite their gratitude, Schwab and the others had no intention of giving anything more than lip service to Gary's "friendly cooperation."

Bethlehem Steel was now in a strong position to face any domestic competitor, even U.S. Steel.

10

Tariff and Labor Controversies

Schwab was confident that Bethlehem Steel could match any domestic competitor, yet he claimed that the company could not withstand unrestricted competition from foreign steelmakers. He believed that Bethlehem's survival was intimately tied to high protective tariffs on steel—that in the absence of tariffs, Bethlehem could be undersold in the eastern United States by European rail producers. U.S. Steel was not as threatened: it sold its rails primarily in the area west of Pittsburgh and thus was not vulnerable to foreign steel. Its chief protection was the high cost of inland railroad freighting; sending steel products 100 miles by rail cost as much as shipping them 3000 miles across the Atlantic.

Schwab also believed that there should be no import tariff on iron ore. He did not want to rely on expensive American ore bought from sources controlled by competing steel companies; instead, he insisted that Bethlehem must develop independent ore supplies. But few undeveloped sources of ore were available in America, so Bethlehem had to depend on the Juragua mines in Cuba. In 1908, however, Bethlehem's Cuban sources were jeopardized; a member of Congress proposed that an import duty of 30¢ a ton be placed on iron ore. This would have benefited U.S. Steel, whose primary source of ore was located within the United States, in the Mesabi range, near Lake Superior.[1] Bethlehem, on the other hand, would have suffered considerably from such a duty.

Schwab and the rest of the eastern manufacturers decided to fight the proposal. Schwab, as head of an association of pig iron and steel manufacturers operating in the eastern United States, became the intermedi-

189

ary between the steelmen and Senator Boies Penrose, the powerful Pennsylvania Republican. (In return for Penrose's support, Schwab contributed heavily to the Senator's political campaigns.) As Schwab noted, "I am sure, my dear Senator, that the best interests of all the Eastern people, who are your constitutents, will be served by free ore, and, indeed, not only served, but preserved. Unless we get this, we will be in no position to compete with our powerful antagonists in the West." [2]

Schwab was not a consistent protectionist. He supported free trade for foreign ore despite the fact that such ore would reduce the market for domestic ore producers—but he defended protective tariffs for finished products, such as steel rails. He was only for free trade when it suited his own interests.

In 1908 the steel industry faced a renewed threat: reduction in protective tariffs. The Chairman of the House Committe on Ways and Means, Sereno E. Payne, a New York Republican, announced that hearings would be held on "Schedule C—Metals," in late November. All interested parties were invited to present their views. The hearings, however, were unsystematic and unfocused. The witnesses seemed to be going through the motions, knowing that their words would not alter the opinions of either advocates or opponents of the tariff. Each speaker appeared to be engaged in an act of public piety—some were there to condemn the tariff as the mother of trusts and the robber of the workingman's pay; others, with equal fervor, argued that the tariff was the cause of America's industrial supremacy and that it safeguarded the American worker against the menace of cheap foreign labor. These were exactly the same arguments which had been raised since 1870, when the first tariff on steel was passed.

Both sides used terms such as "fair profits" and "reasonable prices" repeatedly, without defining them or finding a commonly accepted meaning. The Congressmen and their witnesses wandered through a maze of shifting numbers—they talked of the costs of ore, coke, and limestone, of conversion costs per ton, of freight rates and wage rates, of selling prices and profit margins. But to no avail. These figures varied from year to year and from region to region; there were no fixed reference points, no standards against which changing prices and profits could be judged. [3]

Schwab's advocacy of continued high protective tariffs on foreign rails was made the more difficult by Andrew Carnegie. In early December 1908, in an article in *The Century Magazine*, Carnegie declared that tariff protection was no longer necessary, that he had once supported John Stuart Mill's "infant industry" argument for the tariff, but that now America's infant industry had grown to manhood: "The infant we have nursed approaches the day when he should be weaned from tariff milk and fed upon the stronger food of free competition." [4]

Then, on December 15, 1908, Schwab testified before the committee. He challenged the logic of Carnegie's article, claiming that American steel products could not compete with European exports because of the higher wage scales which prevailed in America. Europe's advantage in paying lower wages had once been offset by the high degree of inefficiency and waste which had characterized many European firms. But now conditions had changed, said Schwab. A "renaissance" had occurred in German steel-making, and producers throughout Europe were copying the cost-saving innovations of the German pace-setters. By combining high efficiency with low wages, Europe could inundate America with its rails—hence the need for tariff protection. [5]

Schwab's argument contained a number of dubious assumptions. In describing Europe's alleged advantage, he treated low wages and inefficiency as if they were causally unrelated. If it were somehow possible for European steel mills to achieve rising productivity without any increase in wages, then what prevented American companies, including Schwab's, from achieving the same state? Schwab always maintained that wage levels were determined by "market forces" or "the law of supply and demand," which, presumably, were beyond his control. But if that were so, how was it that the European steel producers could exempt themselves from the laws of the market? None of the committee members posed these questions to Schwab, but they did raise others which challenged his arguments for tariffs.

Schwab was shown a letter he had written to Frick on May 15, 1899. In it he had boasted that Carnegie Steel could produce rails at a net cost of $12 a ton, while the most efficient steel companies in Europe had costs of $19 a ton. Consequently, he had predicted, Carnegie Steel could do an increasingly profitable business selling abroad, even underselling European producers in their home markets.

> What is true of rails is equally true of other steel products. As a result of this, we are going to control the steel business of the world. You know we can make rails for less than $12 a ton, leaving a nice margin on foreign business. Besides this, foreign costs are going to increase year by year, because they have not the raw materials, while ours are going to decrease. The result

of all this is that we will be able to sell our surplus abroad, run our works full all the time, and get the best practice and costs in this way.[6]

Clearly, this gave no evidence that the American steel industry needed to fear foreign competition—quite the contrary. Schwab had boasted of superior efficiency, lower costs, rising sales in foreign markets, and soaring profits. How, he was asked, did he reconcile this optimistic letter of 1899 with his pessimistic prophecy of 1908?

Although the committee members doubted that Schwab could explain away his earlier optimism, he managed to do so. He told them why he had written the letter. Frick had asked him for a statement of Carnegie Steel's profit prospects which could be shown to a prospective buyer of the company. In order to place the best possible construction on Carnegie Steel's sales outlook, in his letter Schwab had only considered actual mill costs, while excluding charges for interest and depreciation, which would have substantially raised the $12 per ton cost. Schwab explained that "that letter was written as [sic] an enthusiastic and optimistic young man seeking preferment in a great company." [7]

Now, he claimed, conditions had changed significantly, and the prices he had quoted in the 1899 letter had only antiquarian relevance. Although domestic rail sales were rising at a slow rate, costs of production, particularly the charges for labor and transportation, were soaring. Tariff protection was necessary, Schwab insisted, so that the American producers could maintain the high wages they paid to American steel workers. If tariff protection were removed, then those wages would have to be reduced proportionately.[8]

But how could Schwab claim that American rail sales were barely profitable? For some years, ever since the formation of U.S. Steel, rail prices had been stabilized at $28 a ton, a substantial rise from the $17 range in which they had sold in the years immediately preceding 1901. Did this not signify that American producers were now making handsome profits? "Well, it was time we were making some money," Schwab said; it was necessary for steel producers to make a 25 per cent gross profit on their products in order to cover the high turnover in equipment caused by rapid technological obsolescence. When these huge outlays were taken into account, only a very thin margin of profit remained.[9]

Schwab did not mention the fact that manufacturers introduced technological innovations for the very purpose of reducing costs. Additional capital outlays are offset by lower production costs; if that were not the case, then the investments would be uneconomical and the producers would not make them. Schwab seemed to be saying that he wanted to make profits in order to pay for new equipment, whereas the actual situation was that he and the other steelmen introduced new equipment in order to increase profits.

Schwab was also asked whether a price-fixing conspiracy existed among American rail producers: every one of them charged $28 a ton. He denied any collusion. The steel producers charged a uniform price because they all recognized the futility of price-cutting:

> I, for example, as a rail manufacturer, feel that if I were to vary that price of $28 for rails, which seems to have been recognized by all rail manufacturers as a fair price, and giving a fair profit,—if I were to vary that 10 cents a ton to-day, I would precipitate a steel war, to use such a word or expression, that would result in running my works without any profit. Everybody, by tacit and mutual understanding, feels the same thing about that. I would not vary the price of my rails under any circumstances, not if I knew it was to get 100,000 tons in orders, for the reason that my competitor next door would put the price down . . . and we would be in a position where we would be running without any profit at all.[10]

But Schwab knew quite well that secret price-cutting was rampant throughout the industry. The small firms undercut the "official" price established by the industry's giant, U.S. Steel, until they obtained enough orders to exhaust their productive capacity. This fact, one which William B. Dickson of U.S. Steel had noted with such alarm, Schwab chose not to mention. What he was trying to achieve by withholding this information from the government and thus allowing the charge of price-fixing to stand virtually unchallenged is not clear.

When Schwab learned that the House Ways and Means Committee had invited Andrew Carnegie to testify, he became alarmed. He feared that Carnegie might persuade the committee that tariff protection was no longer necessary. He was so worried, in fact, that he tried to get the committee to retract its invitation. Schwab urged Pennsylvania Republican Congressman John Dalzell, his ally on the committee, to suggest this, but Dalzell was not receptive. Several days later, on December 19, 1908, Dalzell wrote to Schwab, explaining why Carnegie's appearance was necessary:

> If the committee should formulate a bill in which tariff duties should be retained to some extent upon the metal schedule, the public would say we had

193

only heard one side of the question; that while you and others claim that duties are still necessary, we had refused to hear parties who deny this. In self-defense, therefore, nothing remained for the committee but to insist upon Mr. Carnegie's making his statement of the case so that when the bill was finally formulated, the public might have an opportunity to judge as to whether or not we were justified in the conclusions that we arrived at. I knew at the time that you telephoned me just how radical Mr. Carnegie's views were.

Dalzell assured Schwab that when Carnegie did testify it would become clear to everyone "that his conclusions were not founded upon figures and were drawn during a period when conditions were entirely different from what they are now." [11]

In his testimony before the committee Carnegie did not contradict Schwab's claim that some form of tariff protection might still be necessary. In fact, he lent his prestige to Schwab's contention that eastern rail producers were seriously threatened by European exports. The only steelmakers for whom tariff protection was unnecessary were those located far inland from the Atlantic seacoast; Judge Gary, Carnegie noted, could well afford to say that tariff protection might no longer be needed. [12]

Schwab and the others were unsuccessful in persuading Congress to maintain a high tariff. In 1909 it passed the Payne-Aldrich Act, which substantially reduced tariff protection for American rail producers. The old duty rate, established in the 1897 tariff act, had been $7.84 per ton; the new rate was $3.92. [13] However, the impact of the reduced tariff on Bethlehem was not severe. By then the company had diversified, and rails represented only a small fraction of its total sales. Over-all profits were not seriously eroded.

By 1910, Bethlehem Steel was flourishing. It once had been a sluggish company which depended heavily on military orders; now it was diversified and well-managed. Rising sales of both commercial and military products, combined with increased productive efficiency, led to rising profits. Bethlehem had become a formidable rival to U.S. Steel, and Schwab enjoyed renewed prestige as an industrial leader. The man who only six years earlier had been widely denounced as a wrecker and brigand was now acclaimed by the same newspapers as a builder and innovator. But once again—as had happened so often—his hard-won successes and achievements were suddenly overshadowed by a new crisis: this time by a strike at Bethlehem Steel in 1910.

The workers at Bethlehem Steel—both the unskilled Hungarians at the base of the labor pyramid and the skilled machinists at the top—had a common grievance: they felt that wage rates were too low and working hours too long. They were particularly irritated by Schwab's refusal to pay time and a half for overtime. When orders surged in and extra tonnage was required, the superintendents simply ordered the men to work overtime at the regular hourly or tonnage rates. Since the company had paid time and a half for overtime prior to the 1907 depression, the workers were asking for a return to the *status quo ante*, but with one change: now that the company was approaching a record level of profits, the men demanded the complete elimination of Sunday work.[14]

Early in February 1910, a delegation of three workers, men representing over 700 mechanics in machine shops #4 and #6, presented their demands to C. A. Buck, the general superintendent of the Bethlehem plant. Buck's response was direct and unequivocal: he fired the three men. A year before, Buck had fired another such delegation which had made the same demands. This time, however, the workers pressed for their demands; they were confident that the company could not afford to refuse them and thereby risk a strike.

The machinists sprang to the defense of the men who had been fired. On February 3 they held a general work stoppage in the two largest machine shops. Buck immediately got in touch with Schwab, and Schwab summoned the disgruntled workers to a mass meeting the next day in the Municipal Hall of South Bethlehem. He promised that he and Grace would listen to their grievances and attempt to reach an amicable settlement. A spokesman for the machinists presented their demands: the company should end Sunday work, pay extra for overtime, and rehire the delegates who had been fired.

Schwab was conciliatory. He was willing to rehire the three men, and he was willing to examine the grievances of all the others—but on one condition: the strikers must first return to work. Only then negotiations could begin. "If we cannot agree," Schwab said, "then the strike is your right. Don't be foolish; do what is right. If we cannot work without fighting, we should separate." [15]

When the strikers reconvened the following day to consider Schwab's offer, their ranks had swelled. The first men to strike had been the skilled machinists. Now other workers, including unskilled laborers and apprentices, had joined the strikers; they also hoped to have their own grievances heard. At the meeting the men complained about their hours and wages. The skilled machinists earned between 15¢ and 27¢ per hour for a normal six-day workweek (including Sunday) of ten hours and twenty-five minutes daily, plus five hours and twenty minutes on Saturday. These were the best terms; men with lesser skills worked longer hours for less pay per hour. The machinists complained that recently they had been ordered to report for work at 6:40 a.m. and to work through until 8 p.m., an extension of three hours a day without time and a half for the overtime.

Prior to the strikers' second meeting, the company issued a statement. Again it was conciliatory. "The question of overtime has never been brought to the attention of officials for arbitration, settlement, or discussion," but it would be considered as soon as the strikers returned to their jobs. At the strike meeting, four speakers addressed nearly 1200 men. One of them suggested it might be best for the men to return to work and thereby test the company's good faith; a strike should be a last resort. The second argued that their only hope for improved conditions would be to organize a branch of the International Association of Machinists. The third urged them to be cautious: the strikers should carefully calculate the extent of their resources for waiting out the strike period; unless they had enough savings to tide them over, they might soon be driven to ask for their jobs back—a humiliating situation. The fourth advised them to abandon the strike, go back to work, and proceed to organize all the workers into a single powerful union; then the next strike would produce a total work stoppage which would paralyze the plant. "Go back and do not forget to organize. In time of peace, prepare for war." [16]

At a press conference held earlier that day, Schwab had minimized the strikers' grievances. The company's payroll for January 1910 had been the largest in its history: its workers had been paid over $458,000 in wages and its foremen and managers had received $117,000 in salaries. This new plateau, Schwab said, was due to the growing volume of orders the company was receiving. He mentioned with obvious pride, an allocation of $12,000,000 for new plant and equipment which the company was planning to spend in the near future. [17]

Schwab had hoped to convince the strikers that Bethlehem Steel's workers had a bright future, but his remarks did not have the desired effect. Instead, the strikers concluded that the company could not dispense with their services even for a short period, and, therefore, it would be impelled to make concessions. They believed that Schwab would not jeopardize the profits which armor and ordnance sales brought him, nor risk financial penalties and forfeitures if Bethlehem failed to deliver on contract deadlines.

It is unfortunate that Schwab, in his brief statement to the press, failed to discuss certain facts which might have strengthened his position vis-à-vis the strikers. Since the men had not yet taken a formal strike vote, this was his best, and, perhaps, his last, chance to defuse the potentially explosive situation. But he did not.

The strikers apparently believed that Schwab, if only he *chose* to, could unilaterally grant their demands, that his assent would not affect the company's costs of production, or its prices, sales, and profits, or, ultimately, their own jobs. He could have made the point that if wages were increased or the workday shortened without any corresponding increase in the productivity of every worker, then someone would have to bear the added costs. Two possibilities existed: either Schwab could have passed along these new costs to his customers in the form of higher prices or he could have absorbed the costs himself by accepting reduced profits. But neither possibility was feasible or fair.

If Bethlehem had raised its prices it would have placed itself in a position of great competitive disadvantage, and it would have lost sales to steel mills which had not raised theirs. And even if all of the domestic companies had raised their prices, there would still have been the threat of foreign steel, which could then have undersold American steel in its home markets—unless foreign steel were excluded by an increase in the tariff, in which case American consumers would have borne the cost of the wage hike.

The strikers had not considered the effect of higher prices on the market for steel products, and Schwab could have explained it to them. If the company could have boosted its prices *without losing most of its customers*, then surely it would have done so, if only for the sake of higher profits. And from those profits it could have paid higher wages. But price increases would necessarily have triggered a slump in the company's sales, and that in turn would have caused a decline in the number of workers needed. Therefore, while some workers would have benefited from a wage increase, others would have been laid off, either temporarily or permanently—a possibility mentioned nowhere by the strikers in their increasingly inflammatory rhetoric.

Then, too, the strikers claimed that their wage demands could be financed if Schwab and the other Bethlehem stockholders would accept reduced profits. That was a belief which Schwab could have challenged successfully. In 1907, Bethlehem and all its competitors had discontinued paying time and a half for overtime, and, at the same time, Schwab had also stopped paying dividends on Bethlehem's stock, both common and preferred. (Dividends were not resumed until 1916, nine years later.) It was not that the company was in trouble; rather, Schwab was plowing all available profits back into the business, financing expansion and continuing the program of diversification. Thus, there simply were no idle hoards of cash which could be used to meet the workers' demands. And even if Schwab himself had been willing to relinquish his entire salary, $50,000, a sum which was colossal compared to a workman's wages, that would only have covered an increase of $6.25 a year for each of Bethlehem's 8000 workers, which would have amounted to a raise of less than one-quarter of a cent per hour.

Of course, Schwab never suggested that he serve without salary; it would have been an injustice. He was the one indispensable man at Bethlehem, even from the strikers' point of view. He was the entrepreneur, the *Arbeitgeber*—the work-giver, the man who created jobs. Under his leadership, Bethlehem Steel's work force had doubled, and it would double again within the next five years.

It is tragic, therefore, that some of his contemporaries viewed him as an exploiter of the workingman. But he himself allowed this distorted view to persist: he failed to challenge the belief which made him the target of the strikers' wrath and resentment—the belief that he, *arbitrarily*, was refusing to grant their demands.

On the afternoon of February 5, the workers held another meeting. At that meeting they voted to strike.

For the strike to be effective, the strikers would have to force a general shut-down of the plant. Management knew that if only a few hundred machinists out of nearly 8000 full-time workers were on strike, they could be replaced by men hired from other cities. The striking machinists knew it too. They sought allies, especially among the many unskilled foreign-born laborers.

Representatives of the machinists and the unskilled workers came to an understanding. The machinists agreed to support a campaign for higher wages for the unskilled laborers (who were then earning 12½ cents an hour), while the laborers agreed to come out on strike and remain out until the machinists' wages, as well as their own, had been increased. A few days later, when Schwab attempted to end the strike by offering the machinists a token raise of one cent an hour, they refused to take it; they kept their solidarity pact with the unskilled men.

The strikers knew that they should organize more effectively, so they accepted the assistance of both the International Association of Machinists and the American Federation of Labor. On February 7, F. J. Conlon, vice president of the International Association of Machinists, addressed the strikers. He told the Bethlehem machinists about the working conditions of his union's members in Washington, D.C. They earned $3.76 a day for nine hours' work, "and all because they belong to the International Association." He also told the strikers that they should organize a strong union; it was their only salvation. Conlon must have been convincing, for, minutes later, when David Williams, chairman of the executive committee of the striking workers, urged them to return to work, he was voted down. Williams believed the men were not well enough organized to withstand a lengthy strike, and he advised them to return to work in the hope that their demands could be satisfied through negotiations with Schwab.

After the men voted to remain out on strike, Williams joined them. He said, "Let us go into it right; let us give them a fight; let's get into these unions. We carry the olive branch to them, and they discriminated against us; let us give it to them. We can tie the plant up." [18]

The strikers' resistance was stiffened by an announcement made earlier that same day by General Superintendent Buck. He had announced that since the men had not accepted the company's first offer, they would not be allowed to return to work as a group. Each striker would have to file a new application for employment before he could obtain his old job.

After the strike vote was taken at that meeting, William Duffy, the secretary of the strike committee, sent a message to Schwab, enumerating the strikers' demands:

> The overtime feature of employment having been a detriment to our health, our homes and families, we ask for its abolition. If it must be a feature of future employment, we ask for an additional compensation in the form of time and 1-half for overtime, so that we may procure additional nutriment to give strength to our bodies to perform the task.

Schwab made no reply to this. He did not need to. The strike was still confined to a minority of Bethlehem's work force, and the machine shops and furnaces reported that production was going on at nearly full capacity. Over at the near-by Saucon Valley plant no one was on strike. And, although Schwab knew that a few more men from the main plant at South Bethlehem had joined the strike in the preceding two days, he also knew that the new strikers were outnumbered by the men who had abandoned the strike and had returned to work.[19]

Later that same day, February 7, the strike intensified. Five hundred unskilled workers walked out, and they were followed by 150 molders. Each group of strikers pledged its support to the others. Encouraged by the growing number of strikers, Chairman Williams exclaimed, "We are going to win. We are to have our nights to ourselves and the Sundays also."[20]

Other groups of workmen, aware that Schwab wanted to avoid a general strike and shut-down, presented him with demands of their own, with the clear implication that if these demands were not satisfied, they, too, would go out on strike. The riggers demanded an hourly wage increase from the current level of 12½ cents to 20 cents. When Schwab refused on the ground that this type of settlement would set a dangerous precedent for other workers who were not on strike, 200 riggers joined the strike. When Williams heard of this development, he predicted, "We will win, for every ten men who return to work, we are getting 200 out." J. P. McGinley, a strike organizer sent in by the A. F. of L., claimed that 200 men had joined the new Bethlehem chapter, and an organizer for the International Molders Union said that 300 molders had joined his union.[21]

During the next few days the strikers waited, but Schwab gave no sign that he was willing to compromise or capitulate. Then, in an effort to win public support, the men tried to gain the allegiance of the opinion-makers of Bethlehem, especially its clergymen. Williams reminded his men that Sunday was the Sabbath, the day intended for prayer and rest. "Go to your pastors and priests and tell them that it is to their interest to come to your aid; that the conditions robbing you of the rest needed and the work on Sundays have been fixed by one man—C. M. Schwab."[22]

The strikers did appeal to the clergy, but there was little favorable response. The only allies they found were the Catholic priests, most of whose striking parishioners were unskilled Hungarian laborers. Jacob Tazelaar, an A. F. of L. organizer and the personal representative of Samuel Gompers, said bitterly, "We had expected that the ministers of the Gospel would remain at least neutral instead of coming to the rescue of this greedy corporation."[23]

The strikers did find other supporters, but this often worked to Schwab's advantage, not to the strikers'. One such ally was Mrs. Gertrude Bresslau Hunt of Chicago, a labor organizer. She attempted to turn the strike into a crusade for socialism—a cause which was repugnant to most Americans. Mrs. Hunt told the strikers, "If they cannot run their steel business at a profit to clothe, feed and shelter their employees decently, if they cannot run it decently, then it is time the Government take possession and see that it runs the business for the whole body of the people." Her remarks, which were widely quoted, won considerable public support for Schwab, while the strikers suffered a corresponding loss of sympathy.[24]

During the next two weeks Schwab showed no signs of relenting. Hundreds of strikers had left Bethlehem for jobs which the union organizers had obtained for them in other cities, and Williams was certain that, given the high demand for steelworkers throughout the country, Schwab would not be able to find sufficient replacements. Williams and the other strike leaders continued to counsel moderation. The strikers were warned against being tricked into acts of violence by "Pinkerton labor spies," and they were advised to stay sober.[25]

As the strike entered its third week, it was still free from violence. However, the very presence of groups of strikers outside the various entrance gates of the South Bethlehem plant created friction. As the strike continued and the strikers' financial resources dwindled, tempers grew shorter. More than anything else the strikers hoped to cause a general shut-down of the plant, and for that reason their greatest anger was not directed at Schwab, but at the non-striking workers. Now, in the third week, as workers passed in or out of the plant gates they were jeered at and chased by strikers. Fearing that violence might erupt, Eugene Grace sent a letter to the Sheriff and the Board of Burgesses of Bethlehem, warning that if any damage were done to the plant by strikers, or if any non-strikers were intimidated or injured, the company would hold the town liable for the damages. Grace demanded that the town double its police protection around the plant.[26]

That protection was sent to Bethlehem even before Grace made his request. Chief of Police Hugh Kelly had already appealed to the Sheriff of Northampton County for additional men, and the Sheriff had relayed the request to Governor Edwin S. Stuart of Pennsylvania. Both the Chief of Police and the Sheriff claimed that the situation in South Bethlehem was potentially menacing, that the recent minor skirmishes might turn into major riots, and that it was impossible to deputize a sufficient number of men from the local area who could be impartial. At first the

Governor refused to intervene, but he reversed his decision after he received a second appeal from the Sheriff. He ordered the State Constabulary to send twenty-five men into Bethlehem.

The appearance of the mounted state police only intensified the conflict. As they rode through town some people hissed and jeered at them. One striker was alleged to have yelled "scab" at a state constable; he found himself under arrest. The constables tried to keep crowds from forming, and they began to use their riot clubs. An innocent bystander, a striker who was patronizing the barroom at the Majestic Hotel, was accidently killed by a stray bullet from a constable's gun. This shooting, which some people suspected was intentional, won new public sympathy for the strikers.[27]

On February 27, Schwab announced that the following day the company would resume its normal work shifts at full production. Schwab was confident that the strike would end: he believed that the presence of the constables would ensure that anyone who wanted to work could enter the plant gates safely, and that this would reduce the ranks of the strikers. In fact, however, the presence of the constables only strengthened the strikers' resistance. At a meeting on the same day, Williams told his men, "I want you all to promise not to go to work while the 'Cossacks' are in town." The strikers then voted unanimously that they would remain out.

Schwab then revealed a new stratagem. He announced that unless the strikers returned to work immediately, Bethlehem Steel would sublet $2,000,000 in contracts to the Carnegie Steel Corporation, a division of U.S. Steel. If he were forced to take this action, then even if the men returned to work there might not be sufficient jobs for all of them.[28] Schwab's statement turned out to be an empty threat, but the strikers, fearing that it might be true, sought allies in Washington.

They launched a direct appeal to the Washington embassies of countries which then had orders with Bethlehem Steel—there were nearly two dozen of them. The strikers urged these governments to cancel their contracts. They offered two reasons: first, that the company had subjected its workmen to intolerable conditions which had driven them to strike, and, second, that work on foreign contracts was being done by unskilled and inexperienced laborers, and their output was inferior and defective.[29]

The strikers believed that this new tactic would force Schwab to relent. They sent Schwab a new statement of their demands, with no significant changes, expecting that he would now agree to a settlement. In a brief statement Schwab replied that he would not negotiate with the strikers; they were not employees of his company.[30]

On March 10, Williams predicted victory within two weeks. He claimed that Schwab was having no success in hiring skilled workers to replace the strikers, that fewer and fewer men were deserting the ranks of the strikers to return to work, and that the A. F. of L. had promised to call a general strike throughout Pennsylvania in sympathy with the Bethlehem strikers. Williams was also encouraged by the introduction in Congress of a resolution demanding a federal investigation of the wages and working conditions at Bethlehem Steel.[31]

Congress authorized the investigation, and President Taft gave the assignment to Charles Nagel, Secretary of Commerce and Labor. He, in turn, delegated it to Ethelbert Stewart, one of the most experienced investigators in the Bureau of Labor. When Schwab learned of this, he said that the company had nothing to hide and he issued a public statement assuring the inspector that he would be given full access both to the company's premises and to its payroll records.[32]

Then, on March 30 Schwab called a meeting of the businessmen of Bethlehem. He told them that he was incensed by the letters which the strikers had sent to members of Congress and to the various foreign governments. The meeting was supposed to be secret, but one of the businessmen who had attended it revealed that Schwab had given them a warning: unless the local businessmen came to the company's aid and defended his reputation, he would have no alternative but to "close down the entire plant and keep it closed." Schwab had insisted that the businessmen must take sides, that they must declare their allegiance either to him or to the strikers, and that if they sided with him, they must do so openly.

Schwab had told the businessmen that only five or six hundred men were still on strike—6500 men were at their jobs, with the plant running at full capacity, even though nearly a thousand strikers had left town for jobs elsewhere. He had charged that the strike leaders were outsiders, "socialistic agitators," and he had expressed both anger and astonishment that they had been allowed to use the Municipal Hall of Bethlehem as their platform for abusing him.[33]

Schwab's ultimatum produced its intended effect. The businessmen approved a resolution condemning the strike leaders for trying to ruin the economy of the area. They also passed a motion which declared to

Congress and the foreign governments that the local businessmen of Bethlehem supported Schwab unanimously.[34] One section of the businessmen's resolution read:

> We condemn the action of these agitators in forwarding to members of Congress and representatives of foreign Governments requests to withhold contracts for work from the Bethlehem Steel Company as a most vicious means of obtaining personal ends. We are confident that the efficiency of the Bethlehem Steel Company has not been impaired by the action of the few hundred employees led by professional agitators.[35]

A special delegation of businessmen was dispatched to visit Democratic Congressman A. Mitchell Palmer, who represented the district. For the moment Palmer refused to commit himself to either side. After hearing the pleas of the businessmen, he told them, "I withhold any expression of opinion . . . until I can learn the facts as found by the investigation which was acted [sic] for by the men and acquiesced in by the company. . . ."[36]

Then the businessmen's delegation visited President Taft. He told them that he opposed boycotts, and that, contrary to the demands of the strikers, the U.S. Army and Navy would not cancel their ordnance and armor contracts with Bethlehem unless the company failed to meet its delivery deadlines. In Taft's opinion, the fact that the company was embroiled in a labor dispute was not a sufficient ground for the government to repudiate its contracts.[37]

After the Bethlehem businessmen met and passed their resolution to support Schwab, the strikers appointed a group of delegates to reply to the resolution. The day after the businessmen's delegation visited President Taft, Williams and Tazelaar came to see him. They were accompanied by Congressman Palmer, who was beginning to reveal that his sympathies lay with the strikers. Taft listened attentively, but he avoided taking sides; he asked the strikers to send him a written statement of their grievances.[38]

Throughout the month of April both Management and Labor waited confidently for the release of the report of the federal Bureau of Labor. Each side expected that the facts would support its claims. On May 4, the report was submitted to the Senate; then it was released to the press. It was factual, neutral, and dispassionate. It consisted primarily of elaborate statistical tables of hours and wages. It neither praised nor criticized the company. But the strikers were pleased, even so; it did document their claims about low wages and long hours. They forgot that those were facts which the company had never really denied.[39]

Schwab, however, was incensed over one crucial omission in the report. It did not state that these same hours, wages, and conditions existed throughout the American steel industry and were not unique to Bethlehem Steel. He complained directly to Dr. Charles P. Neill, the Commissioner of Labor in Washington. Schwab insisted that Neill issue a public statement making it clear that the conditions described were not confined to Bethlehem. Neill did make a statement, but it was cast in a form which Schwab must have found less than satisfying: "Those are conditions of labor which may well be called shocking, but they are not confined to the Bethlehem Steel works." [40] Neill added, "Mr. Schwab himself conceded that the present hours of labor in the blast furnace industry are excessive; but says that competitive conditions impose these hours upon his plant so long as they are common to the steel industry." [41]

Commissioner Neill's statement was widely reported in the press, and it evoked an angry response from Schwab's ally, Senator Boies Penrose. Penrose complained to Charles Nagel, the Secretary of Commerce and Labor:

> For him [Neill] unofficially to go out of his way . . . to brand all the steel industries of Pennsylvania and of the United States as maintaining a condition of "slavery" certainly does not seem to be in line of his official duty and only furnishes campaign material to the Democratic party and the so-called "insurgents" when they assail the tariff bill in the approaching campaign. I would respectfully suggest to you the propriety of restraining Mr. Neill's activities in this connection.[42]

Nagel, politely but firmly, refused. He would not reproach Neill or repudiate his report. "Having listened to an interview between Mr. Schwab and Mr. Neill I am bound to say that I can not believe the report to be colored. . . . Mr. Schwab does not hesitate to deplore the conditions and to declare that he would welcome a change, the introduction of which the conditions of competition alone prevent." [43]

Following the release of the report, the strike continued. Schwab was eager to end it. He was also eager to see Congressman Palmer defeated in his bid for re-election because he had sided with the strikers. Palmer's rival for the Democratic nomination was J. D. Brodhead, who had once served as Congressman from the district and was now attempting a return to office. Schwab told Brodhead that he was willing to settle with the strikers, and if Brodhead acted as intermediary he would receive publicity and praise. While Williams, the strike chairman, and Tazelaar, the A. F. of L. representative, were away in Washington, trying to get Commissioner Neill to arbitrate, Brodhead met secretly with a committee of nine strikers. He offered them Schwab's proposals. And, on May 18, the committee agreed to end the strike.[44]

The strikers had achieved very little by their settlement. The only men who were permitted to return to their jobs were those who had neither damaged the company's property nor attacked its reputation. There were no wage increases; there was no recognition of any union. The settlement stated that "at all times workmen . . . shall be at liberty individually or collectively as workmen of the company and not as representatives of organized labor to approach the president or officers upon any subject of a general nature." This was the exact state of affairs which had existed prior to the strike. The only gain the strikers had made was that overtime and Sunday work were to be optional—but, since there was no increase in wages, any man who did not work overtime and Sundays would face a substantial loss of total income.[45]

By accepting Schwab's terms the strikers admitted that they could gain little by prolonging the strike. During the early weeks, when their strength was greatest, they had been unable either to close down the plant or to cause Bethlehem to lose any contracts. Now they were forced to recognize that the company was operating at almost full capacity and that only five to six hundred men were still on strike. They simply did not have enough support to continue.

The strikers lost for several reasons.

Almost a thousand workers had left Bethlehem during the first weeks of the strike to seek jobs elsewhere. But, as the Strikers' Executive Committee conceded, "Very few strikers can get work in other cities if the firm knows he [sic] comes from South Bethlehem. Whether the Bethlehem Steel Company is responsible or not, the fact exists that a black list is used against the employees of the Bethlehem Steel Company."[46] That black list undoubtedly discouraged other skilled steel workers from leaving their jobs at Bethlehem.

The company sought workers in other cities to replace the men who had gone, and, despite the unions' efforts to discourage such workers from accepting jobs, Bethlehem did succeed in hiring some, though no one knows exactly how many.[47]

The general work stoppage on February 3 had involved all of the machinists and apprentices in Machine Shops #2 and #4. Those were the shops where guns, shells, and shrapnel were made. Machine Shop #2 employed 225 full-time machinists and 92 apprentices, while #4 was manned by 162 machinists and 50 apprentices.[48] If they all had remained out on strike, work on Bethlehem's ordnance contracts, which all had deadlines, would have been halted, and the company might have had no recourse but to grant the strikers' demands. However, the apprentices could not afford to remain on strike. An apprentice earned 8.9 cents per hour, which amounted to $6.00 a week. After four years of service he received a cash bonus of $100 and a wage increase to the journeyman's rate, 17.5 cents per hour. Soon after the general work stoppage the company officials advised the apprentices that they would forfeit their bonuses unless they returned to work.[49] If an apprentice could save $1.00 a week out of his wages, to him the bonus would be the equivalent of two years' savings—an amount he could not jeopardize. After the company's announcement, many of the apprentices went back to work—enough of them to enable the machine shops to resume operations. This helps to explain one of the strikers' major demands: that the number of apprentices be reduced to a ratio of one apprentice for every five machinists.[50] When the apprentices returned to work and took over the jobs of the striking skilled machinists, the strikers charged that the apprentices were inexperienced, that they were turning out inferior work. But the strikers were unable to produce any evidence to substantiate this accusation, and the company suffered no loss of contracts as a result of it.

The strikers suffered another blow after Schwab met with the businessmen of Bethlehem. Responding to Schwab's threat to close down the plant, the merchants and retailers discontinued sales on credit to the men, even to those who were customers of many years' standing.[51]

Beyond all this, the strikers lacked the financial resources to sustain their strike. The Machinists' union and the A. F. of L. did support the strikers, but that support was primarily fraternal and advisory, rather than monetary. When the men joined the newly organized Bethlehem chapters of the various unions, they did not receive financial help. F. J. Conlon of the Machinists' union said, "Of course, none of these new members are entitled to benefits in our association, not having been associated with us the required time." [52] The total amount of outside financial aid the strikers received was $6700, and, as David Williams remarked bitterly, that "was not equal to one dollar a month a man." [53]

Williams and Tazelaar continued to hope for ultimate victory through federal mediation, but most of the strikers had neither the will nor the financial means to continue their resistance. So, when Schwab initiated a modest conciliatory gesture, the strike collapsed.

On the surface, Schwab had won. The strikers had obtained little, and what they did get had cost them 108 days' pay. Yet Schwab could have offered them exactly the same terms at the outset, possibly avoiding a lengthy strike. And, in the end, while he had an outsider arrange the settlement for him, he in fact *did* negotiate with the strikers before they returned to work—which he had said he would never do.

Bethlehem Steel had obtained one thing from the strike—adverse publicity. And Schwab's statements only served to undermine his own reputation and the company's. In 1908 he had argued that the protective tariff was indispensable because it safeguarded high wages for American workingmen, including those who worked for his own company, presumably. But in 1910 his response to the report of the Commissioner of Labor was that the low wages paid by Bethlehem Steel were characteristic of the entire industry. It was an ill-considered comment, for by that single statement he undermined his own earlier arguments for tariff protection and he publicly acknowledged what labor leaders had been claiming for more than a generation—that wages in the American steel industry, although they were higher than those paid by European steel mills, were inadequate. [54]

The federal government, after focusing a glaring searchlight on Bethlehem Steel, decided to take a hard look at the steel industry in general. As Professor David Brody observed, "the entire industry stood implicated. The Bethlehem investigation bore immediate fruit in Washington. The Senate passed a resolution for a full-scale inquiry into the steel industry whose eventual result was a four-volume report of unprecedented thoroughness." [55] This report documented Schwab's claim that working conditions and wage levels at Bethlehem Steel were characteristic of the industry. Further, as Professor Brody said, "The campaign for the eight-hour day on government contract work gained impetus." [56]

In view of the increased federal scrutiny of the steel industry, Schwab's success in ending the strike was at best a phyrric victory. His statement to Commissioner Neill that low wages were characteristic of the industry was taken by labor spokesmen as proof of the urgent need to unionize the entire industry. The steelmen found it increasingly difficult to justify tariffs as a safeguard of high wages for American steel workers. In fact, when another reduction of the tariff was proposed in 1912 and Schwab again spoke out in defense of protectionism, he no longer stressed his customary claim that high tariffs protected high wages. Instead, he shifted his emphasis: he claimed that "the steel industry has not gotten an adequate return on the capital invested in the last few years." He predicted that if the tariff were lowered again it would erase whatever margin of profit still existed for American rail producers, and he issued another of his periodic threats:

> If this bill becomes a law I shall sell out my stock in the Bethlehem works and retire from business. I want it to be understood that my interest in making the Bethlehem works the largest and best in the country is not necessary to me from a financial viewpoint. I do not desire to get richer. I have no heirs to leave my money to, and my greatest pleasure, and just now my only pleasure in life, is to build a big business in Pennsylvania. But with this bill a law I shall drop it all and take no further interest in the steel business.[57]

A year later the debate over tariff revision was still raging. Schwab did not want to testify in person and run the risk of being questioned about low wages; instead, he submitted a brief statement to the House Ways and Means Committee. He pointed out that Bethlehem's average yearly wage in 1911 was $781, "which is more than twice the average [of that] in any European country." Schwab then warned the committee that, since Bethlehem and other eastern manufacturing firms in other industries were particularly vulnerable to German exports, the "inevitable results of tariff reduction would be lower wages and loss of employment to hundreds of thousands of workmen, upon whom are dependent several millions of the population of our country." [58] Schwab had also perceived a growing menace from Belgian rail producers; in 1912 he had warned that their exports alone could "put the Bethlehem Steel Company out of business." [59]

Despite Schwab's grave prophesy of unemployment and falling wages, and despite his threat to retire from business, Congress passed a bill, the Underwood Tariff, in 1913. And, as a result, there was an upsurge of European rail imports.

What Schwab could not foresee—nor could anyone else—was that World War I would break out in August 1914. Once the war began,

German and Belgian rail competition ended abruptly. The war certainly had an impact on Bethlehem; the demand for its products accelerated steadily—first it got orders from Great Britain, then from the United States government. That sharp increase in demand eclipsed any concerns Schwab had about protective tariffs, low wages, or inadequate profits. For the next four years Bethlehem Steel made record profits and continually expanded its productive capacity to accommodate new orders. And for Schwab personally, the war years brought two new opportunities to demonstrate his abilities as an industrialist.

11
Wartime Challenges

On October 20, 1914, the British Admiralty sent a secret message to Schwab, who was then in New York. They wanted him to come to London at once. Schwab guessed that the British were planning to give Bethlehem orders for shells and shrapnel. He immediately telephoned Archibald Johnston, Bethlehem's vice-president in charge of foreign sales and its top ordnance expert, and asked him to come to New York at once, but not to tell anyone where he was going.[1]

That same night, at 1:00, the two men set sail for England aboard S. S. *Olympic*. The huge luxury liner had crossed the Atlantic many times, but on this voyage it was carrying very few passengers. Germany had announced that her submarines would sink the *Olympic* on sight, and hundreds of people had canceled their reservations.[2] On the sixth day out the British Navy alerted the captain that four German U-boats were pursuing his ship. The Admiralty ordered him to head north for Glasgow, instead of continuing straight into the English Channel. As the *Olympic* passed near Loch Swilly, the British naval base on the northern coast of Ireland, she began to receive distress signals: they were from H.M.S. *Audacious*, which had hit a mine and was taking on water.

The *Olympic* stopped to rescue the crew of *Audacious*, 900 men and two nurses, and then made two unsuccessful attempts to tow the stricken ship. She started to make another try, but then she sighted a German submarine. Rather than risk the loss of a second British ship, Admiral Sir John Jellicoe, Commander of the Home Fleet, ordered the *Olympic* to seek refuge in Loch Swilly. The German U-boat could have given chase, but it was more interested in sinking *Audacious*. It did so with a torpedo assault.[3]

Admiral Jellicoe was anxious to suppress the news of the sinking of *Audacious*, fearing that it would have a demoralizing effect on the British public, so he ordered the *Olympic* held in custody at Loch Swilly. But Schwab sent word to Jellicoe that he had urgent business to transact with the Admiralty in London. Jellicoe agreed to release Schwab if he would remain silent about the fate of *Audacious*. All of the other passengers, including Archibald Johnston, were detained; they were released in Belfast a few days later.[4]

When Schwab arrived in London, he met the Admiralty's chief negotiators: Herbert Kitchener, Secretary for War; John Fisher, First Sea Lord; and Winston Churchill, First Lord of the Admiralty. Schwab had been right: the Admiralty needed shrapnel and shells. He received the largest order in Bethlehem Steel's history; the final contract amounted to $135,000,000. It included orders for ten-inch howitzer guns, naval landing guns, and three-inch shrapnel and shells which were to be produced at the record-breaking rate of 50,000 daily. But the last item on the British list, by far the most important, was one Schwab had not anticipated. Britain then had more than fifty submarines patrolling her home waters. The Admiralty wanted twenty more.[5]

Schwab could not make a definite commitment to build submarines for England; he was not certain that production facilities were available. On November 2 he sent a coded cable to H. S. Snyder, Bethlehem's vice-president in charge of shipbuilding, telling Snyder that he would receive a second cable, in a special Admiralty code, in a few hours. The second message told Snyder about the submarine order and instructed him to arrange an urgent meeting in New York with L. Y. Spear and J. W. Powell, the owners of the Electric Boat Company, to whom part of the work would have to be subcontracted. They met the following day. After the conference, Snyder cabled Schwab that the twenty submarines the British needed could be built, ten at the Fore River yards in Quincy, Massachusetts, and ten at the Union Iron Works in San Francisco.[6]

The Admiralty's negotiators pressed Schwab for details about the submarines' construction and the kind of engine they would have. He suggested the H-boat, a type which Bethlehem's shipyards had already produced successfully, and they agreed. The question of delivery time was all-important, and Schwab offered the Admiralty a daring proposal.

Bethlehem usually spent fourteen months constructing a submarine; Schwab said he would guarantee to begin delivery in six months. He proposed that for every week Bethlehem beat its delivery dates, the Admiralty would pay a special premium—$10,000 per submarine. And, as a mark of good faith and to show that his promises were not simply reckless bravado, he agreed to forfeit $5000 per ship for every week they might be late after the delivery deadline.[7]

Lord Fisher was delighted. He wrote to Jellicoe, "We have made a wonderful coup . . . with someone abroad for very rapid delivery of submarines and small craft and guns and ammunition. I must not put more on paper, but it's a gigantic deal done in five minutes! That's what I call war!" [8]

On November 7, Schwab sent word to Snyder that the deal had been closed. The schedule for delivery was four boats in five and a half months, six more in eight months, and the final ten in ten months. Schwab's cable concluded:

> Above in brief is substance of agreement, whole matter largely left to our integrity and competence. Must do everything possible for them. This cable is your authority to proceed at once with every available amount of energy and make an unprecedented record. Organize regardless of expense to make good. . . . This is an opportunity for energetic men to make good showing as my idea is to use at least half of any bonus earned to distribute among people responsible. Will sail Wednesday. Would like to have meeting soon after returning as possible with all heads and in meantime assemble all material make all contracts and start without one moment's delay. I am personally deeply interested and we must make good.[9]

Aware that time was critically short for England, Snyder immediately ordered the production of the steel needed to build the submarines. In ten days, a record time, the steel was produced and delivered to the shipyards. Meanwhile, Schwab and Johnston sailed for America, carrying with them a British Treasury note for $15,000,000—they had insisted on that sum as an advance on the contract.[10]

They reached New York on November 20. When reporters asked Schwab why he had gone to England, he told them that he had merely been checking up on a routine matter—the financing of a large steel order for the government of Chile. But within a few days this explanation was discredited; stories about his top-secret negotiations with the Admiralty began to appear in American newspapers.[11] Someone, perhaps a German spy in the British Admiralty, had leaked details to the press, in the hope that either President Woodrow Wilson or Secretary of State William Jennings Bryan would prevent the contract from being executed. That hope was fulfilled: after consulting with Bryan, Wilson decided that execution of the contract would be an unlawful breach of neutrality, since a neutral power would be shipping weapons of war to a belligerent.[12]

On December 1, Secretary Bryan telephoned Schwab; he asked him to Washington to discuss Bethlehem's contract with Britain. Schwab, accompanied by Johnston and Snyder, met with Bryan the following day. There was no discussion. Bryan announced that President Wilson would consider it to be a violation of American neutrality if Bethlehem shipped submarines to England. But, he told Schwab, Wilson had no objection to the submarines being built, so long as they were not shipped.[13]

Armed with this ambiguous advice, Schwab, Johnston, and Snyder went to New York, where they conferred with Spear and Powell, the engine builders. Schwab then decided to charter a train to Montreal, where he hoped to find a Canadian shipyard which could assemble the submarines from components built by Bethlehem and the Electric Boat Company. In Montreal there was a shipyard owned by the British firm of Vickers. It contained everything he needed: it had the type of equipment required for rapid assembly of the submarines, and most of its facilities were idle. But no one there had the authority to negotiate with him. Schwab immediately returned to New York, and, on December 5, again accompanied by Johnston, he boarded the *Lusitania*, headed for England. The owners of Vickers were there, and they were the men he had to deal with.

They reached London on Sunday, December 13, but before they went to Vickers they paid a brief courtesy call at the Admiralty. Churchill, who had learned of the Wilson-Bryan problem, began an angry monologue. Pacing the floor, puffing furiously on his cigar, he denounced Schwab. It was bad faith, Churchill roared, for him to have made a contract that he could not fulfill. Schwab had deceived the Admiralty, and that deception might even result in Britain losing the war. Precious weeks had been lost, and during those weeks other contractors in other countries might already have been at work on the submarines. Furthermore, Churchill said, Schwab had acted unethically when he had accepted $15,000,000 from Great Britain under such fraudulent conditions.

Churchill did not allow Schwab to reply; he continued to scold until Lord Fisher urged him to be silent. "Schwab, what have you got to say?" Fisher asked. To Fisher's delight and Churchill's astonishment, Schwab said that not only was it unnecessary to cancel Bethlehem's contract, it was not even necessary to extend delivery time. All Bethlehem needed, Schwab explained, was the Montreal shipyard owned by Vickers. He told them he was going to see the Vickers people, and he asked Fisher and Churchill to put in a good word for him with the owners.

Fisher was elated. "If that's all you want," he said, "start back at once and take the yard." There was no need for Schwab to visit Vickers. Fisher promised that he would make all the necessary arrangements. Schwab then cabled Snyder to proceed at full speed. He and Johnston sailed home on the *Adriatic* in time to celebrate Christmas in America.[14]

Bethlehem immediately sent two of its own shipyard presidents—Joseph Tynan of the Union Iron Works of San Francisco and William G. Coxe of the Harlan & Hollingsworth yards at Wilmington, Delaware—to Montreal to supervise construction of the submarines. Schwab ordered American shipyard workers to be transferred to Montreal and, in order to preserve the secrecy of the entire operation, Bethlehem even operated its own food facilities there. Not one of the Vickers officers was retained; all were "retired" at full pay for the duration. Schwab fulfilled his pledge of speed. Bethlehem earned nearly $1,500,000 as bonuses on the Montreal submarines, and, according to Johnston, that entire sum was paid out in bonuses to the construction workers, foremen, and managers.[15]

By the summer of 1915, ten submarines built in Montreal were in active service. Fearing a German attack upon England, the Admiralty Board assigned six of them to a defensive patrol in British home waters. The other four were involved in the Dardanelles campaign, shelling enemy installations on shore.

Schwab was so successful at supplying Great Britain with submarines and munitions that Germany tried to buy control of Bethlehem Steel. In 1915 Count von Bernstorff, the German Ambassador to America, sent a representative to Schwab, offering him $100,000,000 for his controlling interest in Bethlehem.[16] Eugene Grace recalled Schwab's response:

He and I sat with those representatives one evening in which they offered what was an unbelievable sum for his controlling interest in Bethlehem. Without batting an eye, he said: "Gentlemen, I have given my word to the

British that during the period of this World War while we are doing such an enormous business with them, I would maintain my personal control of the Bethlehem interest, and there are no sums that you can offer me that will make me go back on my word." [17]

During this period, Schwab, his wife, and his mother were subjected to personal attacks. He and members of his family received crank letters from German-Americans who threatened his life unless he stopped producing munitions for the British. One of the letters sent to his wife described him as a murderer, while another prophesied that "the curse of God will follow your husband throughout his life." [18] Schwab's business associates shared his family's anxiety. They feared for his life. Finally they persuaded him to accept personal bodyguards, a rotating squad of New York policemen under the supervision of a Captain Titus. Schwab reluctantly agreed to this protection, but thought it unnecessary. He took childish delight in eluding his protectors whenever possible. [19]

When the British government learned of Germany's bid for Bethlehem Steel, it decided to make a counter-offer. England's ambassador to America, Sir Cecil Spring Rice, sent for A. Barton Hepburn, who was a financial adviser to the British Embassy as well as president of the Chase National Bank. There was a certain irony in Rice's choice of an emissary: only seven years earlier Hepburn had turned down Schwab's application for a $1,500,000 loan because Bethlehem's future had seemed too precarious. Now he was instructed to offer Schwab one hundred times that amount, $150,000,000, for Bethlehem Steel, to guarantee that the company would fulfill its contract with England.

Schwab rejected Britain's offer, just as he had turned down Germany's. He told Hepburn, "There is not enough money in Germany or Great Britain combined [sic] to buy the Bethlehem Steel Corporation until it has executed its obligations to the British Government." [20] In 1918, in recognition of what Schwab had done for England, Lord Fisher paid tribute to him. In his autobiography, Fisher called Schwab a "Champion Pusher," and he said, "If any man deserves the gratitude of England, Mr. Schwab does." [21]

Not all of Schwab's wartime activities turned out as successfully. In 1916 Bethlehem became embroiled in a new controversy with the United States government, and again it was over armor plate. This time, however, it was not the quality of the armor which was called into question, but the price. To protect his investment and future profits, Schwab launched a costly campaign to persuade members of Congress and the public to take his side against the Secretary of the Navy, but in the end he was defeated.

The seeds of his defeat had been sown in 1913 by the American Anti-Trust League, an organization of small businessmen who lobbied for vigorous enforcement of the Sherman Act.[22] In 1901, the League's secretary, Henry B. Martin, had tried to persuade Attorney General Philander C. Knox to establish rewards from $10,000 to $200,000 for anyone providing information which led to a conviction for violation of the antitrust laws. But Knox gave him no encouragement.[23]

In 1913, however, Wilson's Democratic administration took office, so Martin felt that he might find a more receptive audience. The League deluged the President's officials with information which it claimed would prove that Bethlehem Steel and the two other armor manufacturers, Midvale and one of the subsidiaries of U.S. Steel, were engaged in a conspiracy to fix prices, and that the conspiracy had been going on for years. Not only was this a gross violation of the Sherman Act, the League said, it was also a threat to the safety of American naval personnel, since the armor of the ships they sailed in was inferior, defective, and obsolete, and would give them no protection from enemy fire.

The first open attack on the manufacturers was launched on the floor of the House of Representatives. Armed with information supplied to him by the American Anti-Trust League, Congressman James M. Graham of Illinois charged that the Navy Department, and especially the Bureau of Ordnance, were the pawns of the Steel Trust and its allies. The League made similar charges in a letter it sent to President Wilson.[24]

The League offered no details to support these accusations, but it promised that evidence would be forthcoming if the government would agree to pay sizable rewards for information leading to the conviction of the armor manufacturers for fraud and conspiracy in restraint of trade. To lend weight to the charge, the League resurrected the armor plate scandal of 1893, telling Wilson that the "record shows that the Carnegie Steel Company has furnished bad armor before, and that company has been convicted of defrauding the Government of $450,000 in one armor contract." [25]

On April 9 Wilson ordered Secretary of the Navy Josephus Daniels to investigate the new charges. But Daniels had already begun an inquiry;

he had received a similar request from Senator Hoke Smith, the Georgia Democrat. Smith had already accused the armor-makers of overcharging the government. Daniels's preliminary investigation seemed to exonerate the companies:

> Whether or not the price at present paid is in excess of the actual cost of the material plus a reasonable profit is a matter on which the Department is not at present prepared to express an opinion and can only state that this price is less than is paid by most foreign governments for their armor and is less than American concerns are receiving for armor plate sold abroad.[26]

It was President Wilson's request which caused Secretary Daniels to continue his inquiry.[27] Regardless of whether the prices charged by the companies were high or low, Daniels was disturbed by the persistent pattern of identical bids on armor contracts for the Navy. He said that he could not understand "how identical bids could be arrived at if there is no combination, conspiracy, or collusion existing between them whereby prices are fixed." Yet he himself offered a plausible explanation:

> The Navy Department, in considering bids for armor and for gun steel, has, wherever the bids of the several companies differed, made a division of award to the several bidders on the condition that the material should be furnished at the price named by the lowest bidder. Both Congress and the Navy Department have, therefore, in a manner fixed the prices of these articles and this fact may explain the identity of bids submitted by the several firms engaged in the business.[28]

Since that was the Navy's policy, the bidding pattern of the armor producers was perfectly rational: instead of the firm with the lowest bid being awarded the entire contract, the order was divided equally among all bidders. Clearly, the companies saw no point in underbidding each other. When a firm offers a potential customer a lower price than that offered by a competitor, it is trying to obtain an order which otherwise might go to that competitor. However, if competing firms know that when any one of them enters a lower bid they will all simply decrease their profits without getting any additional business, then it is understandable that they all find it advantageous to adopt a mutually acceptable bid price.

Secretary Daniels found this situation unsatisfactory. Of course, there was an obvious solution open to him: he could declare "winner take all"—he could award the entire order to the lowest bidder. But he did not. He realized that if one firm consistently underbid its rivals, they might withdraw from armor production entirely, leaving the Navy dependent upon a single supplier. Then, if the nation went to war, it would not be able to obtain all the armor required. The rival firms would no longer possess the facilities to produce armor.

Daniels believed that only one other alternative existed. He told Wilson:

> If . . . it shall appear on further investigation that there is a combination, conspiracy, or collusion amongst the steel manufacturers to such an extent that the government can not secure steel products at a reasonable price and on bids which are actually competitive, I am inclined to believe that the only method by which the Government can compel actual competition will be by establishing its own facilities for the manufacture of armor plate, gun forgings, and other steel products.[29]

So Schwab was once again confronted with the threat of a government-owned armor plant. But he did not take it seriously; the federal government had cried wolf too often. The agitation would die down, he assumed, and "business as usual" would again prevail. However, Schwab misjudged the strength of Secretary Daniels's determination. And he underestimated the crusading zeal of his old antagonist, Senator Benjamin R. Tillman. Although Tillman was now ailing and aged, he was more than willing to reopen the battle he had waged in the 1890's: he still wanted the government to build its own armor plant.[30]

On May 28, 1913, Senator Tillman introduced a resolution authorizing the Secretary of the Navy to find out how much it would cost to build a federal armor plant, what the cost per ton of armor produced by such a plant would be, and how long it would take to construct the necessary facilities.[31] The Senate passed the resolution.

The Bureau of Ordnance, headed by Admiral N. C. Twining, then looked into the matter. Its report stated that a plant with a 10,000-ton capacity could be built for $8,446,000, and that the cost per ton of armor would be $314. Since the price then being paid to private producers was $454 per ton, the government, by producing its own armor, could save $140 a ton, and, on 10,000 tons, there would be an annual savings of $1,400,000, less an interest charge on the money used to build the plant. That would result in a net savings of $1,061,360.[32] The report did not discuss the length of time needed for construction.

When this estimate of costs and potential savings was published it aroused considerable interest, and within weeks the proposal to build a federal plant had gained wide support, particularly from the influential *Scientific American*.[33] In a private exchange of letters with that magazine's editor, Daniels acknowledged that his estimated price per ton included the cost of materials, payroll, and maintenance of plant, but that it did not include depreciation, insurance, or taxes. If those three were added, as they would be in estimating the cost of a privately owned plant, Daniels admitted, there might not be any financial savings in building a government plant. But no matter what the exact amount of savings might be, if any, Daniels favored building the plant. It would give him leverage in dealing with the private manufacturers. At his suggestion, Congressman A. W. Gregg introduced an enabling resolution in the House in August 1913.[34]

Schwab could usually rely upon well-placed friends in Congress to obstruct and ultimately defeat any moves inimical to his business interests. His chief political allies were the Republican Senators and Congressmen from Pennsylvania; Schwab personally contributed heavily to their campaigns. What he did not anticipate and could not control, however, was the overpowering support which the Daniels proposal received from civic and business groups, each eager for the construction of a government plant in its own area. Daniels was bombarded with hundreds of proposals from cities throughout the country, each claiming that it was uniquely suitable as the site for the government plant. Even one of Schwab's oldest and closest friends, Joseph G. Butler, Jr., president of the Chamber of Commerce of Youngstown, Ohio, favored a government plant—especially if it was built in his city.[35] As one newspaper described the situation, "Homer dead had not one-tenth as many cities claiming him as a native son and honored citizen as there are asking that the Government locate the armor plate plant [in the site they suggest] . . . Every place, from Dan to Beersheba, thinks it is peculiarly situated. . . ."[36]

Suggestions poured into Daniels's office—from the Chambers of Commerce, from men who had land to sell, from people who wanted the government to mine for iron ore in their area. To all such suggestions, Daniels made a standard reply: "Until such time as Congress shall make appropriations for the purpose of erecting an armor plate factory, this Department will be unable to make any definite decision as to the place of its probable location." [37] It was a perfectly straightforward and appropriate reply, but one which could easily lead each of its recipients to believe that if only he brought pressure to bear upon the Congress, the prize might be his.

Daniels, meanwhile, had come into direct conflict with the armor-makers. In 1914 he received identical bids of $454 a ton on armor for a battleship. Daniels rejected the bids, as he had the option to do, and called for new ones. Midvale Steel, which had entered into armor production in 1906, submitted the lowest rebid, $400 a ton, and it won the entire order. By awarding the contract on a "winner take all" basis, Daniels had in fact circumvented collusive bidding and restored price competition. But, because he had decided that it would be desirable to build a government plant, he did not pursue that strategy.[38]

Daniels soon had other, more urgent problems to deal with, as did Tillman and Wilson. After World War I began in 1914, the question of a government armor plant faded from public attention. But in 1916, when the Wilson administration proposed a costly military preparedness program, Senator Tillman rekindled the armor controversy.

Tillman, chairman of the Senate Committee on Naval Affairs, reminded Wilson of his pledge to support a bill for a government armor plant, and he stressed the urgency of its speedy passage. "Otherwise," wrote Tillman, "with the immense building program looming ahead of us, the Government will be unmercifully robbed. . . ." Wilson replied, "I stand ready to redeem my promise, but you must be my guide as to how I can help." [39]

The President's willingness to support a government armor plant did not spring from any first-hand enthusiasm for the venture, but rather from political necessity. He had to accommodate both Tillman and those members of Congress who were potentially subject to Tillman's influence. As Professor Melvin Urofsky has observed:

A large number of antiwar, anti-preparedness congressmen might soften their opposition to the Administration's defense program if some of the profits were siphoned off into a federal armor plant. Furthermore, Tillman's goodwill might serve some useful purpose in getting the Southern contingent in Congress, which so far had shown little enthusiasm for the Administration's preparedness schemes, to be more sympathetic.[40]

In February 1916, Tillman's committee held hearings on the armor plant bill. The two chief witnesses against the bill were Eugene Grace of

Bethlehem and Alva C. Dinkey of Midvale. Despite their eagerness to reach an amicable agreement with the Navy on armor prices—first by offering to show their cost and profit figures in confidence to Daniels, then by offering to guarantee a lower price for the next five years—their proposals were rejected. Their adversaries, Tillman and Daniels, were irrevocably committed to the creation of the government's own armor mill.

The Committee on Naval Affairs approved the bill. Its report to the Senate, written by Tillman, contained a bitter attack on the armor manufacturers:

> As long as present conditions continue the armor manufacturers are in a position to force the United States Government, in the language of the highwayman, to "stand and deliver."

> The committee has no desire to criticize unjustly the manufacturers of armor plate. They have done no more than most other men would have done under similar circumstances and temptations. Men in the pursuit of wealth are essentially greedy and hoggish. . . . Give power to any set of men, however excellent and honorable, and sooner or later they will abuse it. Men have been built that way since the beginning of time.[41]

Tillman's sense of indignation was so strong that he did not consider or offer the Senate any guidance on several pertinent questions: by what standard did he condemn the manufacturers as "greedy and hoggish"? What, in fact, would be a fair profit for the armor-makers? If the steel-makers were free to make as high a profit as possible on all the other types of products they manufactured, why should they not be equally free to do so on armor sales? Why should they have to underbid each other, if the Navy divided the orders between all bidders, but at the lowest bid price? Did not high profits act like a magnet, drawing new firms into armor production—witness Midvale's entry in 1906—thus providing the Navy with more suppliers of armor?

Tillman expected his bill to pass easily in the Senate, but he was less sure of its fate in the House, so he turned to Wilson to rally support. With Wilson's help, he said, the legislation could be enacted by March 31. Tillman added:

> Then, we can turn Admiral Harris loose, and he has told me if the red tape is cut in the Navy Department and he is allowed to go ahead, he can have the factory completed and ready to turn out armor in eighteen months; so that by October 1917 we will have the armor problem solved, and the hold-

ups to which the Government has had to submit heretofore will no longer exist.[42]

In mid-March, barely a month after Tillman's committee issued its stinging indictment of the "greedy and hoggish" armor producers, the Senate passed the Armor Plant bill by a 58 to 23 vote. At this juncture Schwab finally realized that he had underestimated the determination and strength of Tillman and Daniels. He consulted public relations expert Ivy Lee, then he began a massive campaign to defeat the armor bill in the House. On March 25, 1916, Bethlehem sent to each member of Congress the first of twelve statements presenting the case against the government's building its own armor plant.[43]

Tillman realized what Schwab was trying to do, and after the Congressmen had received the first two statements, he urged Wilson to intervene. "There is a fearful fight against it, you know, and having passed the Senate so promptly and with such an overwhelming majority, it would be a pity—I would consider it a crime almost—if it doesn't become a law. Promptness will save much money." [44]

When it became clear to Schwab that his messages to Congress would merely delay passage of the Armor Plant bill, but not cause its rejection, he tried to enlist the support of the general public. He ran a series of ads in 3257 daily and weekly newspapers throughout the country. The ads were typographically striking, concise, and forceful in tone. They appealed to reason, to the readers' sense of fairness and justice. All of the ads were signed by Schwab and Grace. One of them read: "We have allowed irresponsible assertions to be made for so long without denial that many people now believe them to be proven facts. We shall make the mistake of silence no longer. Henceforth we shall pursue a policy of publicity. Misinformation will not be permitted to go uncorrected." [45]

Schwab argued that constructing a federal armor plant would be both unfair and unnecessary:

SUPPOSE THIS WAS YOUR BUSINESS! If the Government had asked you to invest your money in a plant to supply Government needs; and after the plant was built, and had become useful for no other purpose, the Government built a plant of its own, making your plant useless and your investment valueless—would that seem fair?

This is precisely what Congress is planning for the Government to do with reference to our investment of $7,000,000 in an armor plant.

223

As an alternative to a government armor plant, Bethlehem now offered to "manufacture armor plate for the Government of the United States *at actual cost of operation* plus such charges for overhead expenses, interest, and depreciation as the Federal Trade Commission may fix. We will do this for such period as the Government may designate." [46]

On April 6, the House Committee voted, 15 to 6, to report favorably on the Armor Plant bill,[47] but in the full House Schwab's campaign slowed its momentum. Tillman was frantic: "Schwab with his hundred millions is bombarding everybody in the House and Senate too, day by day, with special pleadings and lying proclamations against its passage.[48]

When Tillman's target date for passage, March 31, had passed, there were two other bills which still had priority on the House calendar. Tillman begged the President to intervene, to insist that the House give priority to the armor bill. "Pardon my impatience, but I have been fighting these scoundrels for twenty years and know how unscrupulous and greedy they are." But Wilson, having made promises to support the two other bills, was unable to oblige. Although the forces arrayed against the armor bill were "powerful and persistent," Wilson expressed confidence that they could be overcome.[49] He was right. On June 2 the Armor Plant bill passed the House of Representatives by a wide margin. The bill was not immediately made law—in fact, Wilson could not sign it until late August because a joint House-Senate conference committee had to reconcile the many differences between the House and Senate versions of the general Naval Appropriations bill, into which the Armor Plant bill had been incorporated.

Since Bethlehem's attempt to influence Congress had proved futile, Schwab turned to Woodrow Wilson, the court of final appeal. He hoped to persuade the President to exercise his veto power. Schwab wrote to Wilson:

> I cannot but believe that the program to urge a Government armor plant was decided upon without full knowledge of the facts. Various speeches which have been made show that the program was determined upon without the knowledge by many, of our desire at all times to be fair and frank, a purpose we have sought to make additionally clear by our offer to make armor for the Government at any price the Federal Trade Commission might fix.
>
> Is it too much to ask, therefore, that you examine the pages of this book and see if our proposition is not so fair that the Government in full justice to itself will feel warranted in adopting measures which, while giving the Secretary of the Navy authority to proceed with the building of a Government

armor plant, will make such procedure contingent upon the willingness of the manufacturers to supply armor-plate to the Government at such price as the Federal Trade Commission shall decide to be fair alike to the Government and to them? [50]

This last effort was as futile as all the others. Wilson did not even acknowledge Schwab's letter. Schwab was defeated: his campaign to save Bethlehem's $7,000,000 investment had been an expensive failure. The Navy was now free, as Tillman had put it, to "turn Admiral Harris loose" to build a government armor plant.

Schwab's appeals to the American public won wide praise in the press for their candor and reasonableness. Even so, he failed. He had made two assumptions, both of them unjustified. He had thought that ordinary citizens would read his statements, and that they would write their Congressmen, saying that they agreed with Bethlehem's viewpoint. And he had thought that members of the House of Representatives would give his statements a full and unbiased hearing. But 1916 was a presidential election year, and issues other than the Armor Plant bill commanded more attention from both the public and Congress. In late April, for instance, America became embroiled in a new diplomatic controversy with Germany concerning submarine attacks on passenger ships in the English Channel. Moreover, the entire House of Representatives was due to stand for re-election in November, and the Congressmen, understandably, were eager to cut short the debate on the armor plant issue so that Congress could adjourn. Also, some candidates who were up for re-election were reluctant to side with Bethlehem against Wilson, Tillman, and Daniels for fear of having it said that they were lackeys of the armor-makers. By contrast, those who opposed Schwab could claim that they had struck a deathblow against the "war-profiteering" munitions makers. Perhaps if Schwab had waged the same campaign a year or two earlier, when Tillman's crusade had first begun to take shape, he might have succeeded. But Schwab's campaign came too late. By 1916, Tillman and Daniels were too deeply committed to make a graceful retreat, let alone to surrender.

Congress appropriated $11,000,000 for the construction of the government's armor plant, which was to be built on a large tract of land in South Charleston, West Virginia. But in 1917 construction was halted— the nation was at war, and the costs of labor, machinery, and raw materials had skyrocketed. The $11,000,000 appropriated was not enough.

Construction was resumed when the war ended; nevertheless, the final cost exceeded the original estimate by several million dollars. In 1921, after the first armor plates were produced—at a cost nearly double the price per ton charged by private producers—the plant ceased operations.[51]

Hundreds of skilled workmen had been hired to construct and operate the plant. They were summarily dismissed. And the hundreds of homes built for those workers remained empty for the next twenty years. A token force assigned to guard the plant kept a careful vigil over the rusting machines and decaying buildings. Throughout the 1920's the plant remained idle. When Franklin Roosevelt took office in 1933 he was urged to reopen the plant, to help revive the economy of the hapless town which had been saddled with an armor-plated white elephant. One woman told a leading Senator that Roosevelt must act quickly and reopen the plant so that profit-seeking private producers could not "exploit" the government again.[52] Roosevelt ignored all such pleas. The final victory belonged not to Tillman or to Daniels, but to Schwab.

In 1904 Bethlehem's armor sales had produced nearly half of the company's income. By the time of the outbreak of World War I, the armor plant represented only 5 per cent of the company's total investment, and armor sales accounted for only 3 per cent of its annual income. Although Schwab had diversified the company by manufacturing commercial products, the war created an unprecedented demand for Bethlehem's military products. As a result of the war boom, the town of Bethlehem grew. In 1904 it had a population of 13,000; by 1916 the population was 150,000. And during the same period the company's sales grew from $10,000,000 to $230,000,000.[53]

Those who had had the patience to hold onto their Bethlehem stock were well rewarded. In 1904 its common stock had stood at a high of $20 a share; then, during the Panic of 1907, it sank to $8, and Schwab announced that for the indefinite future Bethlehem would discontinue paying dividends and would retain all of its earnings for reinvestment. This news upset Bethlehem's second largest shareholder, Samuel Untermyer, who had purchased a large block of Bethlehem stock after the U.S. Shipbuilding Company controversy of 1903–4. He threatened to sue. But Schwab struck an agreement with him. Untermyer withdrew his suit in return for a place on the Board of Directors, where he could monitor the activities of the company. And he held on all during the next decade, waiting for the payout of dividends.[54] By 1914 the stock had risen to $29.50 per share, but still no dividends were distributed; instead, Bethlehem's earnings were used to finance expansion and to pay off its funded debt. When war began in August 1914, the stock began to climb, and by January 1915 it had reached a new high, $46. It began to soar two months later, and it ultimately reached $600 on October 22 of that year. The long-awaited bonanza came on January 27, 1916, when Schwab declared a cash dividend of $4.50 a share (as well as a 10 per cent wage increase for unskilled workers).[55]

Schwab was widely praised for his development of Bethlehem Steel. *The New York Times,* in an editorial, called Bethlehem "possibly the most efficient, profitable and self-contained steel plant in the country." [56] Even Samuel Untermyer paid tribute to Schwab. He said that Bethlehem "is one of the most progressively managed businesses in the country and has as its head the greatest, most resourceful and far-seeing steel manufacturer, the most remarkable salesman and organizer and the most loyal and enthusiastic official to be found anywhere in this or any other country." [57]

By late 1915 Schwab had become the object of intense public curiosity, and many stories were being published about his spectacular career—more than had appeared at any time since his election to the presidency of U.S. Steel in 1901. After struggling for a decade to make Bethlehem a formidable rival to U.S. Steel and to regain the power and eminence he had held at the turn of the century, he took pleasure in his new public image and in the widespread interest in his career.

Of course, Schwab was not the only industrialist whose success aroused public curiosity. His friend, J. Leonard Replogle, was another celebrity whom reporters pursued. But in vain: Replogle did not care for publicity. He had soared into prominence in 1915, when he had purchased $15,000,000 in stock of the Cambria Steel Company. Many newspapers published the story that Replogle had begun his career in steel at the age of eleven as a waterboy in Cambria's Johnstown plant. He immediately punctured this myth: "I wasn't a water boy; I was an office boy. Not that it makes any difference." He told reporters that he would gladly answer any questions about his business policies and plans, but he begged them to "soft-pedal the flub-dub about me." And when a reporter asked him whether the story of his success could serve as an inspiration for other young men, Replogle replied that boys who require inspiration from others are "pretty hopeless." He refused to offer any formulas for achieving fame and fortune.[58]

Unlike Replogle, Schwab enjoyed his prominence. And he decided to satisfy the public's interest. He would give the world his observations on success. In January 1917 Schwab published *Succeeding With What You Have*, his first and only book. Anyone who read it in the hope of learning something about success or about Schwab must have come away dissatisfied. Its sixty-three pages contain a few pleasant anecdotes and catchy phrases, but it offers only conventional recipes for success and hardly any information about Schwab.

Schwab told his readers that the leaders of American industry "are not natural prodigies. They won out by using normal brains to think beyond their manifest daily duty." Those who are regarded as geniuses, he said, are "merely men who have learned by application and self-discipline to get full production from an average, normal brain."

How does one obtain promotions? Schwab answered, by working "a little harder than any one else on the job one is holding down." He warned against being a clock-watcher and being distracted by the pursuit of pleasures. "Young men may enjoy dropping their work at five or six o'clock and slipping into a dress suit for an evening of pleasure; but the habit has certain drawbacks. I happen to know several able-bodied gentlemen who got it so completely that now they are spending all their time, days as well as evenings, in dress suits, serving food in fashionable restaurants to men who did not get the dress-suit habit until somewhat later in life." [59]

The keys to success, wrote Schwab, are industriousness and perseverance. "Real success is won only by hard, honest, persistent toil." And again, "A man will succeed in anything about which he has real enthusiasm . . . provided that he will take more thought about his job than the men working with him. The fellow who sits still and does what he is told never will be told to do big things." Nor should an ambitious man shy away from physical labor or from expending extra effort: "If a young man entering industry were to ask me for advice, I would say: Don't be afraid of imperiling your health by giving a few extra hours to the company that pays your salary! Don't be reluctant about putting on overalls! Bare hands grip success better than kid gloves." [60]

Schwab had almost no introspective ability. He was unaware of the way his mind worked when he made decisions, and he was unable to tell anyone how he had formed the mental attitudes which had contributed to his success. He only attempted to do so once, and that was in 1935. He said: "There is no secret or mystery about any of my mental processes. I do not sit and figure things out and keep the thing under my hat until I am ready to spring it. I think that whatever plans I have that surprise others and work successfully are due to intuition, or what Andrew Carnegie used to call 'the flash.' I simply get the flash. It is instinctive, and I am not conscious of thought." [61] This accurate self-appraisal was confirmed by several of his oldest associates. Ernest R. Graham said of him, "He had a trip-hammer brain, and I really believe a sort of telepathic power. He arrived at many of his decisions without knowing the mental processes by which they were reached. This is true inspiration. Few men have it." [62] Gary Kerr agreed: "I am also inclined to believe that his mind operates by inspiration. He is inspired by a thought or a plan, and when he sits down to work it out by sober mental processes, he discovers he was right in the first place." [63]

Over the years Schwab had stored in his mind a vast accumulation of factual knowledge and first-hand observations which he drew upon almost without effort. He operated almost entirely by an intuitive or inspirational method; he could reach conclusions and make decisions by a subconscious process of integration—one which his colleagues observed and admired, but which he was unable to teach to anyone else, including those who read his book.

The most remarkable year in Bethlehem's history was 1916. Its earnings in that year—over $61,000,000—exceeded all the gross sales of the first eight, pre-war years that the company had been under Schwab's direction. Even after deducting $18,000,000 for interest charges, depreciation, and depletion, Bethlehem still had a cash surplus of $43,500,000, and that, combined with $19,000,000 from the preceding year which had not been distributed as dividends, gave the corporation a cash reserve of $62,500,000 with which to finance its expansion. [64]

Bethlehem's plants were working around the clock to fill record orders for munitions and ships, but even as they were, the company was not exclusively or even primarily expanding into military production. On the contrary, Eugene Grace was actively planning Bethlehem's withdrawal from war production, anticipating the sudden cessation of the war and the problems of reconversion to peacetime production. Whenever the war did end, much of Bethlehem's equipment would no longer have any value, except as scrap. Thus, while Schwab was occupied with obtaining new orders from the British and American governments, he and Grace were also investigating and negotiating major new acquisitions in the field of commercial steel.[65]

In 1916, Bethlehem bought the Pennsylvania Steel Company, and, as a result, the corporation "doubled its capacity for making pig iron and steel, increased the variety of its products, provided itself with a tidewater plant and made itself the biggest shipbuilding concern in America." Bethlehem not only acquired Pennsylvania Steel's rail-producing facilities, it also took over the properties of Pennsylvania Steel's subsidiary, the Maryland Steel Company, whose shipyard and steel mill were located on the Chesapeake Bay at Sparrows Point, a site so favorable for serving the Eastern market that Bethlehem subsequently developed it into the corporation's largest single facility. A further advantage of the purchase was that Bethlehem obtained Pennsylvania Steel's Cuban ore properties, which were adjacent to Bethlehem's Juragua mines. Another major acquisition, one which Schwab negotiated in 1912, was the Tofo Iron Mines in Chile. Tofo's ore was of high grade—three tons of it were equivalent to four tons of Lake Superior ore. Bethlehem made two other acquisitions in 1916: the American Iron and Steel Manufacturing Company, in Lebanon, Pennsylvania, a firm specializing in nuts, bolts, and rivets; and the iron mines of the Lackawanna Iron and Steel Company, located in the Cornwall region of Pennsylvania. Because of these acquisitions and several others of lesser importance, Bethlehem Steel was prepared to withstand the sudden cessation of war-related orders.[66]

In 1913 and 1914, the steel industry went through a recession. Then, beginning in 1915, the war caused a general upsurge in demand for steel products. Increased orders came from three sources: European governments, which were ordering shrapnel, shells, barbed wire, and rails to carry on the war effort; American companies, which were expanding to fill war orders from Europe and were also ordering structural steel; and the steel industry itself, which was consuming part of its own output in building new steel-making facilities. Supply could not keep pace with demand, and steel prices began to spiral upward.[67] In November 1915, Judge Gary said, "Demand for pig iron and the various lines of steel is at the present time in excess of the producing capacity of the furnaces and mills of the United States. There is nothing to indicate that there will be a decrease of the demand for some time to come." [68]

One consequence of the steel industry's wartime prosperity was a new wave of mergers. While Bethlehem was negotiating to buy the Pennsylvania Steel Company in 1915, a syndicate headed by J. Leonard Replogle purchased a controlling interest in the Cambria Steel Company. Its properties included a rail-producing plant at Johnstown, Pennsylvania, and iron ore sites in the Lake Superior region.[69] Six months later, in February 1916, the Replogle syndicate resold Cambria, at a higher price, to the Midvale Steel Company. Midvale itself had changed owners in 1915; in that year three veteran steelmen, old friends of Schwab and former partners of Andrew Carnegie, purchased it. They were William E. Corey, who had been forced out of the presidency of U.S. Steel in 1909; William B. Dickson, who had resigned as second vice president of U.S. Steel in 1910; and Alva C. Dinkey, Schwab's brother-in-law, who was president of the Carnegie Steel Company, U.S. Steel's subsidiary.[70] During the war years the new Midvale-Cambria group prospered, but it was unable to survive the 1921–22 recession. Bethlehem bought Midvale-Cambria in 1922, and by that purchase it acquired coal and ore properties as well as strategic production sites.[71]

In 1915, however, when Midvale-Cambria was formed, there was no recession in sight, and the prospects for profits seemed limitless. By December 1916 average steel prices (calculated in dollars of constant value) had risen to 240 per cent over the prewar level, and by July 1917 they had increased 370 per cent over it.[72] In order to satisfy the vast demand for steel, construction of new mills was begun and many old, idle mills were reactivated. It became economically feasible to reopen and run plants whose equipment was obsolete or whose production costs in normal times would have been uneconomically high.[73] Yet despite all the efforts that were made to increase production, the feverish demand for steel continued to outrun supply.

The situation grew even worse after April 1917, when America entered the war. Unlike a commercial customer, which might defer its purchases until prices had fallen, the United States government could not postpone its defense orders. In early July 1917, President Wilson told Newton D. Baker, the Secretary of War, that he was prepared to nationalize the steel industry and set prices by Presidential edict unless the manufacturers would agree to offer the government steel at "reasonable prices."

The leaders of the steel industry, meeting in Washington on July 11 with members of the War Industries Board, opposed any plan to stabilize prices on government orders. The steelmen said that they would be caught in a "profit squeeze"—selling prices would be fixed, but the costs of labor and raw materials, which were not subject to controls, would rise. The meeting ended inconclusively, and thereafter Wilson invoked the medieval concept of "just price" as a guideline. But when he tried to explain the concept, his explanation was so vague that it was useless. "What was a 'just price' after all? How were prices to be fixed? Who would administer them? With the market mechanism thus affected, how would goods be distributed?" [74] Wilson left those questions unanswered.

Nevertheless, within two months the leaders of the steel industry had agreed to a price stabilization policy. They would establish and adhere to a schedule of prices to be charged to the government, prices based on "average costs" within the industry and a "fair return" to the companies. [75] Not a single member of the industry's delegation offered any opposition to the *principle* of price controls. Believing that nationalization was the inevitable alternative to controls, they merely held out for the highest possible fixed price. Not one of these captains of industry, these leaders of America's largest steel-producing firms, even tried to make the point that price controls were both unnecessary and undesirable, from the standpoint of the industry and from that of the national war effort. Instead, they haggled over details while allowing the principle of controls to go unchallenged; thus, "war socialism" was adopted in the industry because the spokesmen for capitalism did not understand or did not try to point out the futility of controls and the desirability of relying on the unregulated price system to allocate materials.

Professor Ludwig von Mises has explained that the market system is applicable even in wartime:

> What America needed in order to win the war was a radical conversion of all its production activities. All not absolutely indispensable civilian consumption was to be eliminated. The plants and farms were henceforth to turn out only a minimum of goods for nonmilitary use. . . . The realization of this program did not require the establishment of controls and priorities. If the government had raised all the funds needed for the conduct of the war by taxing the citizens and by borrowing from them, everybody would have been forced to cut down his consumption drastically. The entrepreneurs and farmers would have turned toward production for the government because the sale of goods to private citizens would have dropped. The government, now by virtue of the inflow of taxes and borrowed money the biggest buyer on the market, would have been in a position to obtain all it wanted. [76]

Instead of adopting this program, the government abandoned the price system. From then on, steel prices were fixed, but the manufacturers' costs of production continued to rise. The cost of producing rails rose from $23.02 a gross ton in 1916 to $32.18 in mid 1917, and to $40.78 in early 1918. At these same times, the cost of producing steel plates was $30.95, $44.33, and $53.43.[77]

The price stabilization agreement brought a greater volume of business, but a lower rate of profit. Bethlehem's earnings for 1917 were nearly as large as they had been in the preceding year, but they represented a smaller percentage of return on substantially increased sales. Although Schwab criticized government control of steel prices, he nonetheless believed that the agreement between the War Industries Board and the steel industry had averted nationalization, an alternative he would have liked even less.[78]

Even prior to America's entry into the war and the subsequent price stabilization agreement, Bethlehem's profits had been rising. Schwab was pleased that the company's growth had been primarily in commercial steel rather than in war products. During 1916 domestic business accounted for $117,500,000, and there was an additional $17,500,000 of export orders for steel bars. Munitions sales to overseas customers amounted to $58,500,000—only 30 per cent of the total orders.[79]

In early 1917 a sharp increase in munitions orders from the Allies made it necessary for Bethlehem to raise $65,000,000 in working capital. Fifteen million dollars came from the proceeds of a stock offering; a Class B non-voting common stock was created because Schwab did not want to relinquish his voting control over Bethlehem Steel. The other $50,000,000 was derived from the sale of "Two Year Notes," three-fourths of which were due to be liquidated when British Treasury Notes which Bethlehem had received in 1914 came to maturity. These two issues were expected to cover Bethlehem's financial needs for 1917–18.

When the United States entered the war in April 1917, however, Bethlehem began to receive huge U.S. government orders for munitions, raising the value of orders on hand to a record total of $300,000,000, but leaving the company desperately short of working capital. By August 1917, Schwab felt that Bethlehem had to undertake a new $30,000,000 stock offering; he saw no alternative. In order not to further dilute the common stock, the new issue was to be an 8 per cent cumulative convertible preferred stock.[80] But even the proceeds from the new stock offering were not adequate to meet the company's mounting needs for working capital.

The U.S. government, quite properly, refused to pay in advance for

233

goods a company had not yet produced, but it was willing to make cash advances, subject to one important reservation. Because a contractor might default on deliveries, the government sought to protect itself by requiring collateral equal in value to the cash advanced. The problem which faced Schwab, Grace, and F. A. Shick, Bethlehem's treasurer and comptroller, was to find approximately $15,000,000 in unmortgaged properties or securities to use as collateral. But this was impossible: nothing worth $15,000,000 was available which had not already been pledged as collateral for bank loans or new bond issues.

Schwab's personal fortune was tied up in Bethlehem stock, but the government would not accept common stock as collateral. The only acceptable collateral was land, or bonds, or personal promissory notes. Bethlehem's top executives discussed the problem of collateral, and they found an unconventional solution. Shick later explained it:

> In order quickly to secure such cash advances together with the other executives of the Corporation I endorsed the Corporation's promissory notes in favor of the Secretary of War on behalf of the United States in the amount of $15,579,466.46. By this action the other executives of the Corporation and I, respectively, made ourselves liable for all or any part of said sum.[81]

Bethlehem Steel's executives were so confident of their company's ability to fulfill its government contracts that they did not allow themselves to be deterred by the fact that, even collectively, they did not own $15,000,000 (exclusive of their holdings of Bethlehem common stock). If Bethlehem Steel had defaulted, the government would have held largely worthless promissory notes, but a more serious fate would have awaited the Bethlehem executives: whatever assets they did own would have been seized and sold, and liens would have been placed on their future income until the full sum, plus penalties and interest, had been paid off.

Grace, Shick, and the other top executives were willing to pledge all their personal wealth as collateral, an eloquent tribute to the loyalty and *esprit de corps* Schwab inspired. Those executives had full confidence in Bethlehem—and rightly so. For the firm was so well organized and operated that it could withstand the sudden withdrawal of Schwab in 1918, when he was "conscripted" for government service.

12

Dollar-a-Year Man

When America entered the war in April 1917, she had few troop and cargo ships, and by July of that year the shortage was severe. Industry was straining its productive capcities to meet military requirements, but the shipbuilders were not keeping pace with the nation's mounting needs. And in the meantime, deliveries for Europe were delayed on the docks.

The United States Shipping Board exercised general authority over ship procurement; its function, in consultation with the War and Navy Departments, was to determine the quantity, type, and tonnage of troop and cargo ships needed for the war effort. But the actual work of providing these ships—placing construction contracts and seeing to it that delivery schedules were met—was delegated to the board's subsidiary agency, the Emergency Fleet Corporation.

The Shipping Board was legally empowered to seize all suitable ships from any available sources. At first the board tried to meet its quotas by confiscating German ships in U.S. ports and by "requisitioning" privately owned American ships under construction in U.S. shipyards. But these were insufficient to meet rising military needs, and the Emergency Fleet Corporation had to undertake a massive shipbuilding program.[1]

From May to November 1917, an average of 100,000 American troops were shipped to Europe each month. But then Russia withdrew from the war, and the Germans were free to concentrate their forces on the Western front. In the early months of 1918 they began to make major advances. The United States War Department feared that England and France might be overwhelmed unless American troop shipments were increased substantially, and, in March 1918 the Shipping Board sent orders to the Fleet Corporation from the Army for "1,000,000 tons of troop ships, 2,000,000 additional tons of cargo ships, some additional tankers, colliers, refrigerator ships, and additional hospital ships." This sudden upsurge in Army requirements caught the Fleet Corporation wholly unprepared. It had been hard-pressed when it had had to procure shipping for 100,000 men a month; now it was ordered to provide transportation for 350,000.[2]

The head of the Shipping Board was Edward Nash Hurley, a former businessman and former chairman of the Federal Trade Commission. When President Wilson appointed Hurley in July 1917, he gave him authority to select the head of the Emergency Fleet Corporation. Hurley chose his friend Charles F. Piez, an engineer and businessman.[3]

Early in April 1918, Hurley sent President Wilson a report on the production of the preceding month; it was disappointingly low. Hurley complained of construction delays, shortages of materials, and careless workmanship throughout the industry. He said, "My own feeling has been that the owners of the yards are not fighting as they should be for increased production. It seemed to me that the only way to change the situation was to put the situation squarely before them, letting the public know where the blame lies." Hurley's only recommendation was that the President should denounce the shipyard owners publicly. Hurley himself had done so privately in a telegram he sent them on April 2. In it he had concluded, "the American people want ships, not excuses."[4]

Hurley and Piez found that they were unable to cope with the mounting demands made upon them, so they decided to enlist the services of a leading American businessman to take charge of the shipbuilding program. They hoped that such a man would be capable of both inspiring and directing production. They asked Henry Ford to do the job, but he refused. Charles Schwab was their next choice.[5]

Nearly a year earlier, Secretary of Commerce Franklin K. Lane had suggested that Schwab could solve the shipbuilding problem:

> The President ought to send for Schwab and hand him a treasury warrant for a billion dollars and set him to work building ships, with no government inspectors or supervisors or accountants or auditors or other red tape to bother him. Let the President just put it up to Schwab's patriotism and put Schwab on his honor. Nothing more is needed. Schwab will do the job.[6]

When Hurley proposed Schwab, Wilson did not veto the suggestion, despite the fact that he and Schwab had clashed in 1916 over the Armor Plant bill and that Schwab had made a conspicuous contribution of $100,000 to Wilson's Republican rival for the presidency, Charles Evans Hughes.[7] With Wilson's approval, Hurley telegraphed Schwab on April 5, 1918, asking him to come to Washington on urgent business. Hurley, Piez, and Bainbridge Colby, a member of the Shipping Board, formed a three-man delegation to persuade Schwab to become Director-General of the Emergency Fleet Corporation.

Hurley announced to Schwab that he had been "conscripted" for public service for the duration of the war. Schwab said no. He reminded the delegation that he already was actively supporting the war effort—as chairman of Bethlehem Steel he was working full-time to complete crucial government contracts—and that he, personally, was risking millions of dollars if these contracts were not successfully completed. He also said that he should not be considered for the position because if he became head of the EFC he would be involved in a conflict of interest: the Fleet Corporation would have to make contracts for ships with Bethlehem Steel, the world's largest shipbuilder, and Schwab believed that this would result in charges of financial favoritism. The three men assured Schwab that any practical objections could be overcome. They spent the entire day trying to change his mind; Schwab promised to think over their proposal and to return to Washington for another meeting.[8]

Bainbridge Colby later recalled the circumstances of that second meeting:

> In our determined effort to bring Mr. Schwab into the work I had finally made an appointment with him to meet President Wilson on a certain day [April 16, 1918] at the White House at 2 o'clock in the afternoon. Up to that time we had failed to be convincing, and I well remember the luncheon I had with Mr. Schwab and his associate, Mr. Grace . . . prior to our appointment with the President. Mr. Grace earnestly protested that Mr. Schwab should not take up the work and I quite as earnestly brought forward such counter arguments as occurred to me.
>
> We had reached no agreement and the hour came when we had to keep the appointment with Mr. Wilson. I vividly recall the interview. The President, with whom I had fully discussed the question, and who was entirely in sympathy with the effort to "requisition" Mr. Schwab, came out of an inner room assuming that the matter was settled and that Mr. Schwab was willing to undertake the work. He put out both hands to Mr. Schwab and

spoke in acknowledgement of his sacrifices and his patriotism in a way that would have moved any man. It affected Mr. Schwab, and in that instant his doubts and hesitation were gone and he agreed to be drafted.[9]

Schwab accepted, but with three conditions. All three were met. First, he wanted autonomy; he wanted to run the shipbuilding program without being boxed in by elaborate rules and inflexible procedures. Second, he wanted Wilson's assurance that the President would support him; he asked Wilson, "Will you stand back of me?" and Wilson, putting both hands on Schwab's shoulders, replied, "To the last resources of the United States of America." [10] And, third, he insisted upon a written agreement with the government, stating that he would not have to deal personally with any company of which he was an officer or owner, so that no charge of conflict of interest or financial favoritism could be made against him. This condition was also met; a special committee of the Shipping Board was created to handle all negotiations with Bethlehem.[11]

Then a press release was prepared: "Mr. Schwab will have complete supervision and direction of the work of shipbuilding. He agreed to take up the work at the sacrifice of his personal wishes in the matter. His services were virtually commandeered. His great experience as a steel maker and builder of ships has been drafted for the nation." [12]

When the news of Schwab's appointment as Director-General was announced, many people sent Wilson their congratulations. One described the appointment as evidence of Wilson's "profound sagacity." A second assured the President, "You will get results. He is a genius in steel." Others wrote, "He can do the trick. Your best appointment"—"a grand thing." "Schwab's appointment was a bulls eye shot at Kaiserism. Shoot again." [13] Schwab received similar messages; perhaps the one which pleased him most was a telegram from Winston Churchill:

> I am delighted to learn of your appointment which will enable you to turn your wonderful energy and unique experience to an urgent and vital task. Remembering our work together at the Admiralty at the beginning of the war and the way in which you surmounted every difficulty and successfully completed every undertaking, I feel complete confidence now. All good wishes.[14]

The *Literary Digest* made a survey of press reaction to Schwab's appointment, and it reported unanimous acclaim:

> Conservative newspapers now declare confidently that the ships will be built. Mr. Schwab, it is pointed out, knows steel, knows ships, knows how

to handle labor, and has a prestige which carries weight with every business man in this country or abroad. Even the Socialist New York *Call* joins its "capitalist contemporaries in approving the appointment of Mr. Schwab to this position." It believes he will be a real dictator, that he will have little trouble with the workers, whose aspirations he understands, and that "the output of ships from now on will continually increase when Schwab takes hold." [15]

Soon after he was appointed to head the EFC, Schwab sought out his old adversary, Navy Secretary Josephus Daniels. He wanted to make a truce for the duration of the war and to reassure Daniels of his high-mindedness in accepting his new post. Daniels later told of their meeting. Schwab appeared a somber penitent: "I have made all the money that I want or ever will want. For the balance of my life, particularly while this war is on, I wish nothing except to serve my country. I never want to make another dollar." Schwab sensed that Daniels did not trust him, so he asked him directly: "Do you believe me?" Daniels took refuge in a circumlocution; as he put it, "I did not say 'yes,' but replied, 'I am very glad to know that you, like all good Americans, are patriotically volunteering to serve your country and put that above all private interests.' " [16] This gesture by Schwab was a pathetic act of self-abasement. By making it he sanctioned the belief of Tillman and Daniels, among others, that in wartime one must choose between making profits and being a patriot. He *felt* that these aims were not incompatible, but he could not articulate his feeling—so he deferred to theirs.

Although Schwab told Daniels that profit-making was no longer his own goal, he did believe that profits were a needed incentive for businessmen, during wartime and peacetime alike. In Schwab's view, the maximizing of profits was a businessman's navigational north star, essential in evaluating alternative courses of action. He opposed the idea of either urging or compelling businessmen to put aside their concern for profits and to accede to the government's wartime directives without question. He disapproved of the government's threats to nationalize recalcitrant industries which did not "voluntarily" agree to stabilize prices.

Schwab held that if the government wanted to stimulate effort and efficiency in wartime industries, the best way to do so was to offer incentives, not to threaten stricter controls. He operated the Fleet Corporation just as he had run his steel mills, using the same methods he had employed so successfully throughout his career.

One of Schwab's first decisions was to move the Fleet Corporation's headquarters to Philadelphia. This served two purposes: it allowed him to be closer to the central zone of American shipbuilding, and it kept Washington's officialdom at arm's length. After more than a decade of undisputed authority as head of Bethlehem Steel, Schwab was unwilling to subordinate his judgment to decision-making by committee or to subject himself to the endless conferences and excessive paperwork which he believed were endemic to Washington. He arranged his operations so that he had to spend only about one morning a week in Washington, usually in conference with Hurley.[17]

Schwab followed his own business orientation when he hired the Fleet Corporation personnel. He ignored all considerations of political patronage, agency prestige, or personal friendship as he sifted through hundreds of letters from job-seekers. To applicants who were total strangers, Schwab framed a polite but non-committal reply, stating that he would be happy to consider the applicant as soon as he became more familiar with the needs of the Fleet Corporation. He advised those applicants who were recommended by his friends to write to other federal agencies, where their talents might be put to better use.[18] He refused to hire any prominent public figures. He told Hurley:

> It is a mistake to put men of prominence . . . into working positions. After the appointment which stamps them as having obtained a position of honor in the conduct of the war it is rare that you find them taking detailed charge of the business. I therefore think that wherever appointments are made, we should procure men to whom the salary is a matter of importance.[19]

Just as Schwab sought only men of proven ability to work for the Fleet Corporation, he also selected only companies of proven competence when he placed contracts. He did not encourage the formation of new shipbuilding companies, and he rejected the applications of ambitious amateurs who offered to go into shipbuilding with the financial backing of the government: "our only function is to build merchant ships by contracts placed with shipbuilding companies already in existence."[20]

During his first weeks in his new job, Schwab received several unsolicited reports and complaints about working conditions in shipyards under contract with the Fleet Corporation. One observer wrote that, at the shipyard at Newington, New Hampshire, workmen were told not to work too hard, a carpenter spent nearly an entire workday planing a twenty-foot piece of wood, and a foreman sarcastically told one capable worker that he must not try to build the ship alone. Schwab's eyewitness claimed that on several occasions 100 or 200 men were idle for two hours at a time because production was so uncoordinated and haphazard.[21]

What caused or contributed to such inefficiency? Schwab placed the blame on "cost-plus" contracts, by which builders were paid the full cost of labor and materials, plus a percentage of that amount as profit. Those shipyard owners were being paid sizable sums for the work of superintendence, without financial risk of any sort. They had no reason to insist upon speed or efficiency from their workers or prompt delivery of materials, since the added costs would be borne by the government and their percentage of profit would be calculated on a higher base.[22]

To Schwab, the "cost-plus" system was irrational, an invitation to inefficiency. Within a month of taking over the Fleet Corporation, he ended the system and introduced a plan which tied profits to cost-cutting. On May 15, 1918, Schwab announced that from then on shipyard owners would work only under lump-sum contracts and that they would be responsible for obtaining their own materials without government assistance. This plan relieved the government of the need to supervise shipyard work and placed sole responsibility on the shipbuilders to prevent the waste of manpower or materials. Schwab expected that an efficiently operated shipyard would be able to net about $37,500 as its profit for building one submarine.[23]

Schwab found that the shipyard owners approved of his new plan. He told a friend that he "had no difficulty whatever" in placing contracts in the Great Lakes area. "I made one price for everybody, which price I am told will not only be profitable but satisfactory to all. . . ." [24] Thereafter, when Schwab received reports of shipyard inefficiency, he replied that the new lump-sum plan was safeguarding the taxpayers' interests. Senator Warren G. Harding, acting on information from one of his constituents, complained to Schwab that there had been costly delays at shipyards in the Great Lakes area. Schwab reassured Harding, "all of the ships being built on the Lakes are at a fixed price. If there is any inefficiency or any inefficient labor or management on the part of the shipbuilders the loss falls upon the company and not the Government." [25]

Nor was there a danger, under the new contract plan, that inefficient shipyards could cut corners on safety or quality specifications; all ships were to be inspected three times—first by personnel from the EFC, then by the American Bureau of Shipping, and finally by agents of Lloyds of London before issuance of insurance.[26]

Although a shipyard owner could prevent waste and an inspector could detect safety hazards, neither could speed the construction. Schwab himself took on this assignment. His goal was to accelerate production, raise output per man-hour, and increase the number of ships launched.

In July 1918, Schwab made an extended tour of the West Coast shipping yards. He was greeted like a returning military hero. While shipyard bands played rousing tunes, thousands of workers cheered him. Describing the exuberant welcome Schwab received, one of his aides wrote, "It would be an easy matter to understand how large numbers of workmen could be assembled enroute, but I feel sure . . . [that] the responsive spirit shown by these men was a thing which could not have been ordered." [27]

There was only one yard where he was likely to meet hostility. That was in Seattle. Many of the workers at the Skinner & Eddy Shipyards were members of the I.W.W., the Industrial Workers of the World, a militant radical group which vigorously opposed American involvement in the war. Schwab knew in advance that his reception would be cool, possibly even openly hostile. The owners agreed to assemble the workmen to hear Schwab's exhortation to speed up production and support the war effort, but it seemed almost certain that he would be prevented from speaking by a chorus of catcalls and invective. What could America's munitions king say to socialist workmen? How could he appear to be neither timorous nor imperious, neither pleading nor scolding? Schwab seized the initiative. He began by stating the one fact which undoubtedly was most prominent in the workers' minds, the only fact which they might have expected him to avoid mentioning: "Now boys, I'm a very rich man. . . ." The startled men laughed and applauded, and Schwab went on with his address. When he finished speaking, a delegation of I.W.W. men offered him an honorary membership in their society—and he accepted. [28]

Schwab appealed to the shipyard workers' and managers' patriotism. When he learned that a shipyard strike in Oakland and Alameda had halted production, he fired off a plea to the strikers to put patriotism above their personal grievances. They had placed "small personal differences and interests above their duty to the nation. The news that 4 thousand shipbuilders have stopped work can not but affect adversely the spirit of our boys in France who are at this moment under the fire of German guns." [29] And he encouraged the shipyard workers to view themselves as soldiers in the service of their nation. "We have a great army of workers," he said, "building ships for this emergency. There are 300,000 of us, and we are all fighting for America. You men who swing the cranes are in charge of the big guns. You who drive the rivets are operating the machine guns of the shipyard. Every man who does a full day's work is doing his share to win the war. The gangs at work on a ship are holding a trench and when they launch that ship they go over the top. When they lay a new keel they are digging in and making ready for another long defense." [30]

Believing that all men have a competitive spirit, and that rivalry will drive them to do their best, Schwab promised to publicize the work of particularly able managers and workmen. And he did so, both in the EFC's own publication and in his statements to the general press. He also offered more tangible forms of recognition. He won President Wilson's approval to award "service badges to the men who give four months' faithful service to the Government in the shipyards at building ships, and bars for additional length of service. With these service badges you can walk through the crowds, meet the boys of the Navy and the Army, and hold your heads high." [31]

Schwab established a Competitive Department in the EFC to reward unusual effort in ship production. Rear Admiral F. F. Fletcher agreed to serve as chairman of a three-man board which would evaluate each shipyard's performance. Every month a red flag was awarded to the best yard, a white one to the second best, and a blue one to the third, and medals of gold, silver, and bronze were given to the men whose work had been particularly outstanding at each of the winning plants. [32]

Yet Schwab thought that money meant more to the men than flags or medals. With his own money he established a "bonus of $10,000 which I have offered to the men in the plant which produces the largest surplus tonnage above the scheduled program during 1918." He also persuaded several shipyard owners to match his $10,000 reward. (One of those owners was W. Averell Harriman, the young chairman of the Merchants' Shipbuilding Corporation.) [33]

Schwab was an integrator; he gave cohesion and direction to the activities of men who produced under his leadership what they could not have done on their own. By June 1918, two months after he took charge of the Fleet Corporation, there was already a marked increase in the number of keels laid, ships launched, and ships delivered for active service. And by the fall of 1918, government contracts were regularly completed on schedule, or even ahead of schedule, and the intervals between keel laying, ship launching, and active service were steadily being reduced.[34]

Wilson and Hurley were delighted with Schwab. Even Navy Secretary Daniels revised his estimate of the man. In his diary entry for November 7, 1918, Daniels wrote, "Went to Philadelphia for Emergency Fleet Reception. Schwab. I told him I had forgiven him all his sins." [35]

Schwab wrote to Carnegie in September 1918. He could not conceal his pride:

> I have been connected with the shipbuilding business four and a half months and am getting everything fully organized and in good shape. . . . I feel quite confident that our country will not lose an hour in the war by reason of lack of ships. We put into commission last month three hundred forty thousand dead-weight tons of shipping, or about as many ships as the whole United States was able to complete in any one year prior to the war. I expect to increase this tonnage each month. . . .[36]

Carnegie's reply must have been deeply gratifying. "It gives a record of accomplishment such as has never been equaled, but you know I have never doubted your ability to triumph in anything you undertook. I cannot help feeling proud of you for having far outstripped any of my 'boys.' . . ." [37]

The war ended in November 1918, and three weeks later, on December 3, Schwab sent a telegram to Wilson, asking for permission to resign:

> The changed industrial conditions brought about by the signing of the armistice have brought with them many serious problems affecting industrial operations. I enlisted for the emergency, and now that the emergency is over I desire most earnestly to be released for the great problem of bringing the industries in which I am interested back to normal conditions.
>
> May I have your consent to retire as Director-General of the Emergency Fleet Corporation? [38]

Wilson, then in Paris, replied via his Press Secretary, Joseph Tumulty. "I accept your resignation only because you wish it and because I feel that I must do so in fairness to you." And he thanked Schwab for "a service of unusual value and distinction" [39]—a sentiment which was echoed in editorials throughout the nation's press.[40]

13

Semi-retirement

If Schwab had died late in 1918 just after serving as Director-General of the Fleet Corporation, he probably would have been given an official state funeral. At minimum, he would have been memorialized by Congress for his service to the nation. Woodrow Wilson and Josephus Daniels, as well as politicians and journalists of every political stripe, would have lavished him with eulogies. His wartime efforts had created a collective amnesia about his controversial prewar career; suddenly all the debits and deficiencies were erased from the invisible ledger which the press and the public had maintained on him for the previous twenty-five years. Gone from public memory were such episodes as the armor plate scandal of 1893–94, his gambling adventures at Monte Carlo in 1902, the controversy surrounding the collapse of the United States Shipbuilding Company in 1903–4, the allegations by Senator Tillman and Secretary Daniels that the prices he and other armor-makers charged the government amounted to extortion. All that remained were the assets and credits: his rapid rise in the steel industry, a story reminiscent of a Horatio Alger hero's; his courage in staking the future of Bethlehem Steel on his judgment that the Grey beam was superior to conventional beams, and his refusal to abandon the innovation when the funds needed for its adoption were not available; his supplying Great Britain with the submarines she had needed in 1914. But what earned him the greatest praise was his support of the national war effort; he appeared to have placed patriotism ahead of profit-making. Certainly his well-publicized statement of December 1916 gave this indication:

I believe that Bethlehem is now one of the great assets of the United States, and I may add that the plant will be, in the event that the time ever comes, at the disposal of the U.S. Government to be used as the Government shall see fit, and the U.S. Government shall name the price at which it shall buy the materials produced in that plant.[1]

Furthermore, he had heeded the call to public service, putting aside his business responsibilities when President Wilson "conscripted" him to direct the shipbuilding program. His success as Director-General, coupled with the fact that he had served at his own expense, without any salary except the symbolic dollar-a-year, made him seem doubly deserving of praise from his countrymen. By the moral standards of his era, he had attained heroic stature through public service and self-sacrifice.

Even so, for the next twenty-one years, until his death in September 1939, his wartime service was repeatedly called into question. Behind him were his triumphs; ahead were new accusations, all of which embarrassed and tormented him. What he was never able to understand was that he had brought some of his troubles upon himself—that he had made himself vulnerable to charges of hypocrisy by paying lip service to "ideals" which he repeatedly flouted in action.

His new troubles did not begin immediately. From December 1918, when he left government service, until January 1921 he enjoyed a respite from controversy and criticism. During this period he was free to devote his attention to matters of great personal importance: first Bethlehem's transition to peacetime, then the completion of his opulent country estate.

Schwab, now a civilian, immediately assessed the postwar future of Bethlehem Steel. When the war ended, government contracts for steel and ships were suddenly canceled. Schwab advised Eugene Grace, president of Bethlehem Steel and its subsidiaries, that the contract cancellations gave them a perfect opportunity "to curtail our force and expenditures in every possible direction. You will never have such a good reason or opportunity for doing so and my opinion is that it ought to be done immediately and with all possible vigor." He even proposed that bonuses for shipyard and department managers be eliminated, inasmuch as the plants would not be working at normal or full capacity once the postwar decline set in.

I have no desire to take any steps that might look different from our established principle of bonuses, [but] I feel that during the days when you are unable to have a good basis for the establishment of a bonus, in many instances you had better pay straight salaries until such a bonus may be established. In other words, outside of our general officers, curtail the bonus system as far as possible, unless it can be made on a basis of cost or practice. For example, I think the bonuses on a percentage of beams or rails produced from ingots would be beneficial at any time, or a bonus on cost in any direction, but I think bonuses for general results . . . could well be immediately eliminated.[2]

When Schwab made his first complete postwar review of Bethlehem's cost sheets, he noted one item which disturbed him: each month $36,000 was spent on the operation and maintenance of company automobiles at Bethlehem. "That I think is pretty indicative of a general situation," he wrote. "I would cut off every one, not only for the saving but for the effect. I wonder if you realize that this one item alone at that same rate per year would pay nearly half the dividend on our old preferred stock?"

Schwab knew quite well that many superintendents, managers, and executives would resent sudden and drastic reductions in fringe benefits and bonuses, but he was wholly indifferent to that fact. He told Grace:

Undoubtedly, you may have a lot of men who will be discontented during this period and who may feel and say that they would rather leave the employ of the Company than go on under such reduced basis. My opinion regarding this is that you had very much better lose such men who will probably never be fully contented, and give the younger man who has his mark to make a chance to build into our organization. We have rarely ever made a change of this kind that has not resulted in efficiency and economy.[3]

Schwab realized that when government orders came to an end there would be a collapse in steel prices which would lead to price-cutting by rival firms, and he was eager to prevent ruinous price wars. This could be done if the American steelmakers entered into a sales pooling agreement, but they could not do so legally without the consent of the government. On February 12, 1919, Schwab wrote to President Wilson:

In my opinion, which is shared by most of my associates, the business situation at home is critical and some co-operative methods of conducting manufacture and business is necessary during this transition period, otherwise I fear disaster. Mr. [Bernard M.] Baruch who so ably conducted the War Industries Board and as its Chairman won the confidence of American manufacturers could do more than any other man to help organize some co-operative method of procedure.[4]

247

But President Wilson did not reply. The steel industry sank into a post-war slump; prices fell and profits diminished. Schwab erred when he said that "disaster" was the only alternative to a government-approved and, presumably, government-enforced sales cartel in the steel industry. The postwar recession was both necessary and inevitable; the collapse of prices and profits meant that only the most diversified and most efficiently operated firms would survive the return to peacetime conditions. All the others, especially those which had been created or resurrected during wartime because of the inflationary boom induced by deficit financing of the war, were ventures which simply had no economic justification once the wartime demand came to an end.

Ten months later, Schwab recognized the beneficial aspects of a postwar recession. While he acknowledged that the downturn was severe, he also expressed confidence that recovery would come soon:

> The existing moment is full of difficulties and complexities. Here and there you find prophets of despair. But I want to go on record here as saying that nothing could be healthier for American business than the very condition through which we are now passing. It had to come. I only wish it had come sooner. . . . We are getting relieved of the impurities in our business life. The process is not complete yet. It may take some little time longer. But the patient will in time be cured and when he is cured the great body of American business will emerge with a vigor and an energy the world has never known before.[5]

In the early months of 1921, while the economy was still in the grip of the severe postwar recession, Schwab made an extended tour of the various Bethlehem steel mills and shipyards. Upon his return, he wrote a general appraisal of the situation. It was designed to induce further cost-cutting:

> I don't believe I have ever seen it so bad, with prospects looking so blue in all directions, especially in shipbuilding, etc. However, we have been through dark days and we must go through these. The one thing for us to do is to start anew to introduce economies in every possible direction and save every dollar that we possibly can.

As one means of reducing costs, he told Grace:

> You might have to make some sacrifices that are unpleasant or at least you may have to make some immediate adjustments so that some of the people who have been a long time with us, may be retired with an honorarium or some arrangement that would leave us free to introduce new blood and

especially new methods. It is to be expected that accumulated wealth and luxuries, unfit most men for the serious attention that a crisis like the present demands, and we may have to take steps of a most radical character. . . . At the same time, we must not go so far as to injure our organization.

But, as always, Schwab could find a reason to be hopeful:

Perhaps a few months may see things brightening a little. Our fixed charges are tremendous, and of course I am greatly worried over our ability to meet them. It will probably be all right as long as we have back claims and settlements [on war contracts] which may be coming along during this period of dullness to help us out, but if things continue too long we would inevitably cut our cloth to suit the times.[6]

During the early 1920's Schwab developed diabetes, and thereafter he was required to stay on a sugar-free diet and to receive daily injections of insulin. Whenever he was in New York his friend and personal physician, Dr. Samuel A. Brown, came to "Riverside" to administer the injection. Schwab was, in fact, one of the earliest users of insulin; he had no adverse reaction to it, and, generally speaking, his ability to work was not impaired by his condition.[7]

Earlier, however, during the winter of 1920–21, his doctor and wife urged him to take life easier, and, for a time, he did. But he was only fifty-eight and he had no desire to retire; he was too active and vital ever to tolerate a sedentary existence. But he readily agreed to rest for a few months on his new estate, "Immergrun" (Evergreen), in Loretto.

One day, when Schwab was a boy, he took a walk in the hills surrounding Loretto, and when he returned home he exclaimed, "Mother, Mother, I've found a place in the hills where I'm going to build you a home someday when I'm rich." [8] He built that home in 1898, and it is still standing today. It was an impressive structure for its time, but compared to the palaces Schwab constructed later, it was starkly modest.

By 1914, Schwab had both the money and the desire to build a magnificent estate in Loretto. He chose the site then occupied by his mother's 1898 house for the location of a new forty-four-room mansion. But he did not wish the old house to be demolished; he wanted it moved. However, it was surrounded by an orchard which he was reluctant to destroy. One of Schwab's friends was a Pittsburgh contractor, John Eichleay, Jr., who had devised a process for raising and moving buildings on rollers. Schwab summoned Eichleay. Could he move the house over the treetops? Eichleay said he could, and he did; the old house was relocated a few hundred feet from its original site.[9]

Several years before Schwab actually began construction of the new mansion, he began buying land in the hills of Loretto. The final site covered 4,791,000 square feet—over 1000 acres. When the estate was completed early in 1919, it consisted of eighteen buildings and required seventy full-time employees for its maintenance.[10]

Schwab spent a fortune to make the estate fit his idea of grandeur. From the main house he had a full view of the town of Loretto. Leading up to the house were magnificent parallel stairways, with a waterfall cascading down the center. When the water reached the bottom of the hill, it was pumped electrically to the top to begin its dramatic descent again.[11]

The estate was opulent in every detail: sensuous statues of pagan gods and goddesses ornamented the grounds; the nine-hole golf course was manicured. There were stables for his handsome thoroughbreds, three greenhouses, and ornate garages to house his fleet of cars. Schwab's favorite car was a chauffeur-driven, custom-built Packard, with his initials emblazoned in gold on the doors. His chickens were not kept in conventional coops; they were housed in replicas of French cottages (which today are motels). Schwab had once developed a liking for a village in Normandy, so, price being no object, he built a copy of it. He set aside sixty-six acres on which he had a small community built. The workshops, sheds, and cattle stalls in it resembled prosperous French farmhouses.[12] Schwab provided Loretto with a new reservoir and he had electrical power lines installed, all at his own expense. He also paid for the paving of a four-mile dirt road over which he had driven his father's wagon to deliver the mail.[13]

The Loretto estate was completed in 1919, and Schwab was delighted with it. But he was deeply saddened because Andrew Carnegie was unable to see it. For many years Schwab and Carnegie had hoped that Carnegie would build an estate of his own in the Loretto hills. But in 1919, Carnegie, then eighty-four, lay dying at "Shadowbrook," his mansion in Lenox, Massachusetts. On August 11, the old man asked his secretary to hand him a photograph of Schwab. He looked at it, smiled, and closed his eyes for the last time.[14]

For Schwab, the death of Carnegie was a painful, personal loss. A memorial service was held for Carnegie in Pittsburgh's Carnegie Music Hall in November 1919, and there Schwab spoke movingly about the man he called "my old, my beloved, my greatest friend—indeed, my father." [15]

Death also robbed him of others whom he loved. Except for Frick and his two allies, Phipps and Lovejoy, Schwab had maintained ties with all of the former Carnegie Steel partners. They met annually as the Carnegie Veteran Association. But by now Carnegie's "boys" were aging men. Several of the Veterans died during the early 1920's, among them James A. Gayley, John A. Potter, George Lauder, and J. G. A. Leishman. [16] But nothing grieved Schwab more deeply than the death of Joe, his brother. Joe died on February 17, 1922, the day before Charles Schwab's sixtieth birthday. He was only fifty-eight.

After the war, Joe, in a speculative venture, tried to corner the wheat market. He failed, and that failure destroyed him financially. Even worse, it destroyed his spirit. Joe felt too old and beaten to attempt to rebuild his life, and he began to drink heavily. He ultimately died of alcoholism.

Joe's parents never learned the truth about their son's downfall and death. When they came to visit him on his deathbed at the Hotel Collingwood in New York, his longtime mistress, a Broadway actress, was discreetly absent. His death, they were told, had been caused by diabetes. His body was taken to Schwab's mansion "Riverside," to lie in rest before burial. The presence at the mansion of Joe's estranged wife and his children helped to conceal the true situation. Those who loved Joe kept silent about his final, tragic years. [17]

Two years later, John Schwab died. He was eighty-five. Until the last few months of his life he was a man of remarkable vigor and fitness. While Pauline Schwab came to weigh over 220 pounds, John remained nearly as slender as he had been as a youth. In fact, he had changed very little over the years: he still told the same joke, "I was married in 1861, the year the war began, and it's been going on ever since"; he still lived modestly, and he refrained from drinking and smoking (he told his children, "If God had intended you to smoke, He would have put a chimney in your head"); and he continued to enjoy practical jokes, recalling with special fondness the time Charlie had invited him to visit the steel works and then had told a security guard to throw him out, claiming that he had never seen him before in his whole life. The fact that he was one of the last surviving Catholics to have been baptized by Father Gallitzin was nearly as important to him—and to his friends and neighbors—as the fact that his son was a steel titan. But his son's position had brought him many advantages: he received congratulatory messages from the

Pope and from President Taft on his fiftieth wedding anniversary, he had the opportunity to travel abroad as frequently as he chose, and he could hire a chauffeur to drive him to the quarterly directors' meetings of three near-by banks. (Around the turn of the century, Charles had purchased a controlling interest in three small country banks in Patton, Cresson, and Williamsburg; he had designated his father to represent his interests. John Schwab almost always attended the meetings, but he slept through most of them.) To the end he remained what he always had been: good-humored and nonchalant, proud of his son's achievements, but never pretending to himself or to anyone else that he had been a major influence during his son's formative years.[18]

In November 1922, Schwab said, "I am nominally Chairman of the Board of the Bethlehem Steel Corporation; practically, I am a retired steel manufacturer." [19] He should have said "semi-retired." Although he was no longer involved in the day-to-day management of the business—he had not been since he had become Director-General of the Fleet Corporation in 1918—he did maintain a close watch over Bethlehem, in which the bulk of his fortune was invested. He owned 90,000 shares of preferred and 60,000 shares of common stock in Bethlehem; their combined market value had diminished substantially from the 1915 high of $54,000,000, but they were still worth many millions. And, as Chairman, he received a salary of $150,000.[20]

Eugene Grace was president of Bethlehem, and Schwab was its single largest owner, just as, years before, Schwab had been president of Carnegie Steel and Carnegie had been its single largest owner. But Schwab gave Grace a far greater range of autonomous authority than he himself had ever been given by Carnegie; he had total confidence in Grace's ability and business judgment. Schwab never exercised his veto power, but that did not mean his power was merely titular; rather, he and Grace shared a nearly identical business outlook. The only important differences between them were temperamental and stylistic.[21]

Schwab and Grace were near opposites in personality. Grace was convivial in the company of friends and close associates, but he was shy and soft-spoken when dealing with strangers or reporters. He became considerably more confident and less ill-at-ease by the late 1930's, yet he still disliked being interviewed; he preferred to issue prepared statements instead. Schwab, on the other hand, always wanted to stand at stage center; he delighted in being the object of public curiosity and admiration.

He was ebullient, and eagerly obliged any reporter who asked for an interview. Grace once stalled for five years before granting an interview to B. C. Forbes, one of the most influential financial journalists in America; Schwab was accessible to virtually everyone, even cub reporters on small-town newspapers.

Grace and Schwab also differed in temperament. No one who knew Schwab could recall even a single instance in which he raised his voice in anger, even when his own health or safety had been jeopardized by someone's forgetfulness or carelessness. He believed that people respond best to praise, not criticism, and that when one can find nothing to praise, it is better to keep silent than to condemn. Such self-restraint was alien to Grace's character; he had a short-fused temper. He was brusque and critical when provoked, and he often scolded those who annoyed him. Schwab could never bring himself to fire anyone; when a man proved to be hopelessly incompetent, or dishonest, or an intolerable nuisance, Schwab invariably delegated the unpleasant task of firing him to one of his assistants. Grace fired such men himself.

Schwab and Grace also differed in their attitude toward history and the recent past. Whereas Schwab delighted in talking about the famous men and women he had known and enjoyed reminiscing about his friends and personal experiences in the steel industry, Grace said, "Never look back," and he meant it. He was contemptuous of reminiscence and retrospective analysis. In 1920 he said,

> Don't waste time talking about what you have already done. Use the time in planning or executing something new. Dividends are not earned by reminiscing on the past. . . . when Mr. Schwab returns after some absence and wants me to tell him all about what has been going on, I simply cannot do it fully or satisfactorily. My mind rebels against going back and recovering ground that I covered once and finished up with. . . . You cannot accomplish anything by talking about something you have already done." [22]

As he said, "I live in the present and the future. I have no interest in the past, except as a source of experience to guide me in the future." When his associates encouraged him to talk about his career in steel, he replied, "I don't like to think backwards." [23]

Although the two men were so different in personality and style, their fundamental business attitudes were very much alike, and they had a solid working relationship. There were six major policies on which Schwab and Grace were in full agreement.

253

They both believed in an expansionist strategy—that Bethlehem Steel should keep acquiring ore mines and operating subsidiaries until it became a fully integrated company; they shared the goal of surpassing U.S. Steel, which they believed possessed all of the facilities and none of the vision or methods required for high efficiency and profitability.

Like Schwab, Grace believed that the best way to generate additional profits was to reduce costs; he scrutinized and compared the operating expenses of Bethlehem's various plants with a zeal for detail worthy of a Carnegie or the senior Rockefeller.

They agreed that a giant business firm can only be profitable if its workers and managers have a direct financial stake in the operation; Grace and Schwab rejected any general profit-sharing or stock-purchase agreement which was not directly tied to each man's own work performance. In 1920 Grace said, "The best organization is that which enables the greatest percentage of those in the business to receive in earnings exactly what they *make*, that is, have a system which *measures* each man's results, and then *pay* according to the results attained. Bethlehem has such a system, perhaps the best ever devised." [24] He added: "We have the most elaborate cost-finding and work-measuring system of any plant in America. True, it costs a great deal of money to keep all the records and statistics necessary, but we know that it is worth infinitely more than it costs; it is worth more than any dollars and cents, since it gives our men that satisfied feeling of partnership, that feeling of being remunerated strictly in accordance with the effort put forth." [25]

Grace's favorite slogan, "Always More Production," mirrored Schwab's attitude, at least until the 1929 depression dictated retrenchment and restricted production.

Like Schwab, Grace vehemently opposed unions; he feared that they would encroach upon the prerogatives of management and oppose the Bethlehem system of paying each man the value of what he, personally, had produced rather than a standardized wage. In order to defuse the threat of industry-wide unionization which followed the 1919 strike against U.S. Steel, Schwab and Grace agreed to establish an Employee Representation Plan, which enabled the workers to elect fellow workers as delegates to present their grievances to the company. They found this type of company union preferable to a union controlled and staffed by outsiders—men whom Schwab condemned as "walking delegates from Kamchatka."

Schwab and Grace were equally adamant about excluding "outsiders" from Bethlehem's Board of Directors; unlike many major corporations which included prominent lawyers, educators, bankers, and business-men from unrelated industries on their boards, Bethlehem's board was composed of "insiders"—the company's own operating steelmen and fi-nancial experts—at least until Grace's death in 1960.

Because Schwab and Grace were in agreement on all these policies, Bethlehem could present a united front to its competitors, customers, and workers, as well as to the government and to the press.

Schwab was Bethlehem Steel's roving good-will ambassador. He was an immensely effective speaker, and business, civic, and academic groups were eager to have him address them. Whenever possible, he ac-cepted; he still loved the challenge of capturing an audience's attention and winning its affection.

Unlike Morgan and Frick, who were indifferent to public opinion and cared not one whit whether anyone despised or condemned them, Schwab was highly sensitive about what others thought of him. His ca-reer was punctuated with scandals; he was accused of being a munitions profiteer; his name appeared regularly on lists of the twenty-five most powerful men in America. He was even the target of socialist verses:

> Schwab, Schwab, Charlie Schwab,
> Life and Happiness you rob—
> From the workers in the mills
> To the miners in the hills.

As a speaker he worked to counteract the idea that he was sinister and aloof—to prove that he was charming and approachable—that he was a "regular fellow" despite his wealth and power.

Unlike most businessmen of his era, Schwab was never ill at ease while speaking. He almost always spoke without notes, pouring forth a sonorous stream of pious sentiments, optimistic predictions, and per-sonal reminiscences. Above all, he avoided controversial issues. His au-diences could readily agree with one reporter who lamented:

> To attempt to interview Mr. Schwab and really get anything out of him is as futile a task as a journalist ever undertook. It is like reading the Bible—one only gets fundamental truths. General principles that no one can possi-bly find fault with constitute the entire stock of topics on which he will talk for publication.[26]

If any of his speeches did contain a controversial idea, it was uninten-tional. He simply told his audiences what they wanted to hear. When addressing businessmen, he would stress the necessity of higher profits and lower taxes. When he spoke to workingmen, he would extol the American laborer as a man of "vigor and virility," the cream of "the aristocracy of labor" upon whom industry was dependent for maintaining America's industrial supremacy.[27]

But, uncharacteristically, in a talk he gave in 1918, he seemed to endorse the Bolshevik Revolution. Of course, nothing was further from his mind. He was making an extemporaneous after-dinner talk to the alumni of Public School 40 in New York. Most of the people there were neither wealthy men nor members of distinguished professions, so Schwab merely intended to exalt their status as ordinary workingmen. But this is what he said:

> We are at the threshold of a new social era. This new order of things may work great hardship for many of us. It is going to come upon us sooner than we expect. It is a social renaissance of the whole world. Some people call it Socialism, others call it Bolshevikism. It means but one thing, and that is that the man who labors with his hands, who does not possess property, is the one who is going to dominate the affairs of the world, not merely Russia, Germany, and the United States, but the whole world. . . . This great change is going to be a social adjustment. I repeat that it will be a great hardship to those who control property, but perhaps in the end it will work estimably to the good of us all. Therefore, it is our duty not to oppose, but to instruct, to meet and to mingle with the view of others.[28]

Those remarks were widely reported in the press, and they aroused considerable alarm in the business community. As one of his oldest friends, Charles W. Baker of U.S. Steel, remarked, "undoubtedly his predictions are well founded, but there are some things, even true, that are better left unsaid. . . ."[29] And his former colleague, William B. Dickson, who was regarded as a radical because he proposed to eliminate the twelve-hour workday in the steel industry, told Schwab that he was " 'in bad' with most of your former associates on account of your famous Bolsheviki speech. Perhaps I can add the last bitter drop to your cup of humiliation by informing you that you are now classed with myself as a dangerous anarchist."[30] Schwab's troubled friends apparently forgot that he liked to make momentous prophecies, and that he had offered no detail or explanation, merely a flat assertion, cloaked in murky phraseology. But this experience made him shy away even further from controversial topics.[31]

Few people who heard him speak probably noticed or cared that his talks lacked controversial or original ideas, that he repeated the same time-tested anecdotes, and that (aside from his jokes), his speeches never contained a memorable phrase or quotable passage. What mattered most to him—and what seems to have impressed his audiences most—was his ability to create a mood. He used words as one uses a blanket: he tried to make each listener feel warm and comfortable.

Schwab had no illusions whatever that his speeches were literary masterpieces. His sentences usually were convoluted and confusing; syntax nearly always was drowned in his oratorical stream of consciousness. He often acknowledged that he was not a learned orator, rather "just a plain mill worker" or "a steel puddler from Pittsburgh." But he never used a professional speechwriter to improve the structure or substance of his talks. "It is not what I say that is of interest to the people I address," he said, "but it is the way I look when I say it." [32] In this connection, he complained that radio was spoiling public speaking, that it was necessary for a speaker to see and "size up" his audience. And, he said, "The audience has to see you, and to judge for itself whether you are telling the truth." [33]

Whenever he felt slightly self-conscious about the formlessness and repetitiveness of his talks, he told a self-deprecating joke. Speaking at Princeton University, he concluded by saying: "Boys, I am sure you have never had any one make you such a rambling, such a disjointed speech. I hope your professor of English is not here. A short time ago in Europe I met a lot of foreigners. I had to make a little speech, and I did it and then turned to my friend, Mr. Vickers: 'You speak all these languages. Interpret this to all the people.' He turned to the others and said: 'I have been asked to interpret Mr. Schwab's speech into the various languages, and take pleasure in doing so. I will first render it into English.' " [34]

After the success of his impromptu talk at the University Club dinner in December 1900, Schwab probably felt no need or inclination to prepare a polished speech before he addressed a group of students or businessmen. He relied upon his subconscious mind to supply him with sentiments and anecdotes appropriate for each occasion. Often his talks began along these lines: "I want to state frankly that I came here this evening . . . without a single thought in my mind as to what I am going to say before I arise [sic], but with the hope that I may draw my inspiration from the happy, hopeful prosperous faces I see about me." [35]

257

His most frequently repeated story was designed to narrow the gulf between him and his audience—to make them feel that he had no exalted or inflated image of himself and in that way to create a common bond between them. On dozens of occasions, when the toastmaster or chairman introduced him with a flourish of superlatives which set him apart from his audience, he began:

> When I listened to the introduction of our distinguished Toast Master tonight I cannot [sic] help but repeat a story that I have told often before but which may not be entirely known to yourselves. Some years ago when I got to be a good deal talked about in the newspapers I was riding up the streets of Braddock one day with my colored man in the buggy beside me and I could not help overhearing a woman say to a little child that she was leading along the street, "Look dear, that is Mr. Schwab in the buggy" and the innocent little girl looked up and seeing the colored man and myself asked "Which one?" I always love to tell that story because it brings me to a proper realization of my position." [36]

He believed that in order to capture an audience's attention it was necessary first to win their sympathy. To do this, he often resorted to a ruse. After being introduced, he would walk up to the podium and begin to search through all of his pockets, looking first for his glasses and then when he had found them, he began the search anew, looking for the text of his talk. His listeners probably never realized that his glasses and script were mere stage props which he discarded almost immediately after he finally "found" them. What counted for him was the interval when he searched for them, when most members of the audience began to feel embarrassment and sympathy for him. This ploy created a bond of empathy with his audiences and put them in a more receptive frame of mind for the talks that followed. [37]

His frequent appearances as a speaker won him many new admirers—a factor of considerable importance to a man who drew emotional sustenance from other people's approval. His pleasant memories of such occasions helped to diminish the agony he endured when his honesty and integrity were called into question, as they were, repeatedly.

14

Postwar Conflicts and Challenges

The first criticism of Schwab's wartime service arose in January 1921. A routine audit had been made of the records of the Emergency Fleet Corporation, preparatory to hearings before the House Committee on Shipping Board Operations. The audit was conducted by Colonel Eugene Abadie. On January 20, 1921, Abadie testified that the EFC files contained a voucher for Schwab's personal expenses. The amount involved was $269,543, covering Schwab's expenses for nine months. During his service to the EFC, Schwab had traveled nearly 100,000 miles visiting shipyards and war-bond rallies. All of these trips had been made in his own private railroad car—thus, a major portion of his expenses was for the operation and maintenance of the car and for salaries for its crew. The next largest item was for telephone and telegram messages to Hurley, Piez, and the various shipyard owners.

Of the $269,543 involved, Abadie claimed that Bethlehem Steel tried to recover $100,000 of it by billing the government for it as "cost of ship construction." [1] Since Schwab had served as Director-General without salary or expense account, Abadie's charge, if shown to be true, would have been a serious blot on Schwab's integrity. Schwab learned of the sensational accusation the same day it was made, and he fired off a telegram to the committee asking for an immediate opportunity to deny the charge publicly.

The committee, headed by Congressman Joseph Walsh of Massachusetts, agreed to Schwab's demand; he was asked to testify the following morning, January 21. When Schwab entered the hearing room, he asked someone to point out Colonel Abadie. Schwab walked up to his accuser: "You are Colonel Abadie. I don't believe I know you and I don't recall ever having met you. You have ruined my reputation. You have done me an irreparable injury. Why did you do it?"

Abadie made some brief reply, but reporters standing near by were unable to make it out.[2]

When Schwab came to the witness chair, he began an impassioned denial of any wrongdoing:

> It is absolutely, unmitigatedly, and maliciously false. There is no charge for personal expenses of mine in any manner, shape, or form against the Shipping Board for ship construction to the United States Government or the Emergency Fleet Corporation. Furthermore, no personal expense of mine in any manner, shape, or form while I was director general of the Emergency Fleet Corporation—and they were heavy—were ever reimbursed to me by the United States Government. . . . And I defy anyone to show on the records of the Bethlehem Shipbuilding Corporation any voucher charged against the Shipping Board for my personal expenses. Mr. Chairman and gentlemen of the committee, the statement is so false, malicious, and I might say pernicious that I can hardly find words to characterize its harmful influence.[3]

Fortunately, the resolution of the issue did not depend on Abadie's claims and Schwab's denials. Four days later, Eugene Grace was able to prove conclusively that the government had never been asked to pay the $100,000. Grace revealed that Schwab had not charged his expenses to the United States, but that the entire amount, $269,543, had been reimbursed to Schwab by the Bethlehem Steel Corporation, which had charged it off to general operating expenses. One hundred thousand dollars had been charged off to a subsidiary, the Bethlehem Shipbuilding Corporation, but the Corporation had never asked for reimbursement from the government. In fact, at the time the audit was made for the Shipping Board, none of the bills of Bethlehem or any of its subsidiaries had been presented to the government for payment.[4] Grace was able to document this explanation, and Chairman Walsh offered Schwab a public apology.

Thousands of letters poured into Schwab's office after the Committee cleared him of the charge. Some were from friends and business associates, but the overwhelming majority were sent to him by total strangers—shopkeepers, nuns, farmers, factory workers, even young schoolboys—who had written to express their relief that he had been exonerated. These spontaneous expressions of support were matched by editorials published throughout the country, all consoling Schwab for the suffering he had experienced. The New York *American*, a Hearst newspaper, devoted a half-page editorial to recounting and extolling Schwab's wartime service. The editor, Arthur Brisbane, told Schwab that the tribute had been personally ordered by William Randolph Hearst—the publisher who, two decades before, had devoted a full page to denouncing Schwab's "juvenile egotism" and calling him "only a competent clerk." [5]

A leading magazine, *The Outlook*, asked, rhetorically: "Why is it that more men of first-rate ability do not enter public service?"—and it cited the reckless accusations against Schwab as an answer:

> The looseness of the charge and the inconsiderateness with which it was made are illustrative of the menace which any man who enters American public life faces. . . .
> Men do not like to enter public life, not because its material rewards are small, but because most men have an innate distaste for being hit below the belt. [6]

Schwab was relieved that he had been vindicated, and so soon. Writing to Andrew Carnegie's widow, Louise, he said, "I am glad the horrid nightmare is over and that I have not only been completely exonerated by Congress, but by public opinion. Of course you who know me, knew that such a thing was impossible." [7]

Being in semi-retirement did not shield Schwab from new accusations of wrongdoing. Early in 1922, Bethlehem Steel pressed the federal government for payment of monies still owed to it on wartime shipbuilding contracts with the Emergency Fleet Corporation. Bethlehem claimed it was owed $9,000,000; the EFC's response was that a far lower but still indeterminate amount was owed. On September 1, 1922, the final government audit of EFC war accounts was completed, showing that Bethlehem was entitled to only $3,200,000. Given the wide discrepancy in claims, the dispute was referred for final determination to the EFC's special counsel, William C. Bullitt. Owing to the legal and financial intricacies involved in the rival claims, Bullitt's investigation took nearly three years to complete. [8]

Through early March 1924, the controversy was not disclosed to the public or the press. Then, quite suddenly, it was catapulted to the front page. Congressman James F. Byrnes, a South Carolina Democrat, speaking on the floor of the House, claimed that Bethlehem had made $11,000,000 in "excess profits" on its wartime shipbuilding contracts. According to Byrnes, Schwab's grave misconduct while head of the Fleet Corporation had enabled Bethlehem to reap this wrongful gain. He charged that Schwab had been responsible for awarding shipbuilding contracts to Bethlehem on terms which were more profitable than those given to other companies. Byrnes said, "Necessarily, as a member of the Fleet Corporation, Mr. Schwab had to transact business with his own company. The relationship should have prompted him to be overscrupulous in seeing to it that the Government was protected in all dealings with the Bethlehem Corporation." [9]

Byrnes angrily demanded that the Attorney General, Republican Harry Daugherty, initiate a suit to recover the "excess profits." Byrnes's openly partisan and polemical speech ended with a prediction that the Coolidge administration would suppress the government's rightful claim to the $11,000,000 in return for a generous campaign contribution from Schwab for the forthcoming 1924 election. The government took no action, pending completion of Bullitt's report.

In 1925 Bullitt ruled that the amount still owed to Bethlehem was $5,500,000. Rather than protract the controversy and thus delay receipt of the money, Bethlehem agreed to take the reduced amount. However, the EFC was not willing to accept a compromise settlement. Instead, it repudiated Bullitt's report. It withdrew the report from circulation, and refused to pay the amount Bethlehem had agreed to accept. In fact, the EFC adopted a new stance, claiming that Bethlehem had made excessive profits during the war and therefore that *no money at all* was still owed to it; on the contrary, the EFC, echoing Congressman Byrnes, demanded that Bethlehem "return" $11,000,000.[10] In support of its claim, the EFC filed suit against Bethlehem for recovery of "excess profits"; in retaliation, Bethlehem filed suit against the EFC to obtain the full $9,000,000 it originally had claimed was still owed on wartime contracts.

The litigation, which began in 1925, moved at a glacial pace through the federal courts. Three years went by before Owen J. Roberts was appointed as a Special Master and Referee by the U.S. District Court. Then, two years later, when Roberts was thoroughly familiar with the complexities of the claims and counterclaims, he was appointed Associate Justice of the U.S. Supreme Court and had to withdraw from the controversy. Another nine months elapsed; then William Clarke Mason was appointed to succeed Roberts. Mason had to begin the investigation *de novo*.[11]

The EFC's claim for recovery of excess profits rested upon two contentions: first, that the type of contract it had signed with Bethlehem was invalid and hence could not be enforced on Bethlehem's behalf, and second, that Schwab's dereliction of duty as Director-General had enabled Bethlehem to earn "unconscionable profits."

The form of contract which the EFC repudiated was one which was widely used in business and in many government dealings during the war. Known as the "bonus for savings contract," it stipulated that the government would pay the contractor—in this case, Bethlehem—the "actual cost plus a fixed fee (in no case greater, in some cases less, than 10 per cent of the estimated cost specified in the contract), plus a part (usually one-half) of the saving in the actual cost below the estimated cost."[12] Unlike the "cost-plus" contract, which created an incentive to be wasteful and inefficient in order to receive as profit a percentage of an inflated cost, this type of contract was designed to reduce costs.

Grace, speaking for Bethlehem, claimed that the profits it had earned under these contracts was a tribute to its efficiency in cutting costs, and that the refusal of the EFC amounted to penalizing Bethlehem "for having been successful in keeping down costs,—the very thing which the particular form of contract was designed to accomplish."[13] But the EFC took an entirely different view of the situation. It claimed that the negotiators for the Bethlehem Shipbuilding Corporation had deliberately and fraudulently overestimated the base price in order "to derive excessive, unreasonable, and unconscionable profits" by creating imaginary savings on construction costs.[14]

Following a careful examination of the evidence, the report of the Special Master and Referee, which was issued in December 1935, concluded that "the charge of fraud made by the Government against Bethlehem is without foundation," since the EFC's own witness, G. S. Radford, who had been the contract negotiator for the government, lent no credence whatever to the charge. Instead, Radford agreed with the testimony offered by J. Y. Powell, Bethlehem's negotiator—that, given the uncertainties of wartime, including rising wages and shortages of materials, it was "impracticable to estimate within a reasonable percentage what would be the actual cost of construction."[15] Rather than finding fraud, the Special Master determined that the contracts "resulted from negotiations in which both parties were represented by intelligent, well informed, and experienced officers whose sole object was to make the best trade possible, under conditions which included the uncertainties of wartime contingencies, the results of which were not and could not have been known at the time the contracts were made." [16]

In addition to contesting the form of contract on the ground that it permitted profiteering, the EFC claimed the contract was further invalidated by the fact of Schwab's dual position in 1918: chairman of Bethlehem Steel and Director-General of the EFC. The complaint was not that Schwab had wrongfully used his position to award lucrative contracts to Bethlehem. Schwab, of course, had taken precautions to prevent such charges from ever being raised; when he had accepted the appointment to head the Fleet Corporation, he had obtained a written promise from President Wilson and the U.S. Shipping Board that he would not have to negotiate with Bethlehem.

In its suit against Bethlehem, the EFC did not charge that Schwab had secretly violated this agreement; the complaint was that he had *not* done so! Schwab was accused of dereliction of duty because he had not tried to force Bethlehem to revise its contracts and to accept less profitable terms instead. Frederick I. Thompson, a member of the U.S. Shipping Board, claimed in 1925 that if Schwab had acted honorably he would have attempted to reduce the profits Bethlehem earned on its EFC contracts. Schwab was condemned because the contracts he negotiated with other companies only gave them a net profit of 10 per cent, whereas Bethlehem was permitted to operate under more lucrative contracts— contracts which had been made before he became Director-General. According to Thompson, "They were the first contracts he should have assured himself were within the earning limitation of the profit he set." [17]

The Special Master disagreed. He completely exonerated Schwab, recalling that he, Schwab, was not responsible for the EFC's contract arrangements with Bethlehem, and that he could not have known precisely what profits Bethlehem was making on its shipbuilding contracts since final claims for payment, along with supporting cost sheets, were not filed until long after the Armistice. The report concluded: "The master finds as a fact that Charles M. Schwab had no responsibility as a representative of the Fleet Corporation or of the United States of America in relation to these contracts between Bethlehem and the Fleet Corporation, and that there is no evidence to support the charge that the dual relationship of Charles M. Schwab imposes upon Bethlehem, in good morals, the duty to relinquish all profits from these contracts above 10 per cent of the actual cost of the ships." [18]

The Special Master upheld Bethlehem's claim for additional payment in the amount of $5,380,000—and that finding was upheld in the U.S. District Court.[19] But the government, which was still determined to save part and perhaps even all of the amount awarded to Bethlehem, filed an appeal in the Circuit Court of Appeals, asking for a reversal.[20]

The case was finally decided by the Supreme Court in February 1942—more than two years after Schwab's death. In a five-to-one decision, the majority opinion of the Court, written by Justice Hugo L. Black, held against all of the government's allegations. It ruled that there was no evidence of fraudulent misrepresentation by Bethlehem's negotiators; that under the provisions of the contracts, Bethlehem was entitled to the bonuses; that Bethlehem's profits were not "excessive," since many other firms had made even greater percentages of profits on war orders; that the government's attempts to invoke the common law doctrines of unconscionability and duress as grounds for voiding the contracts were inapplicable; and that there was no justification for the charge that Schwab was guilty of wrongdoing.[21] In a separate, concurring opinion, Justice Frank Murphy pointed out the impropriety of attempting to alter the meaning and application of a contract on the basis of "afterthought and subsequent experience." Noting that contracts must be enforced literally and that the government "had entered into the agreements with full understanding of their terms," Murphy concluded: "The possibility that the Government may be relieved of bargains twenty-four years after agreeing to them is not conducive to mutual trust and confidence between citizens and their government." [22]

Throughout the 1920's, Bethlehem continued to expand its facilities. Beginning as early as 1918, the company spent $20,000,000 enlarging the plate and sheet mills at the Sparrows Point plant near Baltimore. In order to improve the shipbuilding and repair facilities at Sparrows Point, it completed a drydock with a 6,000-ton lifting capacity, and it made similar improvements at the Union shipyard in San Francisco and the Fore River shipyard in Quincy, Mass.[23] Bethlehem also acquired two other major steel-making facilities: it purchased Midvale Steel and Lackawanna Steel, both in 1922.

Bethlehem was in a strong bargaining position because, of all the steel companies, it had suffered the smallest decline in net earnings between 1920 and 1921. Its decline was 29 per cent of net earnings; in the same period U.S. Steel declined 66 per cent, Midvale 143 per cent, and Lackawanna 180 per cent. The last two, of course, had deficits, which explain the owners' willingness to accept Bethlehem's offer to merge.[24] As of December 1921, Midvale and Lackawanna were the third and fourth largest steel producers in America; their steel-making capacities, respectively, were 2,894,000 and 1,840,000 gross tons of ingots, or 5.74 and 3.65 per cent of the country's total capacity. The leaders were U.S. Steel and Bethlehem, the former with 22,700,000 gross tons, or 45 per cent of national capacity, and the latter with 3,050,000 tons, or 6.04 per cent. The union with Midvale and Lackawanna gave Bethlehem a total capacity of just under 10,000,000 tons—over 15 per cent of national capacity.[25]

By acquiring Lackawanna, Bethlehem gained in four ways. Lackawanna was a producer of merchant steel, that is, bars of standard size and shape; these supplemented the special alloy steel bars made by Bethlehem, so the company could now offer its customers a full line. Bethlehem was able to shift some of its rail-making to the Lackawanna plant, thus freeing space for making other products at the main plant at South Bethlehem; this shift also widened the sales area for Bethlehem's rails. Further, Lackawanna's location on the Great Lakes enabled Bethlehem to penetrate the midwestern and Canadian markets, and this improved its competitive position in relation to producers in the Chicago, Pittsburgh, and Youngstown areas. Finally, and perhaps most significantly, the acquisition of Lackawanna's iron ore deposits in the Lake Superior region enabled Bethlehem to reduce its dependence upon ores from Chile and Cuba.[26]

Bethlehem's merger with Midvale took place after only three weeks of negotiations, following the collapse of a proposed merger between Midvale, Republic Steel, and Inland Steel. In 1917, Midvale had assets amounting to $270,000,000, which made it the seventh largest company in the country.[27] The most attractive aspect of Midvale in 1922 was its subsidiary, the Cambria Steel Company, which it had acquired in 1916. Bethlehem acquired Cambria's main plant at Johnstown, Pennsylvania, as well as its ore, coal, and limestone properties and its Great Lakes ore fleet. But it did not purchase Midvale's armor and ordnance plant at Nicetown, Pennsylvania.[28]

Although these acquisitions gave Bethlehem a slight entry into U.S.

Steel's prime sales area, this incursion was peripheral and not intended as an all-out assault. There was a tacit live-and-let-live agreement between Schwab and Gary, and the two largest steel companies were rarely in direct competition. But in 1926 Schwab made a startling discovery which shattered their truce.

In March of that year, Schwab visited Homestead; it was the twenty-fifth anniversary of his departure from Carnegie Steel to head U.S. Steel. During the homecoming celebration Schwab toured the steel works, and, to his amazement, he found that U.S. Steel was in the process of building a mill to produce the patented Bethlehem beam, without his or Grace's knowledge or authorization.

According to Schwab, he confronted Gary, who told him that it was "just an experiment," and that it had "nothing to do with the Grey beam." Schwab found this explanation unacceptable. He and Grace arranged a meeting with Gary and James A. Farrell, U.S. Steel's president, but the outcome of it was no more satisfactory than that of the first encounter. As the meeting ended, Schwab shouted: "Gary, you're the goddamndest liar and double-crosser I've ever known. Your word is no good, and I shall never trust you or have anything to do with you again." [29]

U.S. Steel had decided to build a $31,000,000 wide-flange beam mill after its sales department had reported in 1924 that the absence of such beams from U.S. Steel's product line gave Bethlehem an immense advantage:

> Architects and fabricators generally consider that these sections mark a distinct advance in building construction in the United States involving as they do a great reduction of riveted work, and their introduction has opened up a competitive field which cannot be met with standard sections. . . . The tonnage lost on account of competition with Bethlehem cannot readily be estimated, but it is an ever increasing tonnage and if not lost outright we are obliged to sell our standard sections at unusually low prices in order to compete. [30]

A leading trade journal of the steel industry confirmed the importance of the Grey beam to Bethlehem:

> . . . the exclusive right to manufacture and sell under the Grey patents has meant much more than the benefits derived from selling Grey sections alone. The possession of the Grey right has made Bethlehem the only company in the United States able to supply every structural steel requirement.

Bethlehem not only has the Grey sections, but also a full line of standard sections, plates, bars, and the like. Many customers have been relying upon Bethlehem for their entire tonnage. As a result, Bethlehem has built up a huge structural business in the East.[31]

As a result of his dispute with Gary, Schwab boycotted the May 1926 meeting of the American Iron and Steel Institute. Gary had been its founder in 1909 and its president ever since. Schwab had never missed a meeting, so his absence naturally evoked curiosity and puzzlement. Rumors of a conflict between Gary and Schwab circulated freely among the members of the Institute, but no one knew for certain what had come between the two men. Willis L. King, president of Jones & Laughlin Steel, and a friend of both Schwab and Gary, wrote to Schwab, urging him to attend the forthcoming October meeting:

> The Directors have greatly missed you, and the members of the Institute are wondering at your apparent loss of interest. We would all welcome you back on your personal account, but perhaps, for the more selfish yet important reason that the Institute needs you if it is to accomplish its full benign purpose, and we deprecate any break in the cordial relations which meant so much in the past and will mean so much to the future of the steel industry. So, my dear C.M., if you have suffered injury, why not let the Court give you relief and forget all else.[32]

Schwab replied that he could not be a hypocrite, that he could not pretend that everything was cordial and normal when it was not. He expressed regret that he could not attend and said that the fault lay elsewhere.[33]

Yet at the last moment Schwab changed his mind and brought his houseguest, Queen Marie of Rumania, to address the steelmen. Afterwards, when he spoke, he made no mention of his quarrel with Gary; instead, he stressed the emotional significance of his return. An unabashed sentimentalist, the sight of his old friends and fellow steelmen brought tears to his eyes. "I have missed you more than you have missed me," he said. "One of the delights of my life has been for 25 [sic] successive years to stand before you boys and renew my expressions of love and admiration through the long years. When you reach my age, there is but one thing that counts, and that is the love and appreciation of those with whom you have been associated. I am one of the oldest practical steel men in the United States, with nearly 50 years of service, but the outstanding thing in my life is my love and admiration for you." [34]

In the next few months, no progress was made toward reconciling Gary and Schwab. Early in 1927, however, Gary became ill. Perhaps sensing that he would not recover, he decided to make his peace with Schwab. He asked Schwab and Grace to meet with him and Myron C. Taylor, a director of U.S. Steel. According to Grace, at this meeting Gary admitted that U.S. Steel had been in the wrong, and that he owed Schwab an apology. Schwab and Gary were reconciled, and Grace and Taylor agreed to meet to settle any problems related to the possible infringement of Bethlehem's patent rights by U.S. Steel.[35]

Gary died in August 1927, and U.S. Steel then broke the peace agreement. Bethlehem subsequently filed suit for infringement of patent rights. U.S. Steel's defense was that Henry Grey was not the inventor of the Grey beam and that he, therefore, did not have the right to sell the beam as his own invention. It claimed that Grey did not have a valid patent claim because other men had anticipated and won patents on production processes virtually identical to Grey's. Further, U.S. Steel's attorneys claimed that even if Grey's patent had been valid, it had expired and had not been properly renewed by Bethlehem and hence was now in the public domain. Fearing, perhaps, that none of these claims would suffice, U.S. Steel also claimed that Bethlehem had forfeited the right to sue because it had waited so long to initiate suit after the patent infringement was first discovered.[36]

The conflict never reached a judicial decision. In October 1929, the suit was withdrawn after U.S. Steel agreed to pay royalties for permission to produce the Bethlehem beam.[37]

As a result of the patent fight, Schwab and Grace decided to expand Bethlehem's activities in the area west of Pittsburgh, which was largely dominated by U.S. Steel. An added reason, according to Schwab, was that U.S. Steel had broken its agreements to keep out of Bethlehem's eastern markets.[38]

Schwab tried to penetrate U.S. Steel's sales territory by merging Bethlehem with the Youngstown Sheet and Tube Company, the third largest independent steelmaker in America. Youngstown was particularly attractive because its newest plants, a seamless tube mill in Youngstown and a tinplate and bar mill at Indiana Harbor, could draw upon Youngstown's blast furnaces at South Chicago. If Bethlehem gained control of these new mills, it might challenge U.S. Steel's long-standing dominance of the Chicago steel market.[39]

Youngstown's chairman, James A. Campbell, endorsed the proposed merger when he announced its terms on March 6, 1930: the holders of Youngstown stock could choose either to exchange one of their shares for one-and-one-third shares of Bethlehem's stock, or receive a tentative cash price of $110 per share. On April 8 the tally of proxies revealed that shareholders controlling nearly 70 per cent of Youngstown's stock approved of the merger. Nonetheless, Cyrus Eaton, a minority stockholder, filed suit to prevent the merger. He claimed that while Bethlehem's shareholders were paid no dividends from 1925 to 1928, its executives were paid exorbitant bonuses; he charged that the same fate would befall Youngstown shareholders if they accepted Bethlehem stock in exchange.[40]

Eaton also triggered a new crisis for Schwab. He made public the sensational news that in 1929 alone Eugene Grace had received a bonus of $1,500,000—a fact known to few, if any, of Bethlehem's or Youngstown's stockholders (the precise figure was subsequently revealed to be even higher: $1,623,753). Although the existence of the Bethlehem bonus system was well known—in fact, Schwab boasted that it was the *cause* of Bethlehem's success—the exact amounts paid out to Bethlehem's executives had never been revealed to Bethlehem's shareholders in Schwab's annual reports; the only reported figure was Grace's annual salary of $12,000.

The legal basis of Eaton's suit to block the merger was that Youngstown's directors had failed to acquire and disclose relevant information concerning Bethlehem's bonuses and dividend policies. Eaton argued that the directors should have rejected the merger offer because Grace's huge bonus drained Bethlehem's working capital, making the merger disadvantageous to Youngstown's shareholders. Eaton also claimed that the Bethlehem offer was financially inadequate because it failed to reflect the actual value of Youngstown's properties and assets.[41]

The high point of the trial, which began on June 25, 1930, was the interrogation of Eugene Grace by Eaton's attorney, Luther Day. Grace testified on July 17 that his salary was $12,000, but he refused to disclose his bonus, beyond the cryptic statement: "The factor used to determine my bonus is one and one-half per cent."

Day asked, "One and one-half per cent of what?"

Grace answered: "I don't know."

When Day asked Grace to disclose his salary and bonus agreement, Bethlehem's attorney, Newton D. Baker, objected, on the ground that

Youngstown's stockholders were not entitled to know anything about Grace's remuneration prior to April 30, the day the merger would have taken effect.[42] Baker argued that there had been no fraudulent concealment of the Bethlehem bonus system and that the precise remuneration Grace and others received was not a matter of public concern, that its disclosure would violate the privacy of the recipients and was not relevant to the case since the merger had been approved by a majority of Youngstown's shareholders.[43] Judge David G. Jenkins sustained the objection.

The trial ended in late September, and the court rendered its decision on December 29. It sided with Eaton and thus disallowed the merger. It gave two primary reasons: that the proposed exchange rate of stock undervalued Youngstown, and that the directors of Youngstown had been ill-informed about relevant issues when they recommended approval of the merger to the shareholders. This decision was overturned by the Court of Appeals in 1931, but by then it was too late to salvage the merger. In any case, Bethlehem had lost interest; the stock market had collapsed. Its own stock had fallen from 103 in March 1930 to 27 in October 1931, but Youngstown's stock had taken an even sharper nose dive during the same period, from 140 to 22.[44] While Bethlehem was trying to buy Youngstown Steel as a means of challenging U.S. Steel, it also bought the McClintic-Marshall Corporation, whose fabricating plants were located near U.S. Steel facilities in the areas of Pittsburgh, Buffalo, Chicago, San Francisco, and Los Angeles. The new subsidiary could use Bethlehem beams in constructing buildings and bridges, and thus could offer vigorous competition to U.S. Steel's structural steel division.[45]

When Cyrus Eaton revealed the size of Grace's bonus, some of Bethlehem's shareholders became furious at Schwab, the administrator of the bonus system. On November 1, a group of Midvale Steel stockholders announced the formation of a committee to protect their interests in Bethlehem; they had received stock in 1923, when Bethlehem had bought all of Midvale's facilities except its armor and ordnance works.[46]

On January 13, 1931, four stockholders filed suit in the New Jersey Court of Chancery at Newark. The suit named Schwab and other directors of Bethlehem as co-defendants; it accused Schwab of "favoritism" and "prejudiced judgment" in his administration of the bonus system and of failure to disclose even the aggregate amount of bonuses in Bethlehem's annual reports. The plaintiffs conceded that in 1917 Bethlehem's stockholders had authorized payment of executive bonuses ranging from 3.43 per cent to 8 per cent of net earnings. But they challenged the concept of "net earnings" which Schwab used; bonuses were based on earnings after deducting interest charges and preferred dividends but not depreciation, obsolescence, and depletion. They also contended that the Bethlehem bonuses were vastly in excess of those paid by comparable companies and industries, and even firms which were larger than Bethlehem, such as U.S. Steel, Standard Oil of New Jersey, or American Telephone & Telegraph. The plaintiffs showed that, whereas Grace received $1,600,000 in 1929 and had averaged $814,000 a year between 1918 and 1930, Judge Gary never received more than $500,000 a year at U.S. Steel, George Gordon Crawford of Jones & Laughlin earned only $125,000—the same salary received by Walter C. Teagle, the president of Standard Oil of New Jersey—and Grace's average bonus was more than three and a half times greater than the $250,000 salary of A.T.&T.'s president, Walter S. Gifford.[47]

The plaintiffs demanded that the bonus recipients repay to Bethlehem the $36,493,668 paid out since 1911, and they asked for an injunction to block new bonus payments until Bethlehem supplied a full accounting of the system.[48] This suit forced Schwab into a belated defense of his administration of the bonus plan.

Ivy Lee, who was directing Bethlehem Steel's public relations program, complained to Grace that the policy of not answering criticism of the bonus system was shortsighted and futile. In 1930, at the time when Cyrus Eaton made the first disclosures about bonus payments, Lee recommended that the company issue a pamphlet fully explaining the operation and justification of the bonus plan. His suggestion was overruled, either by Schwab or by Grace, probably by both, since they agreed that Bethlehem was under no obligation to reveal its confidential affairs, such as executive bonuses, to anyone.

In January 1931, Ivy Lee again urged full disclosure. He believed that the company's best defense in the pending suit would be full candor. He cautioned Grace against continuing the policy of silence:

> I really think that in justice to yourself, as well as justice to the company, this subject ought to be ventilated and ventilated aggressively. If we keep our mouths shut and say nothing we will let the public feel that we are disturbed over this thing and are wondering whether our policy was wise.[49]

Piqued at the criticism of the bonus system and irritated by the law suit, Schwab, without consulting Ivy Lee, issued a statement to the press. He assumed full responsibility for what he called the "profit-sharing system." He claimed it was no more than the adaptation to Bethlehem Steel of the policy which had made the Carnegie Steel Company so successful. "The value of the system to Bethlehem," Schwab said, "is clearly shown by the fact that the cost of its executive management is less than that of any other important steel company of which we have definite knowledge, and its value is also reflected in lower manufacturing costs." He also pointed out that the bonus system had been approved originally by the stockholders, who had entrusted its administration to him, and that he personally never received a bonus, so that he would be "entirely free from all possible prejudice" in fixing the amounts of compensation to be paid.[50]

Ivy Lee found Schwab's statement wholly unsatisfactory: "It fell flat," he told Eugene Grace. "I was a little sorry I didn't have an opportunity to talk it over either with you or Mr. Schwab before that statement was put out because I feel that although the public is somewhat agitated over the size of the bonus, there is even more agitation over the theory that it was a secret matter and something that was put over on the stockholders. I don't feel that Mr. Schwab's statement gave adequate attention to this latter phase of the matter." [51]

Opposition to Schwab's administration of the bonus plan did not remain confined to a few shareholders. It became an organized movement, and a group known as the "Protective Committee for Stockholders of Bethlehem Steel Corporation," an outgrowth of the Midvale group, was formed. One of the first actions of the Protective Committee was an attempt to notify all Bethlehem stockholders of the existence of the Protective Committee and to solicit funds for its operation. When Bethlehem refused to make the list of its stockholders available to the dissident minority, the committee won a court order compelling the company to supply the names.[52]

On March 2, 1931, in response both to the Chancery Court suit and to the Protective Committee, Bethlehem Steel sent each of its stockholders a twenty-four-page booklet in the form of a letter from Schwab, explaining and defending "The Bethlehem Bonus System." The letter had another purpose, as well. Schwab wanted to solicit proxies for the forthcoming annual meeting on April 14. At the meeting a motion was to be introduced to approve and ratify his administration of the bonus system since 1917.[53]

Schwab vigorously denied that there had been any concealment, either of the bonus system or of the million-dollar bonuses paid to Eugene

Grace. On the contrary, he had boasted publicly that these bonuses were responsible for Bethlehem's growth and financial success. He cited newspaper accounts of 1915, Bethlehem's Annual Reports of 1916 and 1917, and an October 15, 1930, editorial in *Forbes* as proof that details of the bonus system had been published. However, in the letter Schwab passed over the fact that most members of the Protective Committee became shareholders in 1923, when Bethlehem purchased Midvale Steel, whose stock they had owned. Thus the dissidents legitimately could claim that they had little or no choice about becoming Bethlehem stockholders, that they were entitled to have information about the bonus system supplied to them, and that they were not negligent if they did not each personally engage in extensive research about Bethlehem's financial history. Schwab's defense missed the core of their objection.

In a response to Schwab's statement, the Protective Committee pointed out:

> A careful scrutiny of the annual reports of the corporation for the period in question [1925–28, when no dividends were paid] will show that there is not in any of them the least mention of bonuses nor could the most astute analyst have gleaned from any of the earnings statements or balance sheets accompanying these annual reports anything relating to the compensation of officials, which subject is submerged in one bare item of "operating expenses."
>
> Not only were the bonuses not disclosed but they were studiously concealed. The annual reports deal in the text as well as in the schedules of figures at length and in detail with the pension system of the corporation which involved the expenditure of a few hundred thousand dollars a year but no mention whatsoever is made of bonuses which run into millions annually.[54]

The critics were not objecting to the bonus system as such. Rather, they were challenging the size of the sums paid out, particularly to Eugene Grace. One of the objectives of the critics was to reduce bonuses. Schwab argued forcefully against this proposal:

> If the rates were to be revised downward because the earnings were exceptionally large in any one year, as for example in 1929, the purpose of the plan would be defeated, because the incentive to greater effort would be greatly diminished, if not destroyed. If an individual is promised a given percentage of the profits, and then, when the profits grow (and the purpose of the plan is thus realized), he is told that his share is to be materially reduced, he inevitably concludes that he is not sharing in the profits and that in substance he is receiving only a fixed salary.[55]

One of Schwab's statements particularly exasperated the dissenting stockholders. He said that he had an understanding with the bonus recipients that they would invest their bonuses in Bethlehem common stock. The critics charged that Bethlehem's failure to pay dividends on common stock during 1925–28, despite its earning over $50,000,000 in profits, had depressed the price of the stock. This enabled the bonus recipients, who received $6,800,000 in bonuses during that same period, to buy up a larger number of shares of the stock than they could normally have been able to do.[56]

Schwab anticipated this objection; he challenged the notion that the failure to pay dividends had depressed the price of Bethlehem's common stock. He said:

> If dividends are paid, the stockholders to that extent receive directly and immediately their share of the earnings. If, however, the Board of Directors, believing it to be for the best interest of the Corporation, decides not to pay dividends, but to reinvest the earnings in the business, the amount so reinvested enhances the intrinsic value of the stock. . . . No one can reasonably deny that the reinvestment of those earnings in the business in lieu of paying them out in dividends increased the intrinsic value of the common stock. Nor can anyone seriously urge that the compensation of executives which is based upon earnings should be withheld from them merely because those earnings are thus reinvested.[57]

In short, he claimed that the reinvestment of earnings served to sustain and even possibly to raise the price of common stock.

Strictly speaking, Schwab's reply was correct: if a company passes dividends and instead reinvests its earnings in any way which yields a rate of return greater than zero, the net asset value of the stock is increased. However, Schwab ignored the issue of time-preference. Some stockholders may prefer to get current income through dividends rather than deferring gains until they sell their stock. Those who preferred to get current income would sell if they learned that dividends were being passed. Their action would reduce (at least temporarily) the price of the company's stock, and enable the bonus recipients to buy more shares than otherwise.

Schwab further argued that the payment of bonuses was not and should not have been contingent upon the payment of dividends. Since he was not a bonus recipient, he had no motive to deceive or cheat the owners of common stock. He denied that his actions had been illegal or even ill-advised. He believed Bethlehem had been aided, not injured, by his administration of the bonus plan, and he assumed that this would be evident to everyone.[58]

On March 28, Bethlehem filed its response to the suit, denying every allegation of wrongdoing and fraudulent deception. The company claimed that bonus payments had not been disclosed since 1917 because such a disclosure would defeat the purpose for which the system was designed. The purpose of the bonus plan was to create an incentive for Bethlehem's senior executives to exert their best efforts, but if the bonus given to each recipient were known by the others, it might create jealousies and dissatisfaction within the executive ranks.[59] This claim might have had some truth to it, but it failed to answer a crucial question— why was the *aggregate* amount of bonuses never revealed in the annual reports?

The annual meeting was held on April 14, and there the bonus system was discussed. At one point Schwab could not conceal his annoyance that his administration of the system had been challenged: "I had the feeling that this damn company belonged to me, you know, and I went ahead and did the best I could." [60] Another awkward moment came when a stockholder named Elbert Miller said, "I have been a small stockholder of Bethlehem Steel Corporation for several years and in that time I have attended a majority of the stockholders' meetings, but I never heard of the bonus plan until I read of it in the newspaper accounts of the Youngstown litigation. Why, after fourteen years, are we now asked to ratify it?" Schwab's answer, in its entirety, was, "I can't answer that, it seems to be a legal question." [61] In fact, Schwab could have answered it. He wanted to win immunity from liability before the pending suit in the Chancery Court was settled.

In response to another question, Schwab disclosed that between 1901 and 1908 he had been paid no salary by Bethlehem; that from 1908 to 1920 he had received $50,000 a year; that his salary as chairman had been $150,000 from 1921 to 1929; and that he had received an unsolicited raise to $250,000 from the Board of Directors in 1930 "without my presence or knowledge." [62]

After the balloting was tabulated, it was announced that the overwhelming majority of proxies was in his favor: 30,403 shareholders voted their 2,288,357 shares of common stock on his behalf, out of 3,000,000 shares outstanding, as did 23,464 shareholders representing 734,729 shares of preferred stock out of 1,000,000 outstanding.[63] The Protective Committee had anticipated such a development, and it had already sought an injunction to prevent the vote from releasing Schwab from his liability in the pending suit. Vice Chancellor John H. Backes of the Chancery Court issued a restraining order, allowing the proxy vote to be cast but not allowing it to override the suit of the dissenting minority. The Court stated:

> What the defendants expect to gain by the resolution is not apparent. . . . an overwhelmingly affirmative vote . . . would not avail them, for the vote could have no influence upon the right of even a single shareholder who suffered injury by unlawful conduct. Obviously the resolution [i.e. the vote] of a majority cannot force a complaining minority into suppression of their rights nor find in it immunity, or by means of it frustrate redress for injury.[64]

The restraining order forced Bethlehem to come to terms with the Protective Committee. Settlement of the controversy was announced on July 2, 1931, after protracted discussions in which Samuel Untermyer was an active intermediary between the contending parties. The compromise settlement involved four major changes: first, Schwab's exclusive power to administer the bonus plan was ended, and thereafter it would be administered by those members of Bethlehem's Board of Directors who were not bonus recipients; second, the aggregate amount of bonuses paid would be published in the annual report; third, the by-law providing that 8 per cent of earnings was the maximum available for bonuses was not repealed, but the basis of calculating earnings was changed so that depreciation, depletion, and obsolescence charges were deducted (since these charges had averaged $15,000,000 for the preceding few years, this change meant an approximate reduction of $1,200,000 in the amount available for executive bonuses); and fourth, in anticipation of greatly reduced bonuses, the executives were now to receive larger base salaries, and no bonuses at all were to be paid when dividends were passed on the common stock.[65] At the request of the Protective Committee, Chancellor Edwin Robert Walker dismissed the bill of complaints against Bethlehem, and dissolved the injunction and restraining order.[66]

In this episode Schwab had risked being accused of wrongdoing by withholding information from his stockholders, even though he personally did not stand to gain financially. Schwab believed that the major reason for Bethlehem's success was that it paid large bonuses to its executives, offering them great rewards for productivity, efficiency, and initiative. Since Bethlehem's stockholders might not have approved of such large amounts, Schwab took the easy way out by withholding this information. This policy was apparently rooted in the same attitude which he had held in dealing with armor production at Homestead in 1893. Just as he had been convinced that the government would not have understood or agreed that violations of the armor contract were beneficial to the Navy, so he believed that the stockholders would not have understood that huge executive bonuses were beneficial to them because such incentives would lead to higher profits and dividends in the future. Schwab, the practical steelman, thought that he knew what was best for the company and its stockholders, so he saw no need to consult them about his policies.

In 1916, when Schwab had launched his nationwide series of ads designed to win public support for the defeat of the pending armor plant bill, he had declared: "We shall make the mistake of silence no longer. Henceforth we shall pursue a policy of publicity." His publicist, Ivy Lee, commented: "This campaign is really the first of its kind. I believe that this departure from its former policy [of silence] by the Bethlehem Steel Company marks the beginning of a new era in the conduct of American corporations, particularly in their relations with the public and the Government." *Editor & Publisher* agreed. It wrote: "this change of attitude toward the public on the part of one of the largest corporations in the land . . . cannot fail to be of inestimable benefit to that company in all of its future public relations." [67] Regrettably, Schwab gave the policy of candor far too brief a trial; in fact, he implicitly repudiated it after the Armor Plant bill was passed. He reverted instead to what he evidently thought of as innocent deceptions. Schwab was simply unable to adhere to a policy of candor and direct confrontation.

15

Twilight of a Titan

In late 1927, when Schwab succeeded Judge Gary as president of the American Iron and Steel Institute, he became the senior spokesman for the American steel industry. This post provided him with a semiannual forum in which to express his philosophy of business and his views on the state of the economy.

He foresaw many problems—but only in the context of the steel industry; he had no intimation whatever of the impending general crisis of the economy. He imagined that the stock market boom of the Harding and Coolidge years, fueled by the inflationary policies of the Federal Reserve System, was a sign of solid strength in the economy. But he was alarmed that the steel industry was making lower returns on capital investment than most other industries were, and he believed the cause to be a reckless policy of building new plant and equipment far in excess of present needs. Each firm seemed determined to improve its competitive position at the expense of its rivals, whereas the proper policy, Schwab said, should be one of "live and let live." [1] What was needed, he said, was "an industry-wide agreement which would end the wasteful practice of cross-hauling," thereby giving each firm relative immunity from competition in its own local markets. The agreement also would have to provide for some means of maintaining prices at a profitable level, for now firms were cutting their prices to attract orders in what Schwab called "a wasteful hunger for tonnage." [2]

In 1928 he told his fellow steelmen that if they could agree on uniform prices and avoid price cutting, a new era of prosperity would await the steel industry.[3] And in 1929 he again said that the steel industry was not keeping pace with the general prosperity of the nation. He warned the leaders of the industry that they must resist any temptation to expand plant capacity. Instead, they must use existing capacity to the fullest extent and with the greatest efficiency.[4]

After the stock market collapse in 1929, he told reporters: "Be not afraid. The stock market cannot stop or stem the prosperity that extends throughout this great country of ours." [5] The collapse, he said, was merely another in the long chain of readjustments needed to bring market prices into alignment with "true values." He predicted that recovery would be imminent.

On New Year's Day of 1930 he called in the press to give a glowing message of hope to the American people. He denied that the economic crisis was anything more than a temporary corrective, and he ridiculed the pessimists in the financial world. "I hear men talk about America's situation today as though we had reached some summit. Compared to what this country will be fifty years from now, we haven't even reached the foothills." He was not troubled by the sharp downturn, for he believed the sharper the fall, the more stunning and dramatic the rise, and he looked forward to America's bright future. "Lord, the fun these youngsters of today are going to have in the next fifty years! And I'll be out of it! Like a Civil War veteran while they were fighting in the Argonne." [6]

Throughout the Depression, Schwab offered the ailing economy what he thought was much-needed medicine: frequent doses of well-publicized predictions of a bright future near at hand. He was a prominent "old-timer" who enjoyed talking to the press, and reporters sought him out and treated his colorful statements as important news from a leader of industry. Schwab hoped that his statements about economic recovery might help to restore public confidence; to him, optimism was a self-fulfilling prophecy.

Each time he made a new prediction that the economy would soon take a giant and permanent stride forward, he reminded the public that his opinion was not a mere guess or hunch; rather, it was a conclusion based upon half a century's experience in the world of business. He believed that the businessmen, farmers, workers, and politicians of America were prisoners of their own despair, and that a bright message of hope would channel their efforts away from retrenchment and toward recovery.

In May 1930 he consoled his colleagues with the observation that they could not expect a record year every year. In July he declared that the end of the Depression was at hand. In October he claimed that unemployment would soon vanish; in December he predicted a decade of unprecedented prosperity. Privately, to his colleagues, he admitted to a "short term pessimism" but he firmly believed that recovery was near.[7]

When another year had passed without an end to the Depression, Schwab's optimism remained undaunted. Other steelmen were despondent, but Schwab continued to make cheerful prophecies. He reminded his colleagues that he had lived through the severe depressions of 1893–97, 1907–8, and 1913–14, and that he was able to take a longer, more hopeful view of the present calamity than they were.[8]

Eugene Grace did not share Schwab's optimism. The most urgent objective for the steel industry, he believed, was to reach an agreement to end price-cutting and to stop building new facilities. He felt that Schwab, as the senior statesman of steel, was not taking adequate advantage of his hold on the A.I.S.I. audiences; his influence was too important, Grace believed, to squander on amusing stories and euphoric predictions of prosperity. Grace and Ivy Lee urged Schwab to focus his speeches on the need for industry-wide cooperation. Schwab agreed, and thereafter, every May and October, Lee would draft an address, Grace would amend it, and they would send it to Schwab, who would use it as the basis of his presidential address before the A.I.S.I. membership. Sometimes it was simply handed to him a few minutes before he was due to talk. Applying his near-photographic memory to the text, which he would read only once, rapidly, he would rise to speak, telling his audience: "Now boys, here is a speech written for me by Ivy Lee and Gene Grace, filled with things *they* want me to say to you. Here's what I think of that idea." He would tear the speech into bits of paper and fling them playfully into the air before the astonished eyes of his audience and the cooperatively anguished eyes of Grace and other senior executives of Bethlehem. "Instead," he said, "I want to speak to you from the heart, to tell you my own personal thoughts"—after which he would deliver from memory the speech he had shredded a few moments before.[9]

But since a totally serious speech would have been out of character for him, Schwab continued to spice his remarks with jokes about the Depression, hoping thereby to brighten the increasingly gloomy atmosphere of the A.I.S.I. meetings. He told, for instance, of the immigrant who was taking a citizenship test and was asked, what is the capital of the United States?—to which he answered: "Oh, about one-third of last year." And he told of a steel executive who was helping his little grand-daughter with her arithmetic homework. When she asked him how many mills make a cent, he answered, "Not a damn one." [10]

Many leaders in the steel industry could not comprehend Schwab's optimism. When Leopold Block, president of Inland Steel of Chicago, asked him how he could possibly find anything to be optimistic about, Schwab replied, "What do you want me to do, cause a panic?" [11] People looked to Schwab for bright prophecies, and he was unwilling to disappoint them. His basic attitude, despite occasional moments of pessimism and despondency, was one of hope and confidence. He knew that some people thought he was deliberately evading unpleasant realities, that, like Dr. Pangloss or Pollyanna, he somehow twisted every rude fact to fit a benevolent conclusion. To his critics Schwab stated: "There are some who seem to think that optimism means the inability to see difficulties, but I would rather define optimism as the ability to look at the entire situation and to have enough memory, experience and foresight to realize the good that lies ahead." [12]

In February 1933 he told reporters:

> I am an optimist by nature, and even now, in this serious state of the world when we do not know from one day to another what is going to happen, I remain an optimist.
> The world has never been in worse shape, yet my belief in the future of America as the leading manufacturing nation of the world is unshaken. Something is bound to happen, something unseen and unprophesied, which will put us back on our feet. It always has and it always will.

When asked to be more specific, Schwab cited America's recovery from past depressions. The aftermath of the 1893 depression, he said, was the growth of big business, of giant consolidations and corporations which had benefited America immeasurably. These organizational improvements were followed by major innovations. Whole new industries, operating on previously unknown technology, had come into being: electricity, radio, aviation. They had revolutionized industry and inspired public confidence by increasing employment and profits. "Something like that will appear soon I'm sure." [13]

Meanwhile, in the short run, he placed his hopes on Franklin D. Roosevelt, the newly elected President. Although Schwab was a lifelong Republican, and a substantial contributor to Republican candidates from

McKinley to Hoover, the basis of his allegiance was purely pragmatic. He favored the Republicans because they generally supported high tariffs and low taxes. In 1932 he contributed money to Hoover's re-election fund, and he told the members of the American Iron and Steel Institute: "Many people want a change because of bad business conditions. I believe the best way to better conditions is to elect Mr. Hoover. This is no time to make a change." [14] Nonetheless, he was not alarmed at the prospect of a Democratic victory.

Roosevelt's campaign pledges—that he would reduce government spending and balance the federal budget—made him sound even more safe and conservative than Hoover. Schwab believed that Roosevelt's experience during the 1920's in promoting intra-industry trade associations would be useful during the Depression years. He considered the new President knowledgeable and sympathetic to the problems of business, and so he had no hesitancy in offering Roosevelt his support. "As you know," Schwab wrote, "I am in hearty sympathy with your great and original movement to stimulate business, and I believe you and your associates sufficiently broadminded to accept suggestions." [15]

Schwab was especially pleased with the proposal for a National Recovery Administration. It would suspend the antitrust laws and permit the companies within a given industry to enter into price-fixing and market-sharing agreements. In 1933, as president of the American Iron and Steel Institute, Schwab spoke favorably of the new President, and he promised the loyal support of the steel industry on behalf of proposals which were aimed at economic recovery. [16] One of the requirements of the N.R.A. was that all manufacturers maintain an "open price policy," which meant that prices were to be published in the respective trade journals and then strictly adhered to in contracts. Schwab believed that the N.R.A. had avoided the peril of direct government regulation of business, that, instead, Roosevelt wisely had allowed business to engage in what Schwab called "Industrial self-regulation." [17]

Schwab was disturbed at the comparatively low profits which the steel industry had been earning even during the last years of the 1920's. During the period from 1913 to 1917, Bethlehem's net income per dollar of total revenues had ranged between 9 cents and 20 cents, compared to the decade from 1919 to 1928, when net income had ranged from 3.4 cents to 6.8 cents. [18] He also was distressed by the mounting losses the industry had suffered during the early 1930's. Schwab repeated his belief that in the absence of some program such as the N.R.A., losses would continue indefinitely, because most steel manufacturers were using the period of falling prices and wages to enlarge their plant capacity. Meanwhile, steel mills were operating at a fraction of their capacity, only 15 per cent in early 1933. To make matters worse, steel producers were persistently undercutting each other's prices in order to win whatever orders might still be available. This policy of expansion and price-cutting during depressions—used so successfully by Carnegie in the 1880's and 1890's, and by Schwab and others during the 1907–8 depression—Schwab now called "ruinous, suicidal competition." He hoped that under the N.R.A. the various steel manufacturers would be legally free to adopt a moratorium on all new plant construction.[19]

But in Schwab's view, the recovery of the steel industry, and of American business in general, did not rest solely on the private actions or agreements of businessmen. He urged the federal government to raise tariff barriers because American producers were being undersold at home by foreign producers.[20] Nonetheless, Schwab was wary of the perils of "excessive" government intervention in the economy. He viewed with alarm the proposed legislation to regulate the securities industry; he believed that if it were enacted it would contain so many cumbersome requirements that it would throttle the securities market, thereby endangering old and new firms alike.[21]

In the year following Roosevelt's inauguration, there was a slight upturn in the economy; the crisis of confidence seemed to have passed, and Schwab was jubilant. "The pump has been primed, the necessary stimulus to recovery has been applied, the nation is on its way to prosperity, and what we need now is a sure background for confidence and a real assurance that the period of [New Deal] experimentation is behind us."[22]

Schwab was never troubled by the idea of invoking the aid of government to bolster the profits of the steel industry. He had a simple barometer of proper government policy: low taxes and high tariffs meant prosperity, whereas high taxes and low tariffs threatened the very survival of American business. His concern with political ideology extended no further than the effect of governmental policies on business profits. For that reason, he became progressively disenchanted with President Roosevelt and the New Deal. Three new laws, in particular, made him feel that his interests were being seriously threatened. The Trade Agreements Act of 1934 provided for reciprocal lowering of tariff barriers between nations—but Schwab believed that high tariffs were necessary to protect the dwindling domestic steel market. Then, in mid-1935, the National

Labor Relations Act (the Wagner Act) was passed; it "compelled employers to accede peacefully to the unionization of their plants"—but Schwab was convinced that unions interfered with the prerogatives of management and made extravagant wage demands which jeopardized the profits of the owners. Most alarming of all, from Schwab's point of view, was the passage of the Wealth Tax Act of 1935, a measure widely denounced as a scheme to "soak the rich." Schwab felt that taxes were already too high and that any increases would destroy the lingering traces of initiative and incentive in industry.[23]

Schwab had both fame and great speaking ability, and he might have become the ideological spokesman for business opposition to the New Deal. But he was never inclined to engage in philosophical debate. He felt uncomfortable and intellectually inadequate when discussing economic and political philosophy on any level deeper than after-dinner generalizations. He specialized in acting as toastmaster, trading a few quips with his colleagues, enlivening the proceedings with a few of his vintage anecdotes, and closing with a quotable prophecy of imminent economic recovery. But both his style and his message were increasingly irrelevant to the Depression-ridden 1930's.

Schwab did not noticeably change his style of life during the Great Depression. He felt that there was no reason for him to retrench on the things which gave him pleasure. His aged mother was financially secure. So was his wife; she owned stocks, jewelry, and properties (such as the Loretto estate) in her own name. And since he was indifferent to the size of the estate he would leave at his death, he continued to spend freely.[24]

He still loved gambling and music, and whenever he traveled abroad, which he did at least once a year, he invariably stopped at the roulette tables of Monte Carlo. But at home or en route to Europe, he had to content himself with whist, a form of contract bridge. He was a leading member of the Whist Club of New York, where he played against adversaries who could afford the high stakes which excited him. And, as a sideline, he risked substantial sums in the stock market, most of them based on tips from friends or insiders. He even installed a direct ticker-tape service from New York to his Loretto estate so that he would never be out of touch with the market.[25]

Schwab also spent large sums on musical recitals, hiring the most famous singers and musicians of his day to perform in his home. He greatly preferred private recitals to public concerts because he believed that this brought out the greatness in the artist—"when singers are performing before critics—or before microphones—they are too self-conscious." [26]

Archer Gibson was one of the musicians whose talent Schwab most enjoyed. He paid Gibson an annual retainer of $10,000 to play the organ in his mansion every Sunday and on other special occasions. But he could barely keep Gibson away, pay or not, for Gibson had a passion for the "Riverside" organ, which was the finest in New York City. With his own key to the mansion, Gibson would often let himself in unannounced, ascend the main stairway to the balcony where the organ keyboard was housed in a small room which resembled a Catholic chapel, and play or compose for hours on end. [27]

Schwab loved operatic arias—not those of the somber Wagnerian mode, but rather the lushly melodic and sentimental arias in the works of Puccini and Verdi. He liked only the operatic highlights and seldom attended the Metropolitan Opera for full-length performances; he told one reporter, "I suppose I am too much of a realist to appreciate seeing a man or woman take twenty minutes to die, warbling all the time." [28]

The singer who performed most often at "Riverside" was Marcella Sembrich, the Austrian soprano. Although she had retired from the opera stage in 1909 and, after 1916, no longer gave concerts, each year she sang at "Riverside" on the Sunday nearest Schwab's birthday, February 18: it was her present to him. [29]

Three other singers whom Schwab engaged for his Sunday entertainments were Frances Alda, the New Zealand-born soprano, whom he had met sometime after her New York debut in 1908; Dame Nellie Melba, whom he probably had met for the first time during her singing tours of America in 1918 on behalf of Liberty Bonds; and the contralto Ernestine Schumann-Heink, an Austrian prima donna who sang with the Metropolitan Opera in New York City for many seasons. [30]

Among the operatic tenors who gave private recitals for Schwab and his guests were Enrico Caruso and Herbert Witherspoon. Schwab's friendship with Caruso began in 1903, when they were passengers together aboard the ship carrying Caruso to his American debut, and lasted until Caruso's death in 1921. Schwab probably met Witherspoon during the years they both lived in Pittsburgh—Witherspoon was then singing with the Pittsburgh Orchestra and Schwab was serving as president of Carnegie Steel and U.S. Steel. Another regular at "Riverside" was Victor Herbert, the Irish-born cello soloist. Herbert became the conductor of the Pittsburgh Orchestra in 1898, and after 1904 he began to compose oratorios, comic operas, and songs, some of which were presented under his direction at "Riverside." Fritz Kreisler, the Austrian-born violinist and composer, and Ignace Jan Paderewski, the Polish pianist and composer, also performed there.[31]

While Schwab loved music deeply and had some ability as an organist and singer, it was his sister Mary Jane who possessed a great musical talent. Mary Jane, who was fourteen years younger than Schwab, was graduated from the New England Conservatory of Music; she then continued to study voice and piano at several leading American and European schools, including the Sorbonne. Her family believed she was destined for a brilliant career. While she prepared for her professional debut, she accepted a position as a lay teacher of music in Greensburg, Pennsylvania, at a college run by a religious order, the Sisters of Charity. While there, she developed a mysterious ailment which progressively impaired her ability to sing, or even to speak. Her condition was diagnosed as a concealed goiter, caused by a malfunctioning of her thyroid gland, and she had an operation. It was unsuccessful. As her condition deteriorated, she made a silent vow that if she ever recovered she would devote the remainder of her life to the adoration of God. A second operation was performed by Dr. George W. Crile, a famed surgeon and thyroid specialist in Cleveland; it was successful, even though Dr. Crile had to implant a silver tube in her throat to enable her to breathe.

After her recovery, Mary Jane announced her vow to her family. Her mother, who was deeply religious, was pleased and proud; her brother Charles was not. Since the meaning of life for Schwab was the experiencing of new challenges and adventures, he could not understand her choice—the self-imposed asceticism of a nunnery. He tried to dissuade her from becoming a nun, but failed. In December 1928, she entered a convent of Discalced Carmelites in Altoona. Schwab then offered to build a new convent for the Order if it relocated in Loretto. The Order accepted, and he and Rana each contributed over $100,000 to build the monastery of St. Theresa Lisieux there. Mary Jane became Sister Cecelia of the Blessed Sacrament, and she never again saw any member of her family. Schwab's failure to dissuade her from becoming a nun was one of the most painful disappointments of his life. Fortunately for him, he never learned that, as an added measure of self-abnegation and renunciation, Sister Cecelia gave up every form of music, even the singing of religious hymns; she voluntarily renounced the exercise of her greatest talent. She devoted her time to sewing and praying, and later she served as the mistress of novices. She died in 1954, a few years after celebrating what she called her "silver tubilee," the twenty-fifth anniversary of her second operation.[32]

As Schwab lost his sister Mary Jane, he became closer to his youngest brother, Ed. Born in 1885, Edward H. Schwab was twenty-three years younger than his brother. When Ed attended the University of Notre Dame, his father allotted him only fifty cents a week for spending money—on the theory that a boy who was chronically short of funds could develop few vices. When Ed wrote to "C.M." (as Ed always called him) for a drafting set, he received one by return mail, along with a twenty-dollar bill. Schwab's graduation gift to his brother in 1906 was a three-month, all-expenses-paid trip to Europe for Ed, a cousin, and a friend. Ed would not use C.M. to advance his career. Instead, he accepted a position teaching law at Notre Dame and practicing law in South Bend, Indiana. But he became increasingly dissatisfied with the routine of his life, and after a few years he moved to Chicago, where he became a business broker, buying small firms which were faltering, reorganizing them, and then selling them at a profit.[33]

Edward Schwab's activities were on a small scale compared to those of his brother, but he earned his reputation and income independently, a matter of considerable importance to him. In 1920, after he had established himself, he accepted his brother's suggestion that he relocate in Bethlehem, where several small business firms were in need of reorganization. With Charles's financial backing, Ed took charge of three firms: one which produced spark plugs; a second which made magnetos, spark plugs, and radios; and a third which made the wood cabinets for radios. These firms prospered during the 1920's, but then, like so many other small businesses, they collapsed during the Depression.[34]

In 1929, however, before the onset of the Depression, Ed Schwab moved to New York City to manage his brother's personal finances, including his many side-line business ventures. As the Depression deepened, Ed played an ever larger role in his brother's life. By 1932 Bethlehem Steel's common stock had sunk to $14 a share, reducing the market value of Charles Schwab's 90,000 shares to $1,260,000. In view of the obligations he had assumed, Schwab was financially overextended. One of Ed's chief assignments was to reduce or eliminate those obligations,

which included monthly allowances to twenty-seven friends and relatives. Among those who received Schwab's assistance was his sister,
Gertrude, whose husband's bank in Johnstown, Pennsylvania, had failed
to survive the 1933 "bank holiday." Schwab promised Ed that he would
take on no new obligations, but he was an easy touch, unable to say "no"
to a hard-luck story. Schwab also served as co-signer for over $1,000,000
in personal loans to acquaintances (many of whom defaulted on them),
and he lent large sums to friends to gave him their personal notes and
then blithely treated his loan as a gift. [35]

Despite his own financial reverses, Schwab also made unsolicited gifts.
A football hero whom he had once admired had developed asthma,
which required him to live in a dry climate. When Schwab learned of
the man's predicament, he bought him a 640-acre ranch in Phoenix. But
the land possessed no water—so Schwab also paid the bill for laying irri-
gation pipes. [36]

On another occasion, he received a letter and a check from his aged
aunt, Mrs. Susan Wertzberger, one of his mother's sisters. She enclosed
her life savings, $2000, and asked him to invest it for her so that she
could live on the proceeds for the rest of her life. Fearing that any investment might prove a disaster, Schwab waited a few days and sent her
back a check for $5000, claiming that the difference represented her
profit on the investment. By return mail, he received the check back
again, along with the request that he invest the $5000 in an equally prof-
itable manner. Somehow he managed to persuade her that he could not
work miracles, and she was content with her $3000 gain. [37]

Schwab was distressed by the plight of small businessmen who did
not have the financial resources to ride out the Depression and who
found themselves unable to obtain extensions or renewals of their bank
loans. Schwab wrote to President Roosevelt, admitting that "my letter is
probably telling you nothing with which you are not entirely familiar,"
but he urged Roosevelt to take "some action" to aid small business. Unfortunately, he could not be more specific: "I have no suggestions to offer
you as to how it could be done." [38]

During the Depression, dozens of small businessmen wrote to
Schwab, asking for funds to salvage their enterprises. Some were verging
on bankruptcy and wanted Schwab to buy them out; others hoped that a
man with his millions could afford to lend them a few thousand dollars
until the crisis had passed. He found it difficult to say no, and he became
a co-owner or financial angel of many faltering businesses. Most of them

took comparatively small sums; a few, however, involved major out-lays.[39] He invested $50,000 in the Kullman Dining Car Company, which manufactured roadside diners, one of the fastest growth industries before the Depression. The diners, however, were sold on installments; as long-distance traveling and dining out declined during the Depression, the owners of the diners increasingly defaulted on their payments, and finally Schwab's entire investment was lost. A second investment, one involving $100,000, was made in a business about which he had no direct knowledge or real interest: the Lehigh Valley Silk Mills, located in Bethlehem. When its sales and profits declined, Schwab could find no buyer for his interest. His largest single investment, $250,000, gave him a controlling interest in the West Virginia Zinc Smelting Company; it was a solidly structured business, but one which was unable to prosper during a Depression. In choosing many of his investments, Schwab made the mistake of judging all businesses by the standards he used in the steel industry. He believed that he could make any business yield a profit solely by reducing production costs. As a result, he paid inade-quate attention to marketing and advertising—an oversight which cost him dearly.[40]

On May 1, 1933, Charles and Rana Schwab were the hosts at a party in honor of their fiftieth wedding anniversary. Few of their guests knew that Rana had to be coaxed into attending. She loathed parties in general and this one in particular, for she felt she had little to celebrate.[41]

Their marital relationship had not improved over the years. Schwab wanted companionship, someone to share his pleasures, and Rana was unable (and, in some instances, unwilling) to do so. Rana felt that life as a wealthy industrialist's wife had deprived her of all sense of privacy and personal identity. An omnipresent staff of twenty performed every func-tion in "Riverside" and the Loretto mansion. She had thousands of ac-quaintances, but they were her husband's friends and business associ-ates—an inferior substitute for a few close personal friends. Her only intimate companions were her own relatives: her sisters and nieces. In part her lack of friends was caused by her suspiciousness—she thought that anyone who showed any sign of friendship toward her might be try-ing to manipulate her, as a means of gaining some favor from her hus-band. A millionaire's wife, she felt, could never be certain of anyone's sincerity.

The only unguarded, lighthearted moments of her mature years were those she spent with young children. She was delighted when visitors brought along their children—she felt free to reciprocate a child's spontaneous show of affection. On these rare occasions she could openly express her emotions without fear that the child had any ulterior motive. After such a visit she would send the child a dog or a doll, not as a birthday or Christmas gift, but as her way of saying, "I love you." [42]

Those who knew Rana Schwab all say that her body caused her chronic pain and embarrassment. Inflammatory rheumatism caused her limbs to swell to massive proportions. She loved to eat, and indulged herself, but, inevitably, she became grossly overweight. She was intensely self-conscious in the presence of strangers—she feared they might stare at her. She declined most invitations, even to dinners at the White House, and seldom traveled with Schwab. She envied women who had what she most desired and did not have: children and an alluring figure. The person she most envied was her sister-in-law, Ed Schwab's wife, Edith, who had borne seven children with no noticeable effect on her waistline.

Rana accepted the pain of her chronic illness with stoicism, but she would explode with anger whenever one of her needs or requests was not instantly satisfied. She always carried a cane with a special lead base which she used to summon servants. The cry, "Stone, Stone," accompanied by the relentless banging of her cane, would ring through "Riverside" until the butler, George Stone, heeded her call to draw the drapes, or to pick up something she had dropped, or to open some jar or package which her swollen fingers were unable to manipulate.

Rana spent eight months of each year, from October to May, at the New York mansion; there she passed her time reading, talking to her sisters and nieces—they were frequent visitors and often guests for long periods—and supervising the management of the household, especially scrutinizing the incoming bills to make sure that she was not being overcharged. Her chief hobby was working out jigsaw puzzles; the larger and more complex they were, the more she liked them. When she completed a puzzle, she would scramble up the pieces again and send it to her mother-in-law, Pauline, who was an equally avid devotee.

Every afternoon, weather permitting, Rana went for a drive, accompanied by whomever happened to be visiting her. Her chauffeur drove along the west side of Manhattan and then north for twenty or thirty miles. The route never changed, nor did her observations about the mansions and monuments along the way. She cherished this daily routine; it gave her an occasion to wear one of her fur stoles, and to see something of the outside world without being seen.

But the pleasure of these afternoon outings hardly compared with what she experienced during her visits to the vaults of Tiffany's, the Fifth Avenue jewelers.[43] Schwab regularly gave her gifts of jewelry—canary diamonds, square-cut emeralds, glistening sapphires, bracelets, and broaches—none of which she ever wore; she suspected that each gift coincided with his taking a new mistress. She wore only pearl earrings and two strings of pearls, each pearl the size of a robin's egg; all the rest of her treasure went to Tiffany's for safekeeping. Her visits had a dual purpose. She did not merely deposit new jewels and fondle the old ones; she also clipped the dividend coupons on her portfolio of bonds. Her jewels and securities were her insurance against the possible loss of Schwab's wealth or affection.

By the early 1920's Rana had become reconciled to her husband's extramarital affairs and had recognized the futility of extracting promises of fidelity. She found the situation bearable only because she had no positive proof, merely strong suspicions, reinforced by rumors. As long as "the other woman" was only an abstraction in her mind, the situation was tolerable. She had some consolation; she thought that there was only one woman with whom Schwab had an enduring relationship—herself. This enabled her to believe that she, the "Old Lady," was not threatened by any of the steady succession of young women who briefly captured the "Lad's" attention.

But Rana was wrong; one other enduring relationship did exist. In 1918, while Schwab headed the Fleet Corporation, he met an attractive, full-figured young woman named Myrtle. She became his mistress and remained so until his death in 1939. She was the companion that Rana could not be. Unlike Rana, she was able and eager to attend the theater with him, to accompany him on his periodic cruises to Europe and his visits to the spas, to join him in playing roulette at Monte Carlo, and to serve as a skillful partner at whist. Myrtle felt secure in Schwab's affection in a way that his wife had not for many years. There was only one serious rival for Myrtle's place; she was a contralto at the Metropolitan Opera who was several years younger than Myrtle and considerably more attractive, with dark-dyed hair and a small waist which accentuated her figure. But she was married, a fact which complicated their relationship and thereby secured Myrtle's position.

Rana was wrong in another respect: her marriage to Schwab was never

in jeopardy. He felt no need or desire to divorce her as long as she neither protested nor prevented his extramarital affairs. And he loved her
deeply—not with the passion and emotional intimacy which exists between husband and wife in a close marriage, but rather with the kind of
love he felt toward his mother. In fact, he treated her like a mother—
always expressing concern about her health, her family, her daily activities, always remembering her birthday, always writing to her from
abroad when he traveled, and always sharing with her select highlights
of his own activities. Her acceptance of this motherly role enabled their
marriage to endure.

As Schwab advanced in age, his health grew progressively worse. In
1934, at the age of seventy-two, he resigned as president of the American
Iron and Steel Institute, and at the same time he relinquished almost all
of his business responsibilities. Although he remained chairman of Bethlehem Steel, his role was purely honorific and sentimental; he continued
to receive a salary of $250,000, but it did not represent payment for any
present work. It was a recognition of his past contributions.

But even total retirement did not spare him from new attacks. In 1934,
a Senate committee, headed by Senator Gerald P. Nye of North Dakota,
began to investigate the role played by munitions-makers in America's
entry into World War I. Schwab expressed mild bewilderment that he
and other steelmakers and shipbuilders were being damned for the same
work which had earned them praise during the war years—after all, they
had supplied America and her allies with the ships and munitions needed
to defeat their enemies.[44] As the Nye Committee investigation continued, Schwab and his fellow munitions-makers were labeled "Merchants of Death" and accused of inciting rival nations to war in order to
create a market for munitions.[45]

The two most remarkable aspects of the accusation were, first, that
there was no evidence to substantiate it, and second, that the absence of
evidence did not deter the widespread credence given to the charge. As
Professor John Wiltz has commented: "If such allegations had been
country store or main street gossip that would have been one thing.
They were something else when echoed by the President of the United
States, the Premier of France, two former Secretaries of State, the
League of Nations, *Fortune* Magazine, the *Christian Science Monitor*,
members of Congress, the peace movement, leaders of religion, and even
the *Wall Street Journal* and the Chicago *Journal of Commerce*. Who could

blame people in the mid-thirties for taking seriously this heady business about merchants of death?"[46]

Schwab, the leading American munitions-maker during the World War and often called "the American Krupp," offered no rebuttal. One possible retort would have been to identify the logical fallacy contained in the accusation—*post hoc, ergo propter hoc*. If shipbuilders and munitions-makers build facilities to manufacture war matériel and if a war subsequently occurs, it does not follow that they have *caused* the war; one thing following another in temporal sequence does not prove causal linkage. A second answer to the charge would have been the principle of analogy: if munitions-makers "cause" war, then pharmaceutical manufacturers "cause" disease and educators "cause" ignorance, since in each case profits are made by those who eradicate the malady.[47]

But neither Schwab, nor Grace, nor their publicist, Ivy Lee, made any such response. They all operated on the same premise: never answer accusations unless they are directed at you primarily or exclusively; ignore them if you are merely one menber of a large class that stands accused. Using this myopic strategy, Bethlehem Steel issued no general statement challenging the allegations made before the Nye Committee. For some reason, Schwab was not summoned to testify, perhaps because he was retired and reportedly in ill-health, or perhaps because he was in Europe at the time that Grace and the other Bethlehem executives were scheduled to testify.

Grace told the committee: "We have substantially gone out of the production of what we call 'guns, projectiles, and that class of munitions.' We have converted our plants for peacetime commercial purposes. We are no longer the 'Krupps of America.' We are no longer an ordnance producing concern." This announcement received less attention in the press than did his claim that while it was permissible to conscript men for wartime service, business firms must be allowed to make profits and must not be threatened with nationalization because of a potentially demoralizing effect upon incentives.[48]

When Schwab returned from Europe, reporters asked him for his opinion of Bernard Baruch's testimony before the Nye Committee. Baruch had claimed that a brake could be placed on the sales enthusiasm of munition-makers if taxes were raised so high in wartime that there would be no profits left. Schwab's only comment was: "Mr. Baruch is a splendid man and citizen. I agree that in time of war profits should be heavily taxed."[49] A year later, on Schwab's seventy-fourth birthday,

Baruch's suggestion was again put to Schwab for comment, and his answer had not changed: "Mr. Baruch is entirely right. I think profit ought to be only reasonable, as small as you can make it; but I don't think you can get people to do all that work enthusiastically and well without something." [50]

Schwab was implying that he and his colleagues could not make their maximum effort for the nation's defense unless they had some profit incentive. But this directly contradicted his well-publicized statement of 1916:

> If the United States should become involved in the war, the Government of this country can have any product we manufacture—armor-plate or anything else—*at any price it chooses to pay*; and under such circumstances and *regardless of price* our entire plant will run twenty-four hours a day with every pound of energy we can put behind it. (Italics added). [51]

And in 1918, when he had headed the Fleet Corporation, he had repeatedly exhorted workers to put motives of patriotism above personal gain, to work unselfishly and unstintingly for the national war effort. Moreover, while he opposed the "conscription of capital" (i.e. the taxing away of all profits) and the wartime nationalization of industry, he never once denounced the conscription of soldiers, men who were forced to risk their lives in combat. Either Schwab never recognized the contradictions involved, or he chose to ignore them—but his critics did not. [52]

In May 1936 Schwab made still another casual, ill-considered statement. Returning home from a six-week rest at a German health spa, he told reporters that he thought Germany was in fine condition and that Hitler was "really popular because they credit him with bringing order out of chaos." A writer for a left-wing magazine then denounced Schwab as an admirer of and an apologist for the Nazi dictator. [53]

While Schwab felt that Germany was on the road to recovery, he feared that America was not. He was particularly disturbed by the proposed increase in corporation taxes. "It will be a frightful tax to pay and it will be very hard on industry. It will be a very great burden, and business is bad enough as it is. The vast millions in steel today are making no returns on capital." [54] A year later, Schwab saw one curious indicator that a return to prosperity was not far off: he said, "things can't get much worse." The only remedy he could suggest to speed recovery was the lifting of heavy taxes off what he considered to be the already overburdened shoulders of business. [55]

During the Depression Schwab was paid a salary of $250,000, which aroused considerable resentment among some of Bethlehem's stockholders. In 1934, Bethlehem was operating at only 52 per cent of capacity; it had lost $19,400,000 in 1932 and $8,700,000 in 1933, and was paying no dividends on either its common or preferred stock.[56] At the annual meeting of the stockholders in April 1934, when a stockholder voiced criticism of what he termed Schwab's "excessive salary," Eugene Grace silenced the man by declaring that Schwab's past services to the company more than merited the present salary. During Bethlehem's early years, Grace reminded him, Schwab had not drawn any salary as president, nor had he received dividends on his preferred stock—and he had personally guaranteed the company's loan notes with his own money.[57] This retort only temporarily squelched the attackers.

At the next stockholders' meeting, in 1935, Schwab and Grace faced a minor mutiny. The dissidents owned only 355 votes out of 2,300,000; nevertheless, they were a cohesive group, and they were able to disrupt the meeting. The disgruntled minority, which was led by Lewis B. Coshland, complained that while stockholders were receiving no dividends and workers were earning an average wage of only 67 cents per hour, Bethlehem's three top corporate officers were drawing salaries totaling $488,000. Schwab was receiving $250,000, Grace $180,000, and Robert E. McMath, the secretary, $58,000.[58]

Schwab was told to his face that he had "outlived his usefulness" to Bethlehem and ought to step down. "A few years ago," Coshland shouted, "Mr. Schwab said he was giving up many of his activities on account of his advanced age, but that Bethlehem would always remain nearest his heart. He was wrong in his anatomy. He had held it nearest his stomach, and we have been his meal ticket."

Grace was subjected to a similar assault. A woman who identified herself as a poor widow angrily shouted: "No wonder Father Coughlin preaches about blood money. He knows what he is talking about. There is too much of this. Here we are without a cent, while you men store up millions. Mr. Grace should know there are no pockets in shrouds." [59] The dissidents offered a resolution for stockholder approval which would limit the salaries of Schwab, Grace, and McMath to 20 per cent of Bethlehem's net earnings. The motion was defeated.[60]

But defeat did not discourage the insurgents; in 1936 they renewed their attack on Schwab's salary. At the 1937 meeting, Eugene Grace had to be physically restrained from punching two stockholders when they repeated Coshland's charge that Schwab had "outlived his usefulness" to Bethlehem Steel. Lewis Gilbert, a self-appointed spokesman for minority stockholders in numerous corporations, proposed that Schwab step down from the post of chairman at $250,000 a year and that he accept a newly created post of "honorary chairman" which would carry a $25,000 pension. When put to the vote, Gilbert's motion was rejected, 2,638,909 votes to 260.[61] Management, armed with the proxy votes of absentee shareholders, could effectively silence any dissidents.

Undaunted by defeat, Gilbert reintroduced his resolution at the stockholders' meeting in 1938, but with one modification: Schwab should serve without any salary at all. In reply, Schwab said:

> This moment is the saddest in all my long life. I have devoted my life to the Bethlehem Steel Corporation. I have devoted my entire life and energy to building it up. It is my life-work and I am proud of it. I intend, God willing, to continue to serve Bethlehem during the few years that are left to me.
>
> I do not intend to give way to Mr. Gilbert's suggestion that I serve without pay. For many years I worked and labored for this company. This company owes me a debt it can never repay.
>
> This suggestion of serving without pay is the one sad thought in my sixty years of business experience. Will you, Mr. Gilbert, withdraw it?
>
> —As a personal favor?
>
> Yes.
>
> —As a personal favor, I withdraw the suggestion.[62]

Years later, Gilbert recalled the conversation he had had with Schwab following the annual meeting: "He said that my proposal had caused him suffering and sadness. I began to feel sorry for him. Hesitantly, as I recall, he referred to some financial reverses, manfully admitting that he was a bit hard pressed. It seemed to me as he proceeded that I was perhaps being flinty and unreasonable. (Later, in the train going back to New York, he privately confessed to me that he had once had $40 million, but that it had all gone in the 1929 debacle. I was thus the first to be told the truth.)" [63]

Schwab's losses during the 1930's were not merely monetary. He was rapidly losing his oldest friends. He was deeply nostalgic about the period of his life when he had been struggling to achieve success, and he felt great affection for the men who had been his comrades during that time. Each year, from 1916 to 1933, he was the host at "Riverside" for the reunion of all those who had been partners in the Carnegie Company at the time of its sale to U.S. Steel. But by the mid-1930's, there were only eight survivors of the fifty-one member Carnegie Veterans Association—and Schwab himself was one of two survivors of the "Class of '79," the eight men who entered Carnegie's service in that year. He felt with special intensity the passing of William Ellis Corey, so often his successor at various jobs, but now his predecessor in death.[64]

In 1922, William B. Dickson had written a poem to commemorate the death of Joe Schwab; its sentiment had special poignancy to Charles Schwab after seeing a succession of his friends die in the early 1930's:

> Oft in the stilly night
> Ere slumber's chains have bound me,
> Fond memory brings the light
> Of other days around me.
> The smiles, the tears,
> Of boyhood years,
> The words of love then spoken—
> The eyes that shone,
> Now dimmed and gone;
> The cheerful hearts now broken.
>
> When I remember all
> The friends so linked together—
> I've seen around me fall
> Like leaves in wintry weather,
> I feel like one who treads alone
> Some banquet hall deserted—
> Whose lights are fled,
> Whose garlands dead,
> And all but he departed. . . .[65]

Schwab viewed his years at Carnegie Steel as a lost golden age, a period when he operated at full mental and physical efficiency, a time when those close to him marveled at his limitless enthusiasm and endurance and repeatedly urged him not to overwork or "overthink." By the mid-1930's, his body was ravaged by illness and his spirit increasingly broken by the accusations leveled against him. He felt that there were few people still alive who realized how much he had contributed to the development of the American steel industry. It seemed to him that his greatest achievements were not merely unappreciated, but actually forgotten, or, worse, reviled. He verged on tears when he told B. C. Forbes,

> I really feel that I have contributed something to the development of this country's resources—Bethlehem gives employment to some 30,000 people. It hurts me—it hurts me very much—to be branded as nothing but a greedy, selfish, self-seeking, mercenary, merciless fellow, callous towards workmen and towards everybody else.[66]

The death of his friends meant not only the loss of their companionship, but the loss of what they had always accorded him: recognition and admiration for his productive achievements.

The upkeep on "Riverside" and the Loretto estate continued to deplete Schwab's financial resources. At the same time, he assumed the added burden of supporting the people of Loretto, who had looked to him for aid during the economic crisis. Schwab's estate, which covered over 1000 acres, had always been a major source of jobs for the townspeople. During the 1930's, he liberally added to the employment rolls of his estate. As many as 300 men sometimes were engaged in field work on the estate's grounds and farm, at an hourly wage of 40 cents. These were many more men than were economically warranted, but Schwab felt a sense of *noblesse oblige*. He hired three groups of boys to maintain his nine-hole golf course. One group hand-picked dandelions; the second ran sprinklers over the greens; the third hand-mowed the course. And Schwab deliberately lost sizable wagers when he played with friends who had fallen on hard times.[67]

By 1936, the continuing drain on Schwab's wealth was so great that he tried to sell "Riverside." Because of the huge expenses for its upkeep, he did not expect to find a private buyer. His agents approached the City of New York; they hoped the city would buy the mansion for the official residence of the Mayor. Schwab offered to sell the property, building, and furnishings for $4,000,000, which was only about half of the original cost to him. Perhaps in another decade or in some other city, the offer would have been accepted. It was Schwab's misfortune that the Mayor of New York at that time was Fiorello H. La Guardia, who took pride in the fact that he still lived in the same walk-up tenement in which he had lived before his election. A predecessor, Jimmy Walker, or a successor,

William O'Dwyer, might have welcomed the chance to move into a palace, but La Guardia rejected the idea. He felt that in depression times it was particularly inappropriate.[68]

The primary reason that Schwab wanted to sell "Riverside" was to escape paying property taxes to New York City. The property tax in 1936 was a staggering sum, $110,000. La Guardia told the press that New York City much preferred having the tax revenue to having the mansion itself. What La Guardia did not realize was that Schwab had stopped paying any property tax. He was relying upon the law which permitted a property owner to default for six years, after which the city could seize and sell the property for non-payment of taxes unless the owner elected to pay the past-due taxes and late penalties. Schwab died before the seventh year. New York City never did acquire the mansion; it settled with Schwab's estate for a small fraction of the back taxes.[69]

At the height of his career, Schwab's fortune amounted to $25,000,000, according to a 1973 estimate made by his brother Edward. While this figure may be far too conservative, it is certainly nearer the truth than the legendary fortune ascribed to him in the press during the early 1930's. One widely read account, a two-part "Profile" in the *New Yorker* (1931), popularized a myth when it declared, "It is believed that the bulk of his fortune, estimated at from two to three hundred million dollars, will be bequeathed in foundation form." Schwab declined to explode the myth; instead, as the writer observed, "He is silent on this point." [70]

Schwab was never willing to admit publicly that he was nearly bankrupt, but from 1936 on, he refused any new financial commitments. He had promised a sizable contribution to the building of a library for St. Francis College in Loretto. In December 1936, when the head of St. Francis wrote to Schwab, asking him to redeem his pledge, Schwab was unable to oblige: "As you well know, matters with me financially have been in a bad state and every demand has been met with difficulty. Give me a little time to think over the library situation. . . ." [71]

Schwab had always been devoted to his mother; he loved her, in fact, to a degree which his brother described as being "almost pathetic mother love." Throughout his adult life he telephoned her once or twice a week, and only severe illness ever kept him away from her birthday celebration. Pauline, in turn, felt great pride in Charles's achievements and gratitude for his generosity to her and the things she valued most. He not only had obtained for his parents an audience with the Pope in February 1903, but he also had rebuilt St. Michael's, Father Gallitzin's church in Loretto. His only shortcoming, in her view, was that he did not attend church. When John Schwab died in 1924, Pauline's sons expressed concern about her ability to cope alone, and they asked if there were anything she wanted or needed from them. She seized the occasion to extract a promise from them to attend church each week, but she only succeeded in getting Schwab to attend church with her during the four months a year when he lived on his Loretto estate. And she soon observed that even his being in church did not suppress his customary playfulness: every Sunday, just as the collection plate was being passed, he would hide her purse.[72]

Each fall, from 1924 on, Schwab and his mother attended the Cambria County Fair together. He was its organizer and chief financial sponsor. Schwab was only mildly interested in the exhibitions of livestock and farm produce; for him the chief attraction was the circus. He served as the ringmaster—it was the realization of a boyhood ambition. For his mother this annual experience brought back happy memories of the years when her son was growing up—of the uninhibited young showman who had exclaimed, "I can do something else yet." [73]

In March 1936, a few months after her ninety-third birthday, Pauline developed pneumonia. Her doctors held out no hope for her recovery. As she grew worse, Schwab repeatedly telephoned his brother, Edward, who was at his mother's bedside, seeking reassurance that she was not suffering any pain. The roads were covered with ice, and Schwab did not arrive at her home in Loretto until after her death. When he found his mother laid out on her bed, he asked to be left alone with her for a few minutes to pay his respects. He stayed in her room for over two hours, crying inconsolably, but hesitant to let anyone see the true depths of his loss.[74]

Three years later, on January 12, 1939, Rana Schwab died in her sleep. She had long outlived two doctors who had told her in 1919 that she would be dead within a year, that her heart could not bear the burden of her weight. The news of her death reduced Schwab to tears.[75]

On the day of her funeral, Schwab appeared visibly aged and broken in spirit. His body was slumped, his face was pale and drawn, his eyes were blurred with suppressed tears. He told Alfred McKelvey, the husband of Rana's niece: "I won't see you a year from now; I'll be dead."

When McKelvey assured Schwab that his spirits would rebound after the pain of Rana's death had receded, Schwab disagreed: "A man knows when he doesn't want to be alive, when the will to continue living has gone from him." [76]

After Rana's funeral, "Riverside" was permanently closed; it became the property of the Chase National Bank in settlement for Schwab's outstanding debts. Schwab moved into a small apartment at 290 Park Avenue.

That summer he made his final trip to Europe. He wanted to make one last visit to places which had given him happy memories. In early August, while en route from Paris to London, his plane encountered a storm. The gyrations of the plane made him sick to his stomach, and when they landed he had to be carried off on a stretcher. He was taken to the Hotel Savoy to rest before the return trip to America. [77] But the siege of air sickness was more than a temporary discomfort; it had loosened a blood clot. On August 9, while at the Savoy, Schwab suffered a mild heart attack. Two weeks later, Dr. Edward Gordon of London declared his patient fit to travel; he and a nurse accompanied Schwab to America aboard the S.S. Washington, arriving in New York on August 31. Friends who inquired were told by Jimmy Ward, Schwab's assistant, that he had tolerated the trip well, with no worsening of his condition. After examining Schwab, Dr. Samuel Brown, his personal physician, ordered two more weeks of bed rest, but he expected that Schwab would be up again by the end of the month. Ward wrote to one of the few surviving Carnegie Veterans: "Dr. Brown says that if he does not have another heart attack, he has a better than even chance of pulling through." [78]

But Schwab, who so often had beaten the odds, did not recover; he suffered a second heart attack and died on September 19, 1939.

The executors of Schwab's estate, his brother Edward and Willard A. Mitchell, bore the responsibility of paying Schwab's creditors. Mitchell insisted that all of Schwab's stocks be sold immediately, fearing that the executors would be liable for the difference if the price subsequently slumped. Ed Schwab recommended that the holdings in Bethlehem Steel and the West Virginia Zinc Smelting Company be retained, expecting that the recent outbreak of war in Europe would enhance the market for their products and thus boost their profits and stock price. But Mitchell's adamant stand forced their sale; if either stock had been held only eighteen months longer, until the upsurge in war orders came, Schwab's estate would not have been insolvent.[79] When the estate tax appraisal was completed in 1943, Schwab's assets were $1,389,509, but his liabilities and obligations were $1,727,858—a net deficit of $338,349.[80]

For eight years after Rana Schwab's death, an agent for the Chase National Bank, the new owner of "Riverside," led a succession of would-be buyers through the mansion, but there were no serious bidders. "Riverside" might have become a museum, a foreign embassy, even an historical landmark commemorating the man who built it. But all of these alternatives were economically prohibitive. In 1947 the Prudential Insurance Company of America purchased the site for $1,500,000, and Schwab's palatial home was destroyed by wreckers. Two twin-tower apartment houses were constructed in its place—appropriately enough, with Bethlehem beams.[81]

The Loretto estate is still standing, but hardly in a form Schwab would recognize. The principal heirs of Rana Schwab's estate were her sisters (and, indirectly, her nieces); they hoped that someone would buy the Loretto estate either as a private home or for redevelopment as a resort. But when the property was auctioned off in 1942, the only bid for the mansion and the guest house came from the trustees of St. Francis College.[82] "Immergrun" became a residence for seminarians, young men embarking on a life of poverty, chastity, and obedience.

Today, twenty Franciscan novices sleep in the two master bedrooms which Schwab and his wife once occupied. The front hall of the mansion, through which Schwab's affluent friends once passed, is now dominated by an austere altar and an agonized figure of the crucified Christ, wearing a crown of thorns. A handsome stone building which Schwab had had built especially to house his private ticker tape and telegraph line to his stock broker in New York is now the center from which tens of thousands of religious leaflets and missals are distributed to the laity. And on the grounds of the estate the statues of voluptuous Greek goddesses have been replaced by saintly figures. The contrast could not be more complete. Loretto has returned to the somber religious atmosphere which dominated it in the days of Father Gallitzen.[83]

"Riverside" is gone, and Loretto once again is a sleepy little village. The only monument to Schwab's memory is a stark mausoleum at the crest of the Alleghenies; its sole identifying mark, the word SCHWAB, is barely noticeable.

But the real monument to Schwab's life, and the living testament to his memory, are the steel mills and shipyards, the blast furnaces and bellowing smoke stacks which mark the skyline of America. Perhaps the greatest symbols of Schwab's life can be seen from the center of Park Avenue in New York City, as one looks up at skyscrapers built with the Bethlehem beam.

There Schwab might well invoke the words of Horace:

If you want to see my monuments,
look around you.

Appendix A

The Whipple Notes

In 1935, Sidney B. Whipple, a staff writer for the New York *World-Telegram*, was granted a unique opportunity to interview Schwab at length about his life and career. On his own initiative, Whipple had contacted John C. Long, one of Bethlehem's public relations managers, urging the importance of obtaining Schwab's reminiscences. Long, who regarded Schwab as a figure of major historical importance, enlisted the aid of Schwab's closest assistant, Jimmy Ward, to persuade Schwab to be interviewed.

When Schwab consented, Whipple took a leave of absence from the *World-Telegram*. At a time when other reporters were being laid off or taking salary cuts, Bethlehem paid Whipple the largest salary he had ever earned. Free from other assignments, he was able to spend several hours with Schwab almost every day for nearly a year. But he rarely probed beneath the most superficial level. He relied almost exclusively upon the material which Schwab supplied to him—that is, oral reminiscences and a small number of letters and books. Whipple was a victim of self-induced fear, obsequious and naïvely uncritical. He was barely more than a passive recorder, accepting Schwab's statements at face value and then occasionally embellishing them with remarks which scolded those (like Gary or Frick) about whom Schwab had expressed anything less than total admiration.

Schwab attached one explicit condition to Whipple's assignment: he was never to divulge anything he learned from Schwab as long as Schwab remained alive. The material gathered by Whipple, including

305

several interviews with Schwab's surviving business associates (none of his family members were interviewed), was intended for use solely by a biographer after Schwab's death. But Whipple could not resist. He began to write a biography of Schwab, and he read some of it to one of Schwab's relatives. Whipple hoped that the relative would persuade Schwab to allow Whipple to write his complete biography. When Schwab learned of this breach of trust, he refused ever to see or speak to Whipple again. Schwab ordered John Long to discharge Whipple, reclaim the materials lent to him, and secure all the notes on their conversations.

If there was ever any possibility to employing someone to succeed Whipple—to bring his ill-digested mass of notes into a more systematic form, and to probe further—that possibility was totally destroyed a few months later. A feature story appeared in the New York *World-Telegram* containing details which only Whipple could have supplied. Schwab was furious, and he never again afforded anyone an opportunity to probe his memories.[1]

Appendix B

The Genealogy of an Historical Myth: The Armor Scandal of 1894

Gustavus Myers, author of *The History of the Great American Fortunes* (1907), was the first historian to propagate the myth that Carnegie Steel had sold dangerously defective armor to the Navy. Myers claimed that "the armor mill owners charged their own Government extortionate prices for warship armor plate which, on at least one specific occasion, was found to be worthlessly defective." [1] His only source was the Cummings report. He quoted from it briefly, but he ignored entirely the hearings and documents which accompanied the report and which contradicted his claim.

John Winkler, on the other hand, when he wrote *Incredible Carnegie* (1931), made full use of the hearings and documents, and he quoted at length from the reports of Secretary Herbert, Captain Sampson, and President Cleveland. [2] However, Winkler reproached Cleveland for his "placid disposal of the scandal and mild rebuke of the Carnegie company"—a criticism which seems unjustified in view of the statements made by Herbert and Sampson that no defective armor had been produced. Cleveland did reduce the fine, but the Carnegie Company still had to pay $140,000. That fine hardly supports Winkler's implication that there had been a whitewash.

Matthew Josephson's book, *The Robber Barons* (1934), was widely acclaimed by reviewers for its scholarly accuracy, but it contains nothing original on the armor scandal—except errors. The only two sources Josephson acknowledged were the Myers and Winkler books; he did not go back to the original hearings, documents, and report. Josephson despised Frick: he wrote, scornfully, "Frick, who was now an art patron, wound up by saying that uniformity in steel was unattainable. Each beam or plate is like a poem." He then quoted Frick as stating, "You might as well say that a painter could execute an equally good picture every time. Millet painted but one 'Angelus.' " [3] Josephson did not invent this remark out of whole cloth, but it was Schwab who said it. Frick did not.[4] Josephson adopted this error from a source he did not acknowledge, Burton Hendrick's biography of Carnegie. Subsequent writers assumed that Josephson's account was accurate, and they too ascribed the statement to Frick.

George Seldes's version of the scandal, published in *Iron, Blood and Profits* (1934), appeared to be accurate, but it was not, although he did consult and briefly quote from the hearings and the Cummings report. His book was intended to be a sweeping attack on *all* munitions manufacturers, so he wrote: "The race for profits frequently results in the delivery of dangerous defective materials to the government." But the only example he offered to support this claim was the armor scandal.[5] Seldes carefully avoided any evidence which contradicted his thesis, preferring instead to launch an *ad hominem* attack on Carnegie's lawyer, Philander C. Knox. He also attacked the Navy Department because it had awarded the Carnegie Steel Company contracts for armor even after the scandal— he implied that there was a corrupt relationship between the armor-makers and the Navy.[6]

During the 1930's, at the height of the attacks on munitions-makers as war profiteers, three other brief accounts appeared. The first of these was in *The Merchants of Death* (1934), written by H. C. Engelbrecht and F. C. Hanighen. They relied primarily on Winkler as their source.[7] They also cited Allan Nevins's biography of Grover Cleveland, yet they failed to consult or quote from the source Nevins had used, which was the detailed analysis made by Burton J. Hendrick. Hendrick refuted the charge that defective armor had been produced.[8] Engelbrecht and Hanighen claimed that the armor was "inferior," but they did not answer one vital question: inferior to *what?* There is a vast difference between armor which is defective and armor which does not reach the highest possible standard of excellence. Harvey O'Connor also recounted the armor scandal in his *Steel: Dictator* (1935). O'Connor relied almost exclusively on Winkler, except that he borrowed the bogus Frick comment about Millet's "Angelus" from Josephson.[9] The last version to be published in the 1930's is in Philip Noel-Baker's *The Private Manufacture of Armaments* (1936).[10] Noel-Baker's chief source was a speech made in 1916 by Congressman Clyde Tavenner before the House of Representatives. Tavenner had been trying to persuade the House to curb military expenditures, and in his speech he revived the 1894 armor scandal to embarrass and discredit Schwab, who was then head of Bethlehem Steel, the nation's largest shipbuilder and munitions manufacturer. Noel-Baker accepted Tavenner's statements at face value, and, citing Tavenner, he wrote that "President Cleveland had imposed damages against the Carnegie Company, and that these damages had been paid without demur"—an assertion which demonstrates Noel-Baker's (and Tavenner's) unfamiliarity with the facts.

The myth that Carnegie Steel sold dangerously defective armor plates to the Navy persists to this day. The latest version is in Ben B. Seligman's book, *The Potentates: Business and Businessmen in American History* (1971).[11] Doubtless there are still others.

It is worth noting that these myth-makers were experienced in their calling. Five them them broadcast the legend that J. P. Morgan had earned the first installment of his fortune by selling defective rifles to the Union Army during the Civil War. The Morgan myth was exposed over thirty years ago by R. Gordon Wasson, in *The Hall Carbine Affair*.[12] But a sixth writer, Seligman, continued to perpetuate the myth even after Wasson exposed it. Seligman offered no new evidence; instead he launched a personal attack on Wasson.[13]

These myth-makers shared one conviction: to a man they were hostile to capitalism. They believed that businessmen deliberately foster wars in order to create profitable markets for munitions, and that their greed for profits is so all-consuming that they will not hesitate to endanger human lives and risk their country's welfare. Most of these distorted versions of the armor scandal were written during the 1930's, when America was going through a period of profound disillusionment. World War I had been billed as "the war to end all wars" and the war to "make the world safe for democracy." But after that war, peace was endangered by the rise of dictatorships and militarism in Europe. The dreams of everlasting peace were shattered. American anxiety turned into hysteria, and the quest for explanations gave way to a search for scapegoats.[14] Predictably, the chief candidates were the men who had made huge profits during the war. Schwab was too good a target to overlook; the myth-makers, having found a perfect villain, resurrected the armor scandal of 1894 and then embroidered upon it.

Notes

CHAPTER 1

1. S. J. Wolff, *Drawn from Life* (New York, 1932), 235.
2. Henry W. Storey, *History of Cambria County, Pennsylvania, with Genealogical Memoirs* (New York, 1914), II, 458; Rev. Ferdinand Kittell, *Souvenir of Loretto Centenary, 1799–1899* (Cresson, Pa., 1899), 153, 195; Sidney B. Whipple, manuscript, *Notes on Mr. Schwab's Life* (hereafter cited as Whipple Notes), in Charles M. Schwab Memorial Library, Bethlehem Steel Corporation, Bethlehem, Pa., 7. In 1935, Whipple, a prominent journalist, was hired to record Schwab's reminiscences of his long business career, for the use of a future biographer. When Whipple himself attempted to become Schwab's biographer, he was dismissed, hence the unfinished and unedited state of his notes, nearly 300 pages, based on interviews with Schwab and his surviving business associates and on letters then in Schwab's possession which since have disappeared. Whipple's notes did not extend beyond the material Schwab told or gave him, hence there are many gaps in the coverage, especially regarding the many controversies in which Schwab was involved. (For details, see Appendix A.)
3. Storey, *History of Cambria County*, 464–65; Kittell, *Loretto Centenary*, 195.
4. *Ibid.*, 76–78.
5. *Ibid.*, 33–59, *passim.*; *Mariale 1926*, ed. by Garvey Literary Society of St. Francis Seminary (Loretto, Pa., 1926)—a series of essays on Gallitzin's life and career.
6. Kittell, *Loretto Centenary*, 257.
7. Jesse C. Sell, *20th Century History of Altoona and Blair County* (Chicago, 1911), 456–57; report from Alvin E. Yost to S. H. Yorks, Dec. 8, 1950, Schwab Papers, Bethlehem, Pa. Yost made a field trip to Williamsburg and Loretto to verify details relating to Schwab's family and childhood.
8. Ruth Ayers, "My Boy Charlie: An Interview with Pauline Schwab," Pittsburgh *Press*, July 10, 1932, Magazine Section, 1–2; Whipple Notes, 6, 8; Storey, *History of Cambria County*, 458–59.
9. "My Boy Charlie," Pittsburgh *Press*; Whipple Notes, 8–9.
10. Whipple Notes, 5.
11. *Ibid.*, 11.

12. The Charles M. Schwab Memorial Library possesses hundreds of photographs of Schwab, his family, personal friends, and business associates.
13. "My Boy Charlie," Pittsburgh *Press.*
14. *Ibid.*
15. *Yearbook of the American Iron and Steel Institute, 1918* (New York, 1919), 216.
16. Whipple Notes, 11.
17. "My Boy Charlie," Pittsburgh *Press.*
18. *Ibid.*
19. Whipple Notes, 9–10.
20. *Ibid.,* 7, 9; on Father Bowen, see Kittell, *Loretto Centenary,* 87–88.
21. Based on an 1884 syllabus of courses. For information and material on the history of St. Francis College, I am indebted to the Rev. Vincent R. Negherbon, T.O.R., the President of the College, and to Miss Margaret Tobin, the Librarian.
22. Pittsburgh *Leader,* Sept. 7, 1902.
23. Eugene Grace, *Charles M. Schwab* (Bethlehem, Pa., 1947), 6. A copy of this memorial address can be found in the Schwab Memorial Library.
24. Whipple Notes, 12.
25. *Ibid.,* 10.
26. "My Boy Charlie," Pittsburgh *Press.*
27. Kittell, *Loretto Centenary,* 364.
28. Whipple Notes, 11.
29. *Ibid.,* 11–12.
30. George S. Hellman, *Lanes of Memory* (New York, 1927), 64.
31. Pittsburgh *Leader,* Sept. 7, 1902.
32. *Ibid.,* Sept. 14, 1902.
33. Whipple Notes, 12–13.

CHAPTER 2

1. Fourteen-page ms. essay on Schwab's early years in Braddock, containing details which could have come only from Schwab. Schwab Papers, Bethlehem.
2. Whipple Notes, 12–13.
3. Essay on Schwab's early years in Braddock.
4. Interview with Schwab's cousin, Pittsburgh *Leader,* Sept. 7, 1902.
5. On Jones, see *History of Allegheny County, Pa.* (Chicago, 1889), II, 283–85; *Encyclopedia of Contemporary Biography of Pennsylvania* (New York, 1889), 128–31; John G. Gable, *History of Cambria County* (Topeka, Kansas, 1926), I, 275–79.
6. Whipple Notes, 31–33.
7. Essay on Schwab's early years in Braddock.
8. *Yearbook of the American Iron and Steel Institute, 1929* (New York, 1930), 241, speech by Schwab citing employment records discovered by W. J. Filbert of U.S. Steel; William B. Dickson, *History of the Carnegie Veteran Association* (Montclair, N.J., 1938), 105.
9. Whipple Notes, 16.
10. *Ibid.,* 18.
11. *Bulletin of the American Iron and Steel Association,* Sept. 10, 1901, 131.
12. Whipple Notes, 28.
13. Obituary, Allentown (Pa.) *Call,* Jan. 15, 1939. For the information on the Dinkey family, on which this and the next two paragraphs are based. I am indebted to the late Elizabeth Dinkey (Mrs. Donald Lord Finlayson of Kennebunk, Me.), daughter of Charles Eugene Dinkey, letter to the author, Oct. 14, 1966, and to Rana Ward (Mrs.

J. B. Weiler), daughter of Minnie Dinkey, interview, Delray Beach, Fla., July 26, 1974.

14. Based on interviews with five parties who knew both Charles and Rana Schwab: (1) Rana Ward, niece of Rana Schwab, who lived with the Schwabs for many years in New York; interview, July 26, 1974; (2) Alfred D. McKelvey, former husband of Rana Ward and frequent visitor to the Schwabs' homes between 1935 and 1939; interview, Atherton, Calif., Sept. 12, 19, 1974; (3) Edward H. Schwab, surviving younger brother of Charles M. Schwab, who reached his ninetieth birthday Jan. 23, 1975; interview, Westport, Conn., May 18, 1966, and Fort Lauderdale, Fla., July 26, 1974; (4) Mrs. F. T. Pierson, daughter of Dr. Samuel A. Brown, personal physician to Charles and Rana Schwab; interview, N.Y., N.Y., Nov. 4, 1966; (5) Mr. and Mrs. Arch B. Johnston, Jr., son and daughter-in-law of one of Schwab's closest friends and business associates, Archibald Johnston; interview, Bethlehem, Pa., April 6, 1966, July 16, 1974.

15. Whipple Notes, 16.
16. *Ibid.*, 26.
17. Interviews with Edward H. Schwab.
18. *Bulletin of the American Iron and Steel Association*, April 10, 1901, 52.
19. Whipple Notes, 22.
20. *Ibid.*, 23.
21. *Personality Magazine*, Dec. 1927, 45.
22. Interview with Frank Jerome Riley, New York, N.Y., June 8, 1967, recalling story told to him by Schwab.
23. Every aspect of Carnegie's life and career has been traced anew and is fully documented in Joseph Frazier Wall's definitive biography, *Andrew Carnegie* (New York, 1970).
24. *Ibid.*, 134, 138–43, 197, 295, 306.
25. The following discussion of the technology of iron and steel-making is based on: Louis C. Hunter, "The Heavy Industries," in Harold F. Williamson (ed.), *The Growth of the American Economy* (New York, 1951), 474–81; Peter Temin, *Iron and Steel in Nineteenth-Century America* (Cambridge, Mass., 1964), 125–52; James Howard Bridge, *The Inside History of the Carnegie Steel Company* (New York, 1903), 136–50; Burton J. Hendrick, *The Life of Andrew Carnegie* (Garden City, N.Y., 1932), I, 150–77; W. Paul Strassman, *Risk and Technological Innovation* (Ithaca, N.Y., 1959), 32–36; and Louis M. Hacker, *The World of Andrew Carnegie, 1865–1901* (Philadelphia, 1968), 337–42.
26. Bridge, *Inside History*, 142–43.
27. *Ibid.*, 147.
28. Wall, *Andrew Carnegie*, 500–504.
29. Victor S. Clark, *History of Manufactures in the United States* (New York, 1929), II, 84–85, 155–57, 161–62, 286–91.
30. On Carnegie's expansion policies, see generally Hendrick, *Andrew Carnegie*, I, 178–241, 286–313, and II, 1–53; Hacker, *World of Andrew Carnegie*, 342–62, 385–401; Jonathan R. T. Hughes, *The Vital Few* (Boston, 1966), chapter III, "Carnegie and the American Steel Industry," 220–73, *passim.*; Wall, *Andrew Carnegie*, 316–18, 472–76.
31. Wall, *Andrew Carnegie*, 505.
32. Whipple Notes, 17.
33. Andrew Carnegie, *The Autobiography of Andrew Carnegie* (Boston, 1920), 182–83.
34. *Bulletin of the American Iron and Steel Association*, April 10, 1901, 52.
35. Recollections of David G. Kerr, 1935, appendix to Whipple Notes, 278.
36. Whipple Notes, 17.
37. *Ibid.*, 15.
38. *Ibid.*, 14–15.

39. Clark, *History of Manufactures*, II, 227; Hacker, *World of Andrew Carnegie*, 350–51.
40. Dickson, *History of Carnegie Veteran Association*, 108; Whipple Notes, 42.
41. *Ibid.*, 16, 26.
42. *Ibid.*, 23; Wall, *Andrew Carnegie*, 359.
43. Frank Harris, *Latest Contemporary Portraits* (New York, 1927), 121, based on a 1916 interview with Schwab.
44. *Bulletin of the American Iron and Steel Assoc.*, Oct. 2, 1889, 276–77.
45. *Ibid.*, Oct. 9, 1889, 285; New York *Times*, Oct. 11, 1889; Merle Crowell, "Schwab's Own Story," *American Magazine*, Oct. 1916, 60.

CHAPTER 3

1. *Bulletin of American Iron and Steel Assoc.*, Sept. 7, 1892, 281.
2. Charles M. Schwab, *Succeeding with What You Have* (New York, 1916), 39–41.
3. J. Bernard Hogg, "The Homestead Strike of 1892" (unpublished doctoral dissertation, University of Chicago, 1943), 30–38.
4. Andrew Carnegie, "An Employer's View of the Labor Question," *Forum*, April 1886, reprinted in Edward C. Kirkland (ed.), *The Gospel of Wealth and Other Timely Essays* by Andrew Carnegie (Cambridge, Mass., 1962), 102–3.
5. Wall, *Andrew Carnegie*, 528–30.
6. *Ibid.*, 531.
7. George Harvey, *Henry Clay Frick, the Man* (New York, 1928), 76–105.
8. Wall, *Andrew Carnegie*, 541–42.
9. Hogg, "The Homestead Strike of 1892," 39–47; Wall, *Andrew Carnegie*, 549.
10. Hogg, "The Homestead Strike of 1892," 48–66.
11. Wall, *Andrew Carnegie*, 552.
12. Hogg, "The Homestead Strike of 1892," 82–94.
13. Carnegie to William E. Gladstone, Sept. 24, 1892, Andrew Carnegie Papers, Library of Congress (hereafter cited as ACLC), Vol. 17 #3198–3200; Carnegie to George Lauder, July 17, 1892, ACLC, Vol. 17 #3111fl Carnegie to W. T. Stead, Aug. 6, 1892, ACLC, Vol. 17 #3132.
14. Carnegie to Lauder, July 17, 1892, ACLC, Vol. 17 #3111.
15. Whipple Notes, 74.
16. *Ibid.*, 70; Dickson, *History of Carnegie Veteran Association*, 102; *Bulletin of American Iron and Steel Association*, Oct. 26, 1892; Potter to Carnegie, Oct. 19, 1893, ACLC, Vol. 22 #4169–71.
17. Frick to Schwab, Oct. 22, 1892, Whipple Notes, 65.
18. U.S. Congress, House, Special Subcommittee of the Committee on Naval Affairs, *Investigation of Armor-Plate Contracts*, 1894. Testimony of Charles M. Schwab, July 6, 1894, 619.
19. Whipple Notes, 70.
20. *Ibid.*, 71–72.
21. *Ibid.*, 70.
22. *Ibid.*, 70.
23. Bridge, *Inside History*, 245–46; Harvey, *Henry Clay Frick*, 181; Whipple Notes, 72.
24. *Ibid.*, 71.
25. Robert Seager, "Ten Years Before Mahan: The Unofficial Case for the New Navy, 1880–1890," *Mississippi Valley Historical Review*, LX #3, Dec. 1953, 491–512; George F. Howe, *Chester A. Arthur: A Quarter-Century of Machine Politics* (New York, 1934),

232–40; Leon B. Richardson, *William E. Chandler: Republican* (New York, 1940), 280–93, 380; Allan Nevins, *Grover Cleveland: A Study in Courage* (New York, 1947), 217–23; Mark D. Hirsch, *William C. Whitney: Modern Warwick* (New York, 1948), 297–302, 323–28.

26. *Report of the Secretary of the Navy, 1886*, 10–11, in House Executive Documents, vol. 7, 49th Congress, 2nd session; *Report of the Secretary of the Navy, 1887*, iii–iv, in House Executive Documents, vol. 8, 50th Congress, 1st session; *Report of the Secretary of the Navy, 1888*, iv, in House Executive Documents, vol. 8, 50th Congress, 2nd session.

27. Carnegie to William C. Whitney, Dec. 27, 1886, Whitney Papers, Library of Congress, Vol. 39 #7199; Pittsburgh *Times*, Dec. 24, 1886.

28. Whitney to Hilary A. Herbert (Chairman, House Committee on Naval Affairs), Feb. 28, 1887, Whitney Papers, Vol. 41 #7662; also see Whitney to Senator Eugene Hale, same date.

29. *Report of the Secretary of the Navy, 1887*, Appendix 16, 459–474; Bethlehem Iron Company, *A Statement Concerning the Price of Armor Plate and Congressional Action on the Subject* (South Bethlehem, Pa., 1898), 9–13, copy in Library of American Iron and Steel Institute, New York, N.Y.

30. Carnegie to Whitney, Feb. 1887, Whitney Papers, Vol. 41 #7649.

31. *Report of the Secretary of the Navy, 1890*, 17–18, in House Executive Documents, vol. 9, 51st Congress, 2nd session.

32. *Report of Chief of Bureau of Ordnance, 1890*, 250, appended to *Report of the Secretary of the Navy, 1890*.

33. *Report of the Secretary of the Navy, 1890*, 18–19; U.S. Congress, Senate, Investigation by the Committee on Naval Affairs, *Prices of Armor for Naval Vessels*, Senate Report #1453, 54th Congress, 2nd session. Testimony of Benjamin F. Tracy, Feb. 8, 1896, 143–47, 155–56, and testimony of Andrew Carnegie, 185–91. Wall, *Andrew Carnegie*, 645–46, states incorrectly that Carnegie entered a bid and began producing armor for the U.S. Navy in 1887.

34. U.S. Congress, House, Special Subcommittee of the Committee on Naval Affairs, *Investigation of Armor-Plate Contracts*, 1894. Testimony of C. M. Schwab, 619–21. On Corey, see Dickson, *History of Carnegie Veteran Association*, 66.

35. Hilary A. Herbert to Charles F. Crisp, Speaker of the House, March 26, 1894, in House Executive Document #160, 53rd Congress, 2nd session, 1894, i–ii, xv–xx.

36. *Ibid.*, iii–iv.

37. Frick to Hilary A. Herbert, Dec. 13, 1893, in House Executive Document #160, xxiv; also see Frick's testimony, *Investigation of Armor-Plate Contracts*, July 24, 1894, 691.

38. Frick to Schwab, Sept. 16, 1893, in House Executive Document #160, xxiv; also see Frick's testimony, *Investigation of Armor-Plate Contracts*, 690.

39. Carnegie to Grover Cleveland, Dec. 20, 1893, ACLC, Vol. 23 #4465–67 (this is a rough draft of Carnegie's letter; he kept no copy of the final version).

40. Carnegie to Cleveland, Dec. 27, 1893, quoted in Wall, *Andrew Carnegie*, 651.

41. Captain William T. Sampson to the Secretary of the Navy, Dec. 1, 1893, in House Executive Document #160, xxii–xxiii.

42. Grover Cleveland to Hilary A. Herbert, Jan. 10, 1894, quoted in Herbert's letter to Charles F. Crisp, March 26, 1894, House Executive Document #160, v–vi; excerpts reprinted in Allan Nevins (ed.), *The Letters of Grover Cleveland* (New York, 1933), 343–44.

43. Frick to Hilary A. Herbert, Dec. 13, 1893, House Executive Document #160, xxvi–xxvii.

44. Testimony of H. C. Frick, July 24, 1894, *Investigation of Armor-Plate Contracts*, 689.

45. Nevins, *Grover Cleveland*, 673–74.

46. Wall, *Andrew Carnegie*, 567–68.
47. House Resolution #226, passed May 22, 1894, reprinted in *House Report* #1468, 53rd Congress, 2nd session, 1894.
48. Testimony of C. M. Schwab, July 6, 1894, *Investigation of Armor-Plate Contracts*, 620, 623.
49. *Ibid.*, 626, 636–37.
50. *Ibid.*, 628, 639–40.
51. *Ibid.*, 634–35.
52. *Ibid.*, 628, 632, 639–40.
53. *Ibid.*, 624.
54. Schwab to Frick, Dec. 12, 1893, House Executive Document #160, xxx–xxxi.
55. Schwab's testimony, *loc. cit.*, 647.
56. *Ibid.*
57. *Ibid.*, 631.
58. Captain William T. Sampson to Secretary of the Navy, Dec. 16, 1893, House Executive Document #160, xxxiii.
59. Schwab's testimony, *loc. cit.*, 627.
60. *House Report* #1468, 53rd Congress, 2nd session, 10–11.
61. *Investigation of Armor-Plate Contracts*, 1894. Testimony of Captain Sampson, 29.
62. *Ibid.*, 28–29.
63. Schwab's testimony, *ibid.*, 650.
64. *Ibid.*, 657, 659, 667–68.
65. *Congressional Record, House*, XXVI, part 8, Aug. 23, 1894, 8638–44.
66. Hilary A. Herbert to Charles F. Crisp, March 26, 1894, House Executive Document #160, vi.
67. *Ibid.*; testimony of Captain Sampson, *loc. cit.*, 28–29.
68. John K. Winkler, *Incredible Carnegie* (New York, 1931), 234–35. Italics added.
69. Gustavus Myers, *History of the Great American Fortunes* (New York, 1907; reprinted 1937), 596–97, 600; John K. Winkler, *Incredible Carnegie*, 226–34; Matthew Josephson, *The Robber Barons* (New York, 1934; reprinted 1962), 391–92; George Seldes, *Iron, Blood and Profits: An Exposure of the World-Wide Munitions Racket* (New York, 1934), 236–37, 360–61; H. C. Engelbrecht and F. C. Hanighen, *Merchants of Death: A Study of the International Armament Industry* (New York, 1934), 53–55; Harvey O'Connor, *Steel: Dictator* (New York, 1935), 69–70, 341; Philip Noel-Baker, *The Private Manufacture of Armaments* (London, 1936), 316; Ben B. Seligman, *The Potentates: Business and Businessmen in American History* (New York, 1971), 182.

CHAPTER 4

1. Whipple Notes, 48–49.
2. *Ibid.*, 65.
3. *Ibid.*, 39–40; also see Wall, *Andrew Carnegie*, 625–26.
4. Glenn Porter, *The Rise of Big Business, 1860–1910* (New York, 1973); Samuel P. Hays, *The Response to Industrialism, 1885–1914* (Chicago, 1959).
5. George Rogers Taylor, *The Transportation Revolution, 1815–1865* (New York, 1958), 209–10; Victor S. Clark, *History of Manufactures*, II, 496–500; Fritz Redlich, *History of American Business Leaders: Iron and Steel* (Ann Arbor, Michigan, 1940), 35–47, 91–102.
6. On the desire to avoid competition, and the various means adopted, see Hans B. Thorelli, *The Federal Antitrust Policy* (Baltimore, 1954), 54–85; Eliot Jones, *The Trust Problem in the United States* (New York, 1928), 6–12, 27–29; William S. Stevens (ed.),

Industrial Combinations and Trusts (New York, 1914), 185–87, 211–24; Edward C. Kirkland, *Industry Comes of Age, 1860–1897* (New York, 1961), 201–16; William T. Hogan, S. J., *An Economic History of the Iron and Steel Industry in the United States* (Lexington, Mass., 1971), 236–39.

7. Wall, *Andrew Carnegie*, 332–36.
8. Hendrick, *Life of Andrew Carnegie*, II, 51; Hacker, *World of Andrew Carnegie*, 362. See Carnegie to J. G. A. Leishman, received July 28, 1896: "As for orders take every one offering at any price. Show competitors we are going to run, leave no doubt about it. The lower you go the sooner they will give up. Mr. Peacock [general sales agent] is not responsible for prices, *but he is for scooping the market.*" Italics in original. Schwab Papers, Rare Manuscript Division, Pattee Library, Pennsylvania State University, State College, Pa.
9. *Iron Trade Review*, April 9, 1896, 12, copy in National Archives, Record Group 40 (Dept. of Commerce), file 6518-8-16.
10. Carnegie to Board of Managers, July 11, 1900, quoted in Hacker, *op. cit.*, 407.
11. Wall, *Andrew Carnegie*, 337, 586.
12. On the background and impact of this depression, see Kirkland, *Industry Comes of Age*, 1–12; Rendig Fels, *American Business Cycles, 1865–1897* (Chapel Hill, N.C., 1959), 113–220; Clark, *History of Manufactures*, II, 163–66.
13. *Statistical Yearbooks of the American Iron and Steel Association*, compiled by James M. Swank, quoted by Theodore E. Burton, *Financial Crises* (New York, 1926), 341.
14. Charles M. Schwab, "The Huge Enterprises Built Up by Andrew Carnegie," *The Engineering Magazine*, Jan. 1901, 504–17; Minutes of Board of Managers, July 6, 1897, ACLC, Vol. 43 #8358.
15. Clark, *History of Manufactures*, II, 232–33, III, 38; Hogan, *Iron and Steel Industry*, 551–52.
16. Clark, *History of Manufactures*, II, 235–37, III, 45–46; Hogan, *Iron and Steel*, 239–43.
17. Bridge, *Inside History*, 295.
18. *Ibid.*
19. Hendrick, *Andrew Carnegie*, II, 64; Harvey, *Henry Clay Frick*, 183–84.
20. Whipple Notes, 238.
21. Phipps to Carnegie, Dec. 22, 1894, ACLC, Vol. 29 #5639–40.
22. Carnegie to Leishman, Jan. 4, 1896, ACLC, Vol. 35, quoted by Wall, *Andrew Carnegie*, 662; on Leishman, see *Dictionary of American Biography*, (New York, 1946), Vol. 11, 155–56, and Dickson, *History of Carnegie Veteran Association*, 85–86.
23. Carnegie to Leishman, Feb. 4, 1896, ACLC, Vol. 36, #7090–93.
24. *Bulletin AISA*, Feb. 20, 1897, 45.
25. *Ibid.*, June 20, 1897, 138.
26. John Moody, *The Masters of Capital* (New Haven, 1919), 79.
27. William B. Dickson's unpublished autobiography, Rare Manuscript Division, Pattee Library, Pennsylvania State University, State College, Pa.; also Dickson, *History of Carnegie Veteran Association*, 152.
28. Schwab to Carnegie, Oct. 7, 1897, ACLC, Vol. 45 #8895.
29. Carnegie to Schwab, Oct. 18, 1897, ACLC, Vol. 46 #8979.
30. Based on Minutes of the Operating Department in volumes 45 through 80, e.g. May 7, 1898, ACLC, Vol. 51 #9927.
31. Minutes of Board of Managers, Dec. 12, 1897, quoting letter from Carnegie, Nov. 21, 1897, ACLC, Vol. 47 #9184.
32. Wall, *Andrew Carnegie*, 322, 747–49.
33. Dickson, *History of Carnegie Veteran Association*, 112; *Harper's Weekly*, 46, Nov. 15, 1902, 1709; interview with Edward H. Schwab, July 26, 1974.
34. Carnegie to Schwab, April 15, 1900. Schwab Papers, Pennsylvania State University.

35. *Ibid.*, April 18, 1900.
36. *Bulletin AISA*, Feb. 8, 1893, 42.
37. Carnegie to F. T. F. Lovejoy, Dec. 9, 1895, ACLC, Vol. 35, #6760–62.
38. Further Thoughts on the Minutes of August 31, written by Carnegie on Sept. 18, 1897, ACLC, Vol. 44 #8712–14.
39. Carnegie to Schwab, Oct. 1, 1897, ACLC, Vol. 45 #8828–30. For earlier criticisms by Carnegie of proposals made by Schwab, see Minutes of Aug. 31, 1897, Vol. 44 #8597–8601 and #8689 ff., and Further Thoughts on the Minutes of August 31, Vol. 44, #8712–14.
40. Schwab to Carnegie, Oct. 1, 1897, ACLC, Vol. 45 #8837–41; Schwab to Carnegie, Oct. 5, 1897, Vol. 45 #8871–83.
41. Carnegie to Schwab, Oct. 16, 1897, ACLC, Vol. 46 #8966.
42. Carnegie to Schwab, Oct. 18, 1897, ACLC, Vol. 46 #8977–78, replying to Schwab's letter of Oct. 5, in Vol. 45 #8871–83.
43. Carnegie to George Lauder, Nov. 28, 1897, ACLC, Vol. 47 #9146B.
44. Frick to Carnegie, Dec. 4, 1897, ACLC, Vol. 47 #76.
45. Schwab to Carnegie, Oct. 15, 1897, ACLC, Vol. 45 #8882–3.
46. Carnegie to Frick, Feb. 15, 1897, ACLC, Vol. 41 #8051.
47. Minutes of the Board of Managers, July 27, 1897, ACLC, Vol. 43 #8415–17.
48. *Ibid.*
49. Schwab to Carnegie, March 31, 1897, ACLC, Vol. 63 #12335–36.
50. Whipple Notes, 74.
51. Schwab to Frick, Dec. 3, 1897, ACLC, Vol. 47 #9170.
52. Schwab to Carnegie, Dec. 9, 1897, ACLC, Vol. 47 #9188–89.
53. Schwab to Carnegie, May 26, 1899, ACLC, Vol. 65 #12655; see also Minutes of Operating Dept., Vol. 65 #12666.
54. Minutes of Operating Dept., May 27, 1899, ACLC, Vol. 65 #12666.
55. Minutes of Operating Dept., June 9, 1899, ACLC, Vol. 66 #12741.
56. Minutes of Board of Managers, June 13, 1899, ACLC, Vol. 66 #12768.
57. Schwab to Carnegie, July 15, 1899, ACLC, Vol. 67 #12892.
58. *Ibid.*; also Minutes of Operating Dept., March 5, 1900, ACLC, Vol. 73 #14111.
59. John Maurice Clark, *Strategic Factors in Business Cycles* (New York, 1934), 11; Frick to Henry Phipps Jr., Aug. 1897, ACLC, Vol. 44 #8606.
60. Phipps to Carnegie, Sept. 1, 1897, ACLC, Vol. 44 #8648–52.
61. Minutes of Board of Managers, Nov. 14, 1898, ACLC, Vol. 56 #10904.
62. *Ibid.*
63. Minutes of Board of Managers, March 1, 1898, ACLC, Vol. 49 #9556.
64. Minutes of Operating Dept., Oct. 8, 1898, ACLC, Vol. 55 #10715.
65. American Iron and Steel Assoc., *Annual Statistical Report* (Phila., 1900), 21.
66. Minutes of Board of Managers, July 16, 1900, ACLC, Vol. 76 #14661–65.

CHAPTER 5

1. Schwab's testimony before the Stanley Committee, Aug. 4, 1911, in U.S. Congress, House, *Committee on Investigation of United States Steel Corporation, Hearings*, II, 1285–86, 1288; Whipple Notes, 53.
2. Carnegie to Frick, Sept. 20, 1897, ACLC, Vol. 45 #8734; Whipple Notes, 53.
3. Whipple Notes, 52–53.
4. Clark, *History of Manufactures*, III, 45–47; Hogan, *Iron and Steel Industry*, 265–72.

5. Minutes of Board of Managers, Nov. 14, 1898, ACLC, Vol. 56 #10906–9.
6. *Ibid.*
7. Minutes of Board of Managers, Nov. 15, 1898, ACLC, Vol. 56 #10913.
8. Minutes of Board of Managers, Nov. 22, 1898, ACLC, Vol. 56 #10965; Ida M. Tarbell, *The Life of Elbert H. Gary: The Story of Steel* (New York, 1925), 98–99.
9. Schwab to Carnegie, July 18, 1899, ACLC, Vol. 67 #12970.
10. Minutes of Board of Managers, June 12, 1900, ACLC, Vol. 75 #14504.
11. Minutes of Board of Managers, Jan. 3, 1899, quoting letter from Carnegie, Dec. 30, 1898, ACLC, Vol. 60, #11613.
12. Minutes of Board of Managers, June 6, 1900, ACLC, Vol. 75 #14486.
13. Carnegie to Schwab, June 20, 1900, ACLC, Vol. 75 #14535–37.
14. Minutes of Board of Managers, Dec. 20, 1898, ACLC, Vol. 58 #11341–43; also Schwab to Carnegie, Dec. 12, 1898, ACLC, Vol. 58 #11344.
15. Carnegie to Schwab, Dec. 22, 1898, ACLC, Vol. 59 #11447.
16. On Schoen, see John W. Jordan (ed.), *Encyclopedia of Pennsylvania Biography* (New York, 1914), II, 600–606.
17. Minutes of Board of Managers, Nov. 8, 1898, ACLC, Vol. 56 #10874–75.
18. Minutes of Board of Managers, Nov. 15, 1898, ACLC, Vol. 56 #10911–12.
19. Schoen to Schwab, Dec. 9, 1898, ACLC, Vol. 58 #11271.
20. Carnegie to Schwab, Dec. 10, 1898, ACLC, Vol. 58 #11271.
21. Minutes of Board of Managers, Dec. 13, 1898, ACLC, Vol. 58 #11272.
22. Carnegie to Schwab, Dec. 20, 1898, ACLC, Vol. 58, #11339.
23. Schoen to Schwab, Dec. 21, 1898, ACLC, Vol. 58 #11339.
24. Minutes of Board of Managers, Jan. 16, 1899, ACLC, Vol. 60 #11744.
25. Minutes of Board of Managers, Jan. 31, 1899, ACLC, Vol. 61 #11927–28.
26. Phipps to Frick, Jan. 29, 1899, ACLC, Vol. 61 #11929; Carnegie to Lauder, Feb. 7, 1899, ACLC, Vol. 62 #11992A.
27. Minutes of Board of Managers, Feb. 7, 1899, quoting letter from Carnegie, ACLC, Vol. 62, #12008; Frick to Carnegie, Feb. 7, 1899, ACLC, Vol. 62 #12011–12.
28. Minutes of Board of Managers, Feb. 14, 1899, ACLC, Vol. 62 #12058–62.
29. This and the next two paragraphs are derived from "Grandfather's Talks About His Life Under Two Flags," the unpublished autobiography of Hilary A. Herbert (1903), in Southern Historical Collection, University of North Carolina Library, Chapel Hill, N.C.
30. Francis B. Simkins, *Pitchfork Ben Tillman, South Carolinian* (Baton Rouge, La., 1944; reprinted Gloucester, Mass., 1964), 347–52.
31. Carnegie to Board of Managers, July 29, 1896, ACLC, Vol. 38 #7416–19.
32. *Ibid.*
33. Carnegie to Leishman, Dec. 21, 1896, ACLC, Vol. 40 #7798.
34. Carnegie to Leishman, Dec. 10, 1896, ACLC, Vol. 40 #7768.
35. Minutes of Board of Managers, July 20, 1897, ACLC, Vol. 43 #8399.
36. Schwab to Carnegie, July 27, 1897, ACLC, Vol. 43 #8440–42; also Schwab to Carnegie, Sept. 11, 1897, ACLC, Vol. 44 #8691–92.
37. Carnegie to Schwab, Sept. 24, 1897, ACLC, Vol. 45 #8758–60.
38. Minutes of Board of Managers, May 10, 1898, ACLC, Vol. 51 #9944.
39. Minutes of Board of Managers, May 31, 1898, ACLC, Vol. 52 #10020.
40. Minutes of Board of Managers, May 17, 1898, ACLC, Vol. 51 #9963.
41. *Ibid.*
42. Minutes of Board of Managers, July 5, 1898, ACLC, Vol. 53 #10254–55; also see Schwab to Carnegie, July 7, 1898, ACLC, Vol. 53 #10279–80.
43. Schwab to Carnegie, Aug. 16, 1898, ACLC, Vol. 54 #10494.

44. Schwab to Carnegie, Nov. 22, 1898, ACLC, Vol. 56 #10971.
45. See comments of Frick in Minutes of Board of Managers, March 7, 1899, ACLC, Vol. 63 #12192.
46. Andrew Moreland to Schwab, March 23, 1899, ACLC, Vol. 63 #12270–71.
47. *Ibid.*, #12273.
48. Schwab to Carnegie, March 24, 1899, ACLC, Vol. 63 #12278.
49. Schwab to Carnegie, April 15, 1899, reporting on his meeting in Pittsburgh with Captain O'Neil on April 14, ACLC, Vol. 64 #12436.
50. *Ibid.*
51. Minutes of Board of Managers, May 8, 1899, ACLC, Vol. 65 #12560; also Schwab to Carnegie, May 13, 1899, ACLC, Vol. 65 #12594.
52. Minutes of Board of Managers, May 22, 1899, ACLC, Vol. 65 #12634.
53. Schwab to Carnegie, May 26, 1899, ACLC, Vol. 65 #12656.
54. Minutes of Board of Managers, May 29, 1899, ACLC, Vol. 65 #12675.
55. *Ibid.*
56. *Ibid.* #12676.
57. Schwab to Carnegie, June 2, 1899, ACLC, Vol. 65 #12699.
58. Minutes of Board of Managers, June 13, 1899, ACLC, Vol. 66 #12769.
59. Minutes of Board of Managers, Sept. 28, 1899, ACLC, Vol. 69 #13320–21.
60. *Ibid.*
61. W. R. Balsinger to Schwab, Jan. 25, 1900, ACLC, Vol. 72 #13908–9.
62. Carnegie to Schwab, Jan. 31, 1900, ACLC, Vol. 72 #13939–40.
63. Minutes of Board of Managers, Oct. 16, 1900, ACLC, Vol. 79 #15077.
64. Minutes of Board of Managers, Nov. 27, 1900, ACLC, Vol. 80 #15272.
65. Minutes of Board of Managers, Dec. 4, 1900, ACLC, Vol. 80 #15306.
66. Harvey, *Henry Clay Frick*, 200–217; *Life of Andrew Carnegie*, II, 77–88; Wall, *Andrew Carnegie*, 714–64.
67. Minutes of Board of Managers, June 27, 1899, ACLC, Vol. 66, quoted in Wall, *Andrew Carnegie*, 731.
68. Carnegie to Schwab, Nov. 26, 1899, Whipple Notes, 59–60. (The Whipple Notes contain several unpublished letters from Carnegie to Schwab, all of them originally handwritten and confidential; Carnegie retained no copies or drafts, and the originals which were in Schwab's possession in 1936 have disappeared subsequently.) Carnegie to Schwab, received Feb. 22, 1900, Schwab Papers, Penn State.
69. Harvey, *Henry Clay Frick*, 219–21; Hendrick, *Life of Andrew Carnegie*, II, 93–100.
70. Carnegie to Lauder, Nov. 25, 1899, ACLC, Vol. 70, quoted in Hendrick, *Life of Andrew Carnegie*, II, 96–97.
71. Carnegie to Schwab and Board, Nov. 26, 1899, Whipple Notes, 61.
72. *Ibid.*, 60–62.
73. Carnegie to Lauder, Nov. 25, 1899, *loc. cit.*; Carnegie to Schwab and Board, Nov. 26, 1899, Whipple Notes, 62.
74. Carnegie to Phipps, *ca.* Nov. 26, 1899, ACLC, Vol. 71 #13659.
75. Carnegie to Schwab, Nov. 26, 1899, Whipple Notes, 59–60.
76. Schwab to Carnegie, Nov. 27, 1899, quoted in Wall, *Andrew Carnegie*, 743.
77. Schwab to Carnegie, Nov. 27, 1899, quoted in Hendrick, *Life of Andrew Carnegie*, II, 95.
78. Carnegie to Schwab, *ca.* Dec. 9, 1899 (received by Schwab on Dec. 11, Whipple Notes, 66.
79. *Ibid.*, 238–39.
80. Schwab to Frick, Dec. 3, 1899, quoted in Winkler, *Incredible Carnegie*, 248–50.
81. Carnegie to Schwab, *ca.* Dec. 9, 1899, Whipple Notes, 66.

82. Hendrick, *Life of Andrew Carnegie*, II, 100–113; Harvey, *Henry Clay Frick*, 237–57; Wall, *Andrew Carnegie*, 745–64.
83. *Ibid.*, 753, quoting John Walker's 1928 interview with Burton J. Hendrick.
84. Whipple Notes, 64.
85. Carnegie to Schwab, received Jan. 26 and Jan. 31, 1900, Schwab Papers, Penn State.
86. Julius Moritzen, "The Great Steel-Makers of Pittsburgh and the Frick-Carnegie Suit," *American Monthly Review of Reviews*, 21, April, 1900, 433.
87. Carnegie to Schwab, received Jan. 31, 1900, Schwab Papers, Penn State.
88. Dickson, *History of Carnegie Veteran Association*, 168.

CHAPTER 6

1. Ralph L. Nelson, *Merger Movements in American Industry, 1895–1956* (Princeton, 1959), 71–105; Kirkland, *Industry Comes of Age*, 306–25; Marian V. Sears, "The American Businessman at the Turn of the Century," *Business History Review*, XXX (1956), 382–443; Alfred D. Chandler, Jr. "Beginnings of Big Business in American Industry," *ibid.*, XXXIII (1959), 1–31.
2. Edward S. Meade, "The Genesis of the U.S. Steel Corporation," *Quarterly Journal of Economics*, XV (1901), 531–32; Hendrick, *Life of Andrew Carnegie*, II, 115–19.
3. *Ibid.*, II, 117–18; Wall, *Andrew Carnegie*, 767–68.
4. Carnegie to Schwab, July 11, 1900, quoted in Minutes of Board of Managers, July 31, 1900, ACLC, Vol. 76.
5. Schwab's testimony to the Stanley Committee, Aug. 4, 1911, U.S. Congress, House, *Committee on Investigation of United States Steel Corporation, Hearings*, II, 1312; Schwab's testimony in *U.S. v. U.S. Steel*, May 19, 1913, 4198–99, 4391, 4400–401 (copy of transcript in Columbia University Law School Library); Hendrick, *Life of Andrew Carnegie*, II, 121–23.
6. On the superiority of Carnegie Steel, see Meade, *loc. cit.*, 539–44; on the precarious position of Carnegie's rivals, see Abraham Berglund, *The United States Steel Corporation* (New York, 1907), 66–67; *Report of the Commissioner of Corporations on the Steel Industry* (Washington, 1911), I, 7–13, 78–85, 98–106; Hendrick, *Life of Andrew Carnegie*, II, 119–21.
7. Carnegie to Schwab, June 20, 1900, ACLC, Vol. 75 #14535–37.
8. Schwab to Carnegie, Jan. 24, 1901, ACLC, Vol. 81 #15526.
9. U.S. Congress, House, *Committee on Investigation of United States Steel Corporation, Hearings*, testimony of John W. Gates, 31.
10. *Ibid.*, testimony of Elbert H. Gary, 205.
11. Hendrick, *Life of Andrew Carnegie*, I, 380; II, 76.
12. See, for example, Arundel Cotter, *The Authentic History of the United States Steel Corporation* (New York, 1916), 15, and Gabriel Kolko, *The Triumph of Conservatism, 1900–1916* (New York, 1963), 33. Four-page telegram from A. C. Case to J. McSwigan, copy in Schwab papers, Bethlehem, Pa.
13. For Schwab's recollection of what he said in 1900, see *Investigation of United States Steel Corporation, Hearings*, 1276–77, and Schwab's testimony in the 1913 antitrust dissolution suit against U.S. Steel, 4135–36, and his speech of Dec. 31, 1901 to a group of Chicago bankers, in Whipple Notes, 95–103.
14. *Ibid.*, 86.
15. *Investigation of United States Steel Corporation, Hearings*, testimony of John W. Gates, 30–32.

16. Whipple Notes, 86.
17. *Ibid.*, 87–88.
18. *Ibid.*, 88
19. The original figure named by Carnegie was $400 million, but the final figure was increased to $492 million. Hacker, *World of Andrew Carnegie*, 434; Hendrick, *Life of Andrew Carnegie*, II, 136–37.
20. Carnegie to Schwab, March 7, 1901, Schwab Papers, Penn State.
21. Unless otherwise noted the following section is based on interviews with Edward H. Schwab (born 1885) on May 18, 1966 and July 26, 1974, and with Rana Ward (born 1901), a niece of Rana Schwab and the daughter of Minnie Dinkey and Dr. Marshall Ward, on July 26, 1974.
22. Whipple Notes, 47.
23. *Ibid., passim.*
24. *Ibid.*, 93.
25. Interview with Rana Ward.

CHAPTER 7

1. *Report of the Commissioner of Corporations on the Steel Industry* (1911), I, 131; Hogan, *Iron and Steel Industry*, 473–84.
2. For a sampling of negative reaction, see Mark Sullivan, *Our Times: The United States, 1900–1925* (New York, 1927), II, 351–55.
3. "Morgan's Clique Wields Blocks of Billions," New York *Journal*, April 7, 1901, and Richard T. Ely, "An Analysis of the Steel Trust," *The Cosmopolitan*, XXI (1901), 428–31. I have found only three articles which spoke favorably of the new merger: "The Billion-Dollar Corporation," *Gunton's Magazine*, May 1901, 421–24; "The United States Steel Corporation," *Iron Age*, April 4, 1901, 41; and Charles S. Gleed, "The Steel Trust and Its Makers," *The Cosmopolitan*, XXI (1901), 25–31.
4. Milwaukee *Sentinel*, April 23, 1901.
5. New York *World*, May 15, 1901.
6. *Ibid.*, May 16, 1901.
7. By-laws of U.S. Steel, in U.S. Congress, House, *Report of the Industrial Commission*, XIII, 481–87, House Document #4343, 57th Congress, 1st session.
8. Schwab interview with Burns Mantle, St. Louis *Democrat*, April 9, 1916, quoted in Paul-Louis Hervier, "American Silhouettes: Charles M. Schwab," *The Living Age*, Vol. 298, Sept. 14, 1918, 664.
9. This and the next two paragraphs are based on the by-laws of U.S. Steel, *loc. cit.;* testimony of Elbert H. Gary, *Committee on Investigation of the United States Steel Corporation, Hearings*, 62nd Congress, 1st session (1911), vol. I, *passim.;* and Ida M. Tarbell, *Elbert H. Gary*, 126–51.
10. Allan Nevins, *John D. Rockefeller* (New York, 1940), II, 423–24, based on Nevins's interview with Schwab; Whipple Notes, 216.
11. Schwab testimony, *U.S. v. U.S. Steel*, May 19, 1913, XI, 4172ff.
12. Tarbell, *Elbert H. Gary.*
13. Whipple Notes, 94.
14. Tarbell, *Elbert H. Gary*, 136–37; John A. Garraty, *Right-Hand Man: The Life of George W. Perkins* (New York, 1960), 95–97.
15. Schwab to Perkins, July 3, 1901, George W. Perkins Papers, Special Collections, Columbia University, Box 10, file 1.
16. Schwab testimony, *Committee on Investigation of the United States Steel Corporation, Hear-*

ings, 1298; Ronald Marsching, "Charles M. Schwab: A Business Biography," unpublished senior thesis, Woodrow Wilson School of Public and International Affairs, Princeton University, April 1950, 84–86; Whipple Notes, 39; Tarbell, *Elbert H. Gary*, 174–76.

17. New York *Sun*, New York *Journal*, New York *Herald* and New York *World*, May 9, 1901.
18. See, for example, editorials in Kansas City (Mo.) *Star*, May 10, 1901, and Burlington (Vt.) *News*, May 11, 1901. Many of the newspaper stories and editorials cited in this chapter are drawn from the press scrapbooks for 1901–1903 in Schwab papers, Bethlehem, Pa.
19. Editorial, *The Commoner*, May 24, 1901; Lebanon (Pa.) *Lutheran*, June 13, 1901.
20. *Report of the Industrial Commission*, 57th Congress, 1st session, House Document #76 (1901), Schwab's testimony on the tariff, XIII, 454–58, 464–66.
21. Columbus (Ohio) *Press Post*, May 13, 1901; editorial, St. Louis *Republic*, May 14, 1901.
22. *Report of the Industrial Commission*, Schwab testimony, 454.
23. St. Louis *Republic*, May 14, 1901.
24. Editorial, New York *Times*, May 13, 1901.
25. *Ibid.*; also see New York *News*, May 14, 1901, and Saginaw (Mich.) *News*, May 20, 1901.
26. *Report of the Industrial Commission*, Schwab testimony on labor and unions, 459–62.
27. Philadelphia *North American*, May 13, 1901; Ithaca (N.Y.) *News*, May 13, 1901.
28. See, for example, "Mr. Schwab's Twaddle," Brooklyn (N.Y.) *Citizen*, May 12, 1901, and *Gunton's Magazine*, June 1901, 542.
29. New York *Journal*, May 16, 1901.
30. Minutes of U.S. Steel Finance Committee, April 25, 1901, and U.S. Steel General Statements, Oct. 1902, Perkins Papers, Box 50.
31. Chicago *Times Herald*, May 25, 1901.
32. *Harper's Weekly*, Aug. 2, 1902, 1009–10; *Architectural Review*, Oct. 1902, 537–38; New York *Herald Tribune*, May 11, 1947; Altoona (Pa.) *Mirror*, May 13, 1947; John Wilcox, "Emma Schwab's Folly," *American Weekly*, July 27, 1947; Pittsburgh *Sun Telegraph*, March 6, 1948; Whipple Notes, 116; Andrew Tully, *Era of Elegance* (New York, 1947), 147–68.
33. Whipple Notes, 116.
34. Editorial, *Harper's Weekly*, Aug. 2, 1902.
35. New York *Mail & Express*, Dec. 26, 1901.
36. Boston *Globe*, Jan. 12, 1902; Milwaukee *Journal*, Jan. 15, 1902.
37. New York *Sun*, Jan. 13, 1902.
38. Milwaukee *Journal*, Jan. 11, 1902.
39. Cable from "Wakeful" (Carnegie) to Schwab, Jan. 14, 1902, copy in Perkins Papers, Box 11, file 2; Carnegie to Lauder, Jan. 28, 1902, ACLC, Vol. 88 #106A.
40. Carnegie to Morgan, Jan. 14, 1902, Perkins Papers, Box 11.
41. Schwab to Perkins, Jan. 14, 1902, Perkins Papers, Box 11.
42. Schwab to Perkins, Jan. 15, 1902, Perkins Papers, Box 11, file 5.
43. "When Schwab Made Monte Carlo Stare!," *The World*, March 3, 1901; New York *Evening World*, June 11, 1901.
44. Perkins's cables to Schwab, Jan. 15 & 16, 1902, Perkins Papers, Box 11, file 6; Schwab to Perkins, Jan. 17, 1902, *ibid.*
45. Schwab to Perkins, Jan. 15, 26, 28 & 30, 1902, Perkins Papers, Box 11, file 7; Perkins to Schwab, Jan. 27 & Feb. 13, 1902, *ibid.* Also see John A. Garraty, "Charlie Schwab Breaks the Bank," *American Heritage*, April 1957, 44–47, 103; Garraty, *Right-Hand Man*, 97–100.
46. Carnegie to Lauder, Jan. 19, 1902, ACLC, Vol. 87 #16685.

47. New York *World*, Feb. 9, 1902; New York *Journal*, Feb. 16, 1902.
48. Schwab to Carnegie, Jan. 26, 1902, ACLC, Vol. 87 #16689.
49. Chicago *Advance*, Feb. 6, 1902.
50. *The Nation*, Jan. 23, 1902.
51. Schwab to Carnegie, March 19, 1902, ACLC, Vol. 88 #16757A.
52. Whipple Notes, 90.
53. Schwab to Carnegie, March 19, 1902, ACLC, Vol. 88 #16757A.
54. Schwab to Perkins, July 30, 1902, Perkins Papers, Box 11, file 3.
55. Schwab to Perkins, Aug. 16, 1902, Perkins Papers, Box 11, file 3.
56. Schwab to Carnegie, Aug. 28, 1902 (erroneously dated July 28), ACLC, Vol. 90 #17053A.
57. Pittsburgh *Dispatch*, Sept. 2, 1902.
58. Schwab to Perkins, Sept. 21, 1902, Perkins Papers, Box 11, file 3.
59. Schwab to Perkins, Feb. 5, 1903, Perkins Papers, Box 12, file 1.
60. Perkins to Schwab, Feb. 18 & 27, 1903, Perkins Papers, Box 12, file 1.
61. Schwab to Perkins, March 2, 1903, and Perkins to Schwab, March 4, 1903, Perkins Papers, Box 12, file 1.
62. Perkins to Norman B. Ream, March 15, 1903, Perkins Papers, Box 12, file 1.
63. Perkins to Morgan, June 23, 1902, Perkins Papers, Box 11, file 1.
64. Schwab to Perkins, Aug. 16, 1902, Perkins Papers, Box 11, file 3.
65. New York *American*, April 12, 1903, reprinted in *New Yorker*, April 15, 1903.
66. *Ibid.*
67. New York *Daily Tribune*, Nov. 26, 1905, Sec. II, and Dec. 10, 1905, Sec. II; interviews with Edward H. Schwab and Rana Ward, July 26, 1974.
68. Johnstown (Pa.) *Democrat*, May 6, 1903; Chicago *Chronicle*, May 2, 1903.
69. Whipple Notes, 247; Eugene Grace, *Charles M. Schwab*, 22.
70. Thorelli, *Federal Antitrust Policy*, 279–80.
71. Dickson, *History of Carnegie Veteran Association*, 93.
72. Franklin M. Reck, *Sand in Their Shoes: The Story of American Steel Foundries* (New York, 1952), 13–14, 149.
73. New York *Daily Tribune*, June 1, 1901, 1, and June 2, 1901, Sec. 8, 2; *Harper's Weekly*, 46, Nov. 15, 1902, 1709; interview with Edward H. Schwab, July 26, 1974.
74. New York *Herald*, July 12, 1904, 15; interview with Edward H. Schwab.
75. Dickson, *History of Carnegie Veteran Association*, 153–54.
76. Recollections of David G. Kerr, Whipple Notes, 278; on Kerr, see Dickson, *History of Carnegie Veteran Association*, 81.
77. New York *Herald*, Dec. 16, 1907, 8; Dec. 18, 1907, 3 & 6; Dec. 19, 1907, 3.

CHAPTER 8

1. Henry R. Seager and Charles A. Gulick Jr., *Trust and Corporation Problems* (New York, 1929), 215; Arthur S. Dewing, *Corporate Promotions and Reorganizations* (Cambridge, Mass., 1914), 487. On the background of the U.S. Shipbuilding Company, see Seager and Gulick, 196–215, and Dewing, 464–509; also L. Walter Sammis, "The Relation of Trust Companies to Industrial Combinations as Illustrated by the United States Shipbuilding Company," *Annals of the American Academy of Political and Social Science*, vol. 24 (July–Sept. 1904), 241–70; and Henry W. Lanier, "One Trust and What Became of It," *World's Work*, Feb. 1904, 4445–57.
2. Andrew Carnegie, "Popular Illusions about Trusts," *Century Magazine*, May 1900,

reprinted in Andrew Carnegie, *The Gospel of Wealth*, edited by Edward C. Kirkland (Cambridge, Mass., 1962), 88.

3. New York *World* and New York *Tribune*, June 8, 1901.
4. New York *News*, June 9, 1901; editorial, New York *Journal*, June 9, 1901.
5. New York *Herald*, June 9, 1901; Pittsburgh *Times*, June 15, 1901.
6. Dewing, *Corporate Promotions*, 486–87.
7. Hearings before a Special Examiner, *Conklin et al. v. United States Shipbuilding Company* (hereafter cited as *Conklin v. USSB*), four-volume transcript, in National Archives, Federal Archives and Record Center, 641 Washington St., New York, N.Y., file #4697. Testimony of George W. Perkins, Dec. 8, 1903, 1125; testimony of Charles Steele, Dec. 17, 1903, 1212–23, 1231.
8. Perkins to J. P. Morgan, July 30, 1902, Perkins Papers, Box 11, file 3; also Perkins testimony, *Conklin v. USSB*, 1181.
9. Dewing, *Corporate Promotions*, 498–99.
10. *Ibid.;* also testimony of Daniel Dresser, Oct. 7, 1903, *Conklin v. USSB*, 125.
11. New York *American*, May 27 and 28, 1903; *Journal of Commerce*, May 28, 1903.
12. New York *Sun*, June 12, 1903.
13. New York *Herald*, June 12, 1903.
14. On Untermyer, see *Dictionary of American Biography*, Supplement Two. Untermyer's papers, in the American Jewish Archives, Cincinnati, contain nothing of use.
15. New York *Herald*, June 12, 1903.
16. Statement of Henry Wollman, quoted in *ibid*.
17. New York *Tribune*, June 14, 1903; Schwab testimony, Jan. 7, 1904, *Conklin v. USSB*, 1780.
18. New York *Tribune*, June 30, 1903.
19. *Ibid.*, July 1, 1903.
20. Schwab to Perkins, July 2, 1903, Perkins Papers, Box 12, file 1.
21. Perkins to Schwab, July 5, 1903, *Ibid.*
22. New York *Evening Post*, July 1, 1903; also see New York *Journal of Commerce*, July 4, 1903; Montreal *Star*, July 6, 1903; Rockfield (Utah) *Reader*, Aug. 3, 1903; *Literary Digest*, XXVII, July 11, 1903, 34.
23. Pittsburgh *Gazette*, Aug. 5, 1903.
24. *Ibid.*
25. *Ibid.;* New York *Journal of Commerce*, Aug. 5, 1903.
26. La Crosse (Wis.) *Chronicle*, Oct. 7, 1903; Baltimore *News*, Oct. 14, 1903; Louisville (Ky.) *Herald*, Oct. 14, 1903; also see *Literary Digest*, XXVII, Aug. 15, 1903, 186.
27. New York *Sun*, Oct. 6, 1903.
28. *Ibid.*
29. Dresser's testimony, *Conklin v. USSB*, 131, 179–80, 188, 249–50; also see *Literary Digest*, XXVII, Oct. 24, 1903, 533.
30. Nixon's testimony, *Conklin v. USSB*, 419–1122, *passim*.
31. Smith's Report is reprinted in William Z. Ripley (ed.), *Trusts, Pools and Corporations* (Boston, 1916), 403–38.
32. New York *Commercial*, Nov. 2, 1903, emphasis added.
33. Schwab to Carnegie, Nov. 3, 1903, ACLC, Vol. 99 #18719A.
34. New York *News*, Dec. 23, 1903.
35. Pittsburgh *Press*, Dec. 23, 1903.
36. New York *Times*, Dec. 23, 1903.
37. William D. Guthrie to Schwab, July 16, 1903, in Robert T. Swaine, *The Cravath Firm* (New York, 1946), I, 696–97.
38. Schwab testimony, *Conklin v. USSB*, 1870–87, *passim*.

39. Dewing, *Corporate Promotions*, 504.
40. *Ibid.*, 506–7; Seager and Gulick, *Trust and Corporation Problems*, 214.
41. New York *Times*, Feb. 6, 1904. Schwab's press clipping service, which provided many of the newspaper stories cited in this chapter, ended a few months after he left the presidency of U.S. Steel.

CHAPTER 9

1. B. C. Forbes, New York *Journal* syndicated column, Aug. 12, 1933; Merle Crowell, "Schwab's Own Story," *American Magazine*, Oct. 1916, 12; *Literary Digest*, LVII, May 11, 1918, quoting story from Seattle *Post-Intelligencer*.
2. Whipple Notes, 38, 213.
3. *The Iron Age*, Nov. 24, 1910; also see two articles which stress Bethlehem's excellent plant and equipment: R. Hughes, "From Ore to Armor Plate," *Cosmopolitan*, XXVIII, Feb. 1900, 405–13; "Manufacture of Guns and Armor at the Steel Works," *Scientific American*, vol. 82, May 19, 1900, 306, 312–13; June 9, 1900, 353, 358–59, and vol. 83, July 14, 1900, 17, 24–5.
4. Joseph M. Levering, *A History of Bethlehem, Pennsylvania, 1741–1892* (Bethlehem, 1903), 724–25.
5. Peter Temin, *Iron and Steel in Nineteenth Century America: An Economic Inquiry* (Cambridge, Mass., 1964), 171, 174–75.
6. James M. Swank (ed.), *Classified List of Rail Mills and Blast Furnaces in the United States* (Philadelphia, 1873), 6; *Bulletin AISA*, VII (Oct. 15, 1873), 460, and XII (July 3, 1878), 153.
7. John Fritz, *The Autobiography of John Fritz* (New York, 1912), 174–77, 182.
8. *Ibid.*, 164, 184, 187.
9. H. F. J. Porter, "How Bethlehem Became Armament Maker," *Iron Age*, Nov. 23, 1922, 1340; *Iron Trade Review*, Feb. 20, 1913, 482.
10. Porter, "How Bethlehem Became Armament Maker."
11. For Bethlehem's gun forging contracts with the Navy, dated May 1, 1887, see U.S. Congress, House Executive Document #294, in vol. 36, 51st Congress, 2nd session. For the gun forging contracts with the Army, see U.S. Congress, House Document #151, in vol. 48, 54th Congress, 2nd session, and Eugene G. Grace, "Manufacture of Ordnance at South Bethlehem," *Yearbook of the American Iron and Steel Institute* (New York, 1912), 172–74.
12. *Supra*, Chapter three.
13. James M. Swank (ed.), *Directory of Iron and Steel Works in the United States* (Philadelphia, 1892), 11th edition, 102, and 12th edition (1894), 93; B. F. Fackenthal, Jr., "John Fritz, the Ironmaster," *Proceedings and Addresses of the Pennsylvania German Society*, XXXIV (Oct. 1923), 105–6; Report from Kossuth Niles, Lieutenant, to Secretary of the Navy, Dec. 4, 1896, in *Report of the Secretary of the Navy on the Cost and Price of Armor*, House Document #151, in vol. 48, 54th Congress, 2nd session, 91.
14. Fackenthal, "John Fritz, the Ironmaster," 106.
15. U.S. Congress, House Document #151, in vol. 48, 54th Congress, 2nd session, 29.
16. Frederick W. Taylor, "The Gospel of Efficiency, II: The Principles of Scientific Management," *American Magazine*, LXXI (April 1911), 787–88; Frank B. Copley, *Frederick W. Taylor, Father of Scientific Management* (New York, 1932), II, 46.
17. Papers of Frederick W. Taylor, Stevens Institute of Technology, Hoboken, N.J., files 32, 33 and 119C.

18. Taylor to General William Crozier, April 20, 1910, quoted in Copley, II, 160; Henry L. Gantt, "Compensation of Workmen and Efficiency of Operation—Part III: Task and Bonus," *Engineering Magazine*, XXXIX, April 1910, 17–23.
19. Taylor to Crozier, April 20, 1910, *loc. cit.*
20. *Bulletin AISA*, Aug. 25, 1904; New York *News Bureau*, Oct. 19, 1904. Press clippings on Bethlehem Steel for 1904–10 in the Pliny Fisk Collection, Firestone Library, Princeton University, Princeton, N.J.
21. *Bulletin AISA*, Dec. 10, 1904.
22. This and the next two paragraphs are based on Bethlehem Steel Corporation, *Annual Report*, 1905.
23. These additions are described in detail in "The Bethlehem Steel Company's Recent Extensions," *Iron Age*, Nov. 1, 1906, 1142–46.
24. *Wall Street Journal*, April 4, 1908; Philadelphia *Public Ledger*, March 1, 1909; New York *Times*, March 1, 1909.
25. *Iron Trade Review*, Dec. 28, 1905.
26. Bethlehem Steel Corp., *Annual Report*, 1906; Arundel Cotter, *The Story of Bethlehem Steel* (New York, 1916), 17.
27. *Bulletin AISA*, March 15, 1906.
28. *Ibid.*, July 1, 1906.
29. Recollections of F. A. Shick, June 3, 1949, Grace Biographical Project, Bethlehem Steel Corp., Bethlehem, Pa.
30. Recollections of James H. Ward, May 18, 1949, Grace Biographical Project, *loc. cit.*
31. *Ibid.*
32. Whipple Notes, 257.
33. For breakdown of costs, see *Wall Street Journal*, March 19, 1908; also Bethlehem Steel Corp., *Annual Report*, 1906, 12–13, and 1908, 13.
34. Recollections of James H. Ward, *loc. cit.*
35. *Ibid.*
36. George W. Burrell, "Traces Development of Grey Mill," *Iron Trade Review*, Nov. 4, 1926, 1165; obituary of Henry Grey, *Iron Age*, May 8, 1913, 1147.
37. Burrell, "Traces Development of Grey Mill;" Henry Grey, "A New Form of Structural Steel," *Iron Age*, June 17, 1897, 14; "A New Process for Rolling Structural Steel Shapes," *Engineering News*, XLVI (Nov. 21, 1901), 387.
38. F. Denk, "Mill Design for Rolling Flat-Flanged Beams," *Blast Furnace and Steel Plant Journal*, V (Feb. 1917), 61.
39. "Henry Grey and the New Structural Mill," *Iron Age*, Dec. 31, 1908, 1994.
40. *Wall Street Journal*, Dec. 9, 1905; Philadelphia *American*, Dec. 22, 1905.
41. Whipple Notes, 121.
42. *Ibid.*, 133, 140.
43. Schwab to Harvey Fisk and Sons, Dec. 18, 1906, in Corporate Records Division, Baker Library, Harvard University.
44. Arthur Pound and Samuel T. Moore (ed.), *They Told Barron* (New York, 1930), 84–85.
45. Clark, *History of Manufactures*, III, 11–12, 104.
46. Eugene Grace, *Charles M. Schwab*, 26.
47. Whipple Notes, 133.
48. *Ibid.*, 140; Recollections of F. A. Shick, *ibid.*, 147.
49. *Ibid.*, 133, and part two, 20, 140.
50. Hepburn to Schwab, Aug. 26, 1908, Schwab Papers, Bethlehem, Pa.
51. Carnegie to Schwab, March 22, 1906, Schwab Papers, Bethlehem.
52. Logbook of Loretto, Trip #105, Schwab Papers, Bethlehem.
53. The rumor that Carnegie was giving financial aid to Bethlehem Steel was reported in

the *Wall Street Journal*, Dec. 24, 1908. Carnegie's aid was confirmed in 1951 by Schwab's personal aide; see James H. Ward to John C. Long, Dec. 11, 1951, Grace Biographical Project, *loc. cit.*

54. U.S. Congress, House, *Committee on Ways and Means, Hearings*, House Document #1505, 60th Congress, 2nd session, testimony of Andrew Carnegie, Dec. 21, 1908, 1787–88.
55. Schwab to Grace, Aug. 7, 1908, Grace Papers, Bethlehem.
56. Grace to J. C. Long, April 18, 1947, Grace Papers.
57. Schwab to Grace, undated, *ca.* Aug./Sept. 1908, Grace Papers.
58. Schwab to Archibald Johnston *et al.*, Oct. 21, 1908, Grace Papers.
59. Whipple Notes, 253; interview with Mr. and Mrs. Arch B. Johnston, Jr., April 6, 1966, Bethlehem, Pa.
60. Whipple Notes, 251–52; interview with Alfred D. McKelvey, Sept. 12, 1974.
61. *Wall Street Journal*, Sept. 3, 1908; Grace, *Charles M. Schwab*, 25.
62. Whipple Notes, 158.
63. *Ibid.*, 159.
64. *Ibid.*, 160.
65. *Ibid.*, 160–61; *Historical Sketch of the Development of the Bethlehem Steel Company and Bethlehem Steel Corporation*, pamphlet dated Oct. 2, 1911, 9, copy in Schwab Memorial Library; Desk Diary notation for March 22, 1909: "Closed up matter of Gimbel Building with E. R. Graham and Pliny Fisk," quoted in Whipple Notes, 161.
66. For examples of resistance to innovation, see Clark, *History of Manufactures*, II, 71, 275, concerning hostility to the Bessemer process for making rails and boiler plates; H. F. J. Porter, "Nickel Steel: Its Practical Development in the United States," *Cassier's Magazine*, XXII (Aug. 1902), 483, on early prejudice against steel in any form (as opposed to wrought iron); and W. Paul Strassman, *Risk and Technological Innovation* (Ithaca, N.Y., 1959), 55, on opposition to crucible steel.
67. Recollections of Eugene G. Grace (1935) in Whipple Notes, 257–58, 270.
68. Recollections of Ernest R. Graham (1935), *ibid.*, 268–70.
69. *Wall Street Journal*, Aug. 2, 1909; New York *Times*, Aug. 5, 1909.
70. Letter from Henry Grey, *Iron Age*, Jan. 14, 1909; also see *Bethlehem Review*, Feb. 1949, 6–7.
71. Recollections of Ernest R. Graham, Whipple Notes, 270.
72. *Ibid.*, 168.
73. Recollections of E. R. Graham, *ibid.*, 272.
74. *Ibid.*, 269–70.
75. *Congressional Record*, 64th Congress, 1st session, LIII, Part I, 287 (1916); U.S. Congress, Senate Document #521, 61st Congress, 2nd session (1910), Appendix E, 126.
76. Bethlehem Steel Corp., *Annual Report*, 1910, 9.
77. *Ibid.*; Recollections of F. A. Shick, Grace Biographical Project (1949).
78. Bethlehem Steel Corp., *Annual Report*, 1906; Whipple Notes, 154.
79. J. Stephen Jeans (ed.), *American Industrial Conditions and Competition* (London, 1902), 176–78.
80. Cotter, *Story of Bethlehem Steel*, 19–21.
81. Schwab to Board of Directors, March 15, 1917, quoted in Defendant's Answer in *Berendt et al. v. Bethlehem Steel Corporation*, 45, New Jersey Court of Equity (1931), Supreme Court Vault and Microfilm Library, State House Annex, Trenton, N.J.
82. Charles M. Schwab, *The Bethlehem Bonus System* (Bethlehem, 1931), 3.
83. David Brody, *Steelworkers in America: The Nonunion Era* (Cambridge, Mass., 1960), 24–25, 89; Cotter, *Story of Bethlehem Steel*, 13–14, 20–22; Garraty, *Right-Hand Man*, 110–13.
84. Edward S. Meade, "The Price Policy of the United States Steel Corporation," *Quar-*

terly Journal of Economics, XXII (May 1908), 452–66; Abraham Berglund, "The United States Steel Corporation and Price Stabilization," *Quarterly Journal of Economics*, XXXVIII (Nov. 1923), 1–30.

85. Unpublished autobiography of William B. Dickson, *loc. cit.*
86. Dickson to Wm. E. Corey, Aug. 10, 1904, Dickson Papers, *loc. cit.*
87. *Ibid.*; Dickson to Corey, Feb. 16, 1909, and section on Gary in Dickson's autobiography.
88. *Iron Age*, Feb. 25, 1909, 648. On the Gary Dinners, see National Archives, Record Group 40 (Dept. of Commerce), file #6518-8-16; Maurice H. Robinson, "The Gary Dinner System: An Experiment in Cooperative Price Stabilization," *Southwestern Political and Social Science Quarterly*, VII, Sept. 1926, 137–61.
89. Gary's speech of Oct. 15, 1909, in vol. I of *Addresses and Statements of Elbert H. Gary*, compiled by James A. Farrell, 8 vols., 1927, copy in Baker Library, Graduate School of Business Administration, Harvard University.

CHAPTER 10

1. U.S. Congress, House, *Committee on Ways and Means, Hearings*, House Document #1505, 60th Congress, 2nd session, testimony of Schwab, Dec. 15, 1908, 1627–79, *passim*.
2. Schwab to Boies Penrose, April 22, 1909, Schwab Papers; also see New York *World*, April 19, 1909.
3. Observations based on reading the hearings on Schedule C in *Committee on Ways and Means, Hearings*.
4. Andrew Carnegie, "My Experience With, and Views Upon, the Tariff," *Century Magazine*, Dec. 1908, reprinted in *Committee on Ways and Means, Hearings*, 1766f.
5. *Ibid.*, testimony of Schwab, 1654-55.
6. *Ibid.*, 1628–29.
7. *Ibid.*, 1629.
8. *Ibid.*, 1649.
9. *Ibid.*, 1640, 1669.
10. *Ibid.*, 1650–51.
11. John Dalzell to Schwab, Dec. 19, 1908, Schwab Papers, Bethlehem.
12. *Committee on Ways and Means, Hearings*, testimony of Carnegie, 1781–86, *passim*.
13. Abraham Berglund and Philip G. Wright, *The Tariff on Iron and Steel* (Washington, 1929), 105–9, 131–35; Frank W. Taussig, *The Tariff History of the United States* (New York, 1964), 8th edition, 342, 384, 441.
14. *Report on the Strike at the Bethlehem Steel Works, South Bethlehem, Pennsylvania*, prepared under the direction of the Commissioner of Labor, Charles P. Neill (hereafter cited as *Report*), Senate Document #521, 61st Congress, 2nd session (1910), 7–11, 17–18.
15. Bethlehem *Globe*, Feb. 4, 1910.
16. *Ibid.*, Feb. 5, 1910.
17. *Ibid.*
18. *Ibid.*, Feb. 7, 1910.
19. *Ibid.*
20. *Ibid.*, Feb. 8, 1910.
21. *Ibid.*, Feb. 9, 1910.
22. *Ibid.*, Feb. 11, 1910.
23. *Ibid.*, Feb. 14, 1910; "The Struggle for Industrial Democracy," *The Outlook*, July 16, 1910, 544-45.

24. Bethlehem *Globe*, Feb. 11, 1910.
25. *Ibid.*, Feb. 17, 1910.
26. *Report*, 50–51.
27. Bethlehem *Globe*, Feb. 26, 1910.
28. *Ibid.*, Feb. 28, 1910.
29. *Ibid.*, March 2 & 3, 1910.
30. *Report*, 25–26; *Iron Age*, March 10, 1910, 576; *The Amalgamated Journal*, March 10, 1910, 12.
31. Bethlehem *Globe*, March 10, 11 & 12, 1910.
32. *Ibid.*, March 16, 1910.
33. *Ibid.*, March 31, 1910.
34. *Report*, 84; John A. Fitch, "Bethlehem, the Church and the Steel Workers," *The Survey*, Dec. 2, 1911, 1294.
35. Copy of resolution in National Archives, Record Group 122 (Bureau of Corporations), file 151-10. For the strikers' rebuttal to the businessmen's resolution, see *Congressional Record, Senate*, April 19, 1910, 4957–58.
36. Bethlehem *Globe*, March 31 & April 1, 1910.
37. *Ibid.*, April 6, 1910.
38. *Ibid.*, April 7, 1910. Strikers' statement reprinted in *Report*, 130–31. For Gompers's criticism of Taft's non-committal attitude, see *The Amalgamated Journal*, April 14, 1910, 16.
39. "The Bethlehem Strike," *The Survey*, May 21, 1910, 306–8.
40. *Ibid.*, 308; *Iron Age*, May 19, 1910, 1191.
41. Statement by Charles P. Neill, Commissioner of Labor, in National Archives, Record Group 40 (Dept. of Commerce), file 68114/8.
42. Boies Penrose to Charles Nagel, May 13, 1910, *ibid.*
43. Nagel to Penrose, May 16, 1910, *ibid.*
44. Stanley Coben, *A. Mitchell Palmer, Politician* (New York, 1963), 27; Bethlehem *Globe*, May 18, 1910; *Iron Age*, May 26, 1910, 1239; *Industrial Relations: Final Report and Testimony Submitted to Congress by the Commission on Industrial Relations*, Senate Document #415, 64th Congress, 1st session (Washington, 1916), testimony of David Williams, May 8, 1915, vol. 11, 10957.
46. *Report*, 39.
47. *The Amalgamated Journal*, May 12, 1910, 1.
48. *Report*, 79–81.
49. *The Amalgamated Journal*, March 17, 1910, 1, article by M. F. Tighe; *Machinists Monthly Journal*, March 1910, 248, report by F. J. Conlon; *Report*, 39.
50. *Report*, 32.
51. *Industrial Relations: Final Report, loc. cit.*, testimony of David Williams, vol. 11, 10957.
52. *Machinists Monthly Journal*, March 1910, 248.
53. *Industrial Relations: Final Report, loc. cit.*, testimony of David Williams, vol. 11, 10957.
54. Data comparing European and American wage rates in steel are difficult to obtain and even harder to evaluate because of differences in living standards and price relationships. For the argument that American producers paid higher wages, but that this cost was offset by the greater productivity of the American steelworker, see Berglund and Wright, *The Tariff on Iron and Steel*, 158.
55. Brody, *Steelworkers in America*, 161; Charles P. Neill, *Report on Conditions of Employment in the Iron and Steel Industry*, Senate Document #110, 62nd Congress, 1st session, four volumes (Washington, 1911–13).
56. Brody, *Steelworkers in America*, 161–62.
57. *Iron Age*, Feb. 12, 1912, 419.

58. Schwab to O. W. Underwood, Jan. 7, 1913, in U.S., Congress, House of Representatives, Committee on Ways and Means, *Hearings on the Tariff*, 1913, 1101.
59. *Bulletin AISA*, Aug. 1, 1912, 72.

CHAPTER 11

1. Recollections of Archibald Johnston, Whipple Notes, 208.
2. New York *Times*, Oct. 22, 1914.
3. Archibald Johnston Memo, based on Johnston's diary for 1914, original in possession of Arch B. Johnston, Jr., Bethlehem, Pa.; Gaddis Smith, *Britain's Clandestine Submarines, 1914–1915* (New Haven, 1964), 29.
4. New York *Times*, Nov. 4, 1914; Sir John Jellicoe, *The Grand Fleet, 1914–1916* (New York, 1919), 150–51; Johnston Memo.
5. Gaddis Smith, *op. cit.*, 30–31.
6. Henry S. Snyder Memo, "Data on the Building of Twenty Submarines for the British Admiralty," no date. Original in possession of William B. Snyder, New York.
7. Gaddis Smith, *op. cit.*, 38, Johnston Memo.
8. Arthur J. Marder (ed.), *Fear God and Dread Nought* (London, 1959), III, 66.
9. John K. Mumford, "The Story of Bethlehem Steel, 1914–1918," unpublished manuscript, Schwab Memorial Library, Bethlehem Steel Corp., Bethlehem, Pa.
10. Snyder Memo; Johnston Memo.
11. New York *Times*, Nov. 21 & 26, 1914.
12. Johnston Memo; Gaddis Smith, *op. cit.*, 34; Melvin I. Urofsky, *Big Steel and the Wilson Administration* (Columbus, Ohio, 1969), 98–104.
13. Johnston Memo; Arthur S. Link, *Wilson: The Struggle for Neutrality, 1914–1915* (Princeton, 1960), 61–62.
14. Johnston Memo.
15. *Ibid.*; also Recollections of Archibald Johnston, Whipple Notes, 208.
16. New York *Times*, April 29, 1921, speech by Darwin P. Kingsley. Hepburn revealed this episode during a 1915 luncheon address; see *Iron Age*, May 5, 1921.
17. Grace, *Charles M. Schwab*, 34.
18. Whipple Notes, 197.
19. *Ibid.*
20. *Iron Age*, May 5, 1921; "Germany's Offer to Buy Charles M. Schwab," *Commercial and Financial Chronicle*, May 7, 1921, 1929–30.
21. Schwab to Carnegie, Sept. 5, 1918, ACLC, Vol. 237 #44687–89, quoting a letter just received from Fisher, along with extracts from Fisher's Autobiography.
22. For fragmentary information on the League, "an organization of which too little is known," see Hans B. Thorelli, *The Federal Antitrust Policy* (Baltimore, 1954), 337, 339, 351, 429.
23. American Anti-Trust League to Philander C. Knox, Sept. 6, 1901, in National Archives, Record Group #60 (Dept. of Justice), file 60-138-0(13); testimony of Henry B. Martin before the Committee on Interstate Commerce, in U.S. Congress, Senate Report #1326, 62nd Congress, 3rd session (1913).
24. For Graham's speech, see *Congressional Record, House*, 63rd Congress, 1st session, 3925–30; American Anti-Trust League to Sen. Hoke Smith, Feb. 28, 1913, in National Archives, Record Group 80 (hereafter cited as NA/RG 80), file 10580-148; American Anti-Trust League to Woodrow Wilson, April 7, 1913, Wilson Papers, Library of Congress, file 387.

25. *Ibid.*
26. Josephus Daniels to Sen. Hoke Smith, April 5, 1913, NA/RG 80, #10580-148:1.
27. Wilson to Daniels, April 9, 1913, and Daniels to Wilson, April 11, 1913, NA/RG 80, #10580-148:1.
28. Daniels to Wilson, April 12, 1913, NA/RG 80, #10580-148:1.
29. *Ibid.*
30. Urofsky, *Big Steel and the Wilson Administration*, 121–24; Francis B. Simkins, *Pitchfork Ben Tillman, South Carolinian*, 349–52, 511–13.
31. Copy of Tillman Resolution in Josephus Daniels Papers, Library of Congress, file 417-418.
32. New York *Times*, July 14, 1913; Twining Memo to Daniels, June 7, 1913, eleven pages, plus appendix of contracts and prices and all armor orders since 1887, in Daniels Papers, file 417-418; Twining Memo to Daniels, June 28, 1913, NA/RG 80, #4174-157.
33. *Scientific American*, July 26, 1913, 62.
34. J. Bernard Walker to Daniels, Aug. 1, 1913, and Daniels to Walker, Aug. 27, 1913, NA/RG 80, #4174/169; Congressman A. W. Gregg to Daniels, Aug. 28, 1913, *ibid.*
35. For claims by Baltimore, Chester, Pa., Youngstown, Wilmington, Tuscaloosa, see Daniels Papers, file 417-418; also Bureau of Ordnance memo to Daniels, Jan. 7, 1914, listing all cities for which bills were introduced in Congress as the site of armor plant, *ibid.*; also Joseph G. Butler to Daniels, March 3, 1914, *ibid.*
36. Philadelphia *Evening Telegraph*, ca. Dec. 12, 1913, clipping in Daniels Papers, file 417-418.
37. See, for example, Daniels to Sen. W. O. Bradley, March 13, 1914, NA/RG 80, #4263-189.
38. Bureau of Ordnance memo to Daniels, Jan. 19, 1914, Daniels Papers, file 417-418; *Annual Report of the Secretary of the Navy*, 1914, 8–10, 39–47; Urofsky, *Big Steel and the Wilson Administration*, 125–28.
39. Tillman to Wilson, Jan. 5, 1916, and Wilson to Tillman, Jan. 6, 1916, Wilson Papers, file 387.
40. Urofsky, *Big Steel and the Wilson Administration*, 133.
41. U.S. Congress, Senate, Committee on Naval Affairs, *Armor Plant for the United States, Hearings*, 64th Congress, 1st session (1916); Urofsky, *Big Steel and the Wilson Administration*, 134–37; Tillman's speech to the Senate, in *Congressional Record, Senate*, 64th Congress, 1st session, 2566; U.S. Congress, Senate, Committee on Naval Affairs, Report #115, 64th Congress, 1st session, Feb. 8, 1916, 1–2.
42. Tillman to Wilson, March 9, 1916, Wilson Papers, file 387.
43. Bethlehem Steel Corporation file, Ivy Lee Papers, Princeton University Library, Princeton, N.J.; Ray Eldon Hiebert, *Courtier to the Crowd: The Story of Ivy Lee and the Development of Public Relations* (Ames, Iowa, 1966), 159–62. For a survey of press opinion, see "Threat of the Armor-Plate Makers," *Literary Digest*, LII, Feb. 26, 1916, 491–92, and "For a government Armor-Plant," *ibid.*, LII, April 1, 1916, 886.
44. Tillman to Wilson, March 29, 1916, Wilson Papers, file 387.
45. *The Bethlehem Steel Company Appeals to the People* (South Bethlehem, Pa., 1916), 23, copy in Schwab Memorial Library, Bethlehem, Pa.
46. *Ibid.*, 52.
47. U.S. Congress, House, Committee on Naval Affairs, Report #497 and Report #497, part two (Views of the Minority), April 6, 1916, 64th Congress, 1st session, in *House Reports, 1915–1916*, vol. 2, Miscellaneous.
48. Daniels to Wilson, April 6, 1916, and Wilson to Tillman, April 21, 1916, Wilson Papers, file 387.

49. Tillman to Wilson, April 21, 1916, and Wilson to Tillman, April 25, 1916, *ibid.*
50. Schwab to Wilson, July 15, 1916, NA/RG 80, #26256-271:9.
51. "South Charleston, West Virginia Naval Ordnance Plant," NA/RG 80, Entry 13, Box 3300.
52. C. F. Adams, Secretary of the Navy, to the Chairman of the House Committee on Naval Affairs, Jan. 20, 1932, *ibid.;* Claude A. Swanson, Secretary of the Navy, to Henry A. Wallace, July 10, 1935, *ibid.;* Mrs. Georgina C. Harris to Senator Gerald P. Nye, Dec. 13, 1936, *ibid.*
53. *Iron Age,* Dec. 28, 1916, 1477.
54. Robert T. Swaine, *The Cravath Firm,* II, 72–73.
55. *The Independent,* Jan. 31, 1916, 168; *Iron Age,* Jan. 27, 1916, 278.
56. *Christian Science Monitor,* Feb. 9, 1915; *Iron Age,* April 8, 1915, 826; New York *Times* editorial, April 14, 1915; New York *Herald,* Sept. 26 & 27, 1915; Edwin Wildman, "Charles M. Schwab: The American Krupp," *The Forum,* Aug. 1916, 201–8.
57. New York *Sun,* June 1, 1915.
58. New York *Press,* Nov. 1915, clipping in Schwab scrapbook, Bethlehem.
59. Charles M. Schwab, *Succeeding with What You Have* (New York, 1917), 5–6, 8, 9.
60. *Ibid.,* 32, 15, 14–15.
61. Whipple Notes, 231.
62. *Ibid.,* 274.
63. *Ibid.,* 280.
64. Schwab to stockholders of Bethlehem Steel Corp., Jan. 25, 1917, copy in Marvyn Scudder Financial Library, Graduate School of Business, Columbia University.
65. Schwab, Bethlehem Bonus System, 8–9; statement by Eugene Grace in Breslin Historical Memo (1935), Schwab Papers, Bethlehem, Pa.
66. Hogan, *Iron and Steel Industry,* 543–48.
67. Urofsky, *Big Steel and the Wilson Administration,* 84–86; Clark, *History of Manufactures,* III, 306–10.
68. *The Independent,* Nov. 15, 1915, 283.
69. *Literary Digest,* LI, Oct. 30, 1915, 947–48.
70. Urofsky, *Big Steel and the Wilson Administration,* 94–95; *The Independent,* Feb. 21, 1916, 285–86.
71. Clark, *History of Manufactures,* III, 330; Recollections of Quincy Bent, in Grace Biographical Project, Bethlehem Steel Corp., Bethlehem, Pa.; newspaper clippings in William B. Dickson Papers, Pennsylvania State University, State College, Pa.
72. Bernard M. Baruch, *American Industry in the War* (New York, 1941), 118–19.
73. George Weiss, "What the War Has Done for Steel," *The Forum,* LVII, Jan. 1917, 113–24; Baruch, *American Industry in the War,* 119.
74. Robert D. Cuff, *The War Industries Board* (Baltimore, 1973), 126–27.
75. Grosvenor B. Clarkson, *Industrial America in the World War* (Boston, 1923), 318–21; also see Robert D. Cuff and Melvin I. Urofsky, "The Steel Industry and Price-Fixing during World War I," *Business History Review,* XLIV, Autumn 1970, 291–306.
76. Ludwig von Mises, *Human Action: A Treatise on Economics* (New Haven, 1949), 821–22.
77. Eugene Meyer, Jr., War Industries Board, to G. Carroll Todd, Assistant Attorney General, Feb. 20, 1918, in National Archives, Record Group 60 (Dept. of Justice), file #60-138-0(13).
78. Bethlehem Steel Corp., *Annual Report,* 1918.
79. Schwab to stockholders of Bethlehem Steel Corp., Jan. 25, 1917, copy in Marvyn Scudder Financial Library, Columbia University.
80. *Ibid.,* Aug. 29, 1917.
81. Statement of F. A. Shick, Jan. 28, 1935, in Breslin Historical Memo, *loc. cit.*

CHAPTER 12

1. U.S. Congress, House, *Select Committee on U.S. Shipping Board Operations, Hearings*, 66th Congress, 3rd session (1921), testimony of Edward N. Hurley, 5109-10.
2. Edward N. Hurley, *The Bridge to France* (Philadelphia, 1927), 27-30, 128-29; *Shipping Board, Hearings*, testimony of Charles Piez, 4134; W. C. Mattox, *Building the Emergency Fleet* (Cleveland, 1920), 17-25.
3. On the careers of Hurley and Piez, see *Dictionary of American Biography*, Supplement I (New York, 1944), 446-47, 598-99.
4. Hurley to Wilson, April 3, 1918, Wilson Papers, Box 484A.
5. Hurley, *The Bridge to France*, 135, 137-39.
6. Anne W. Lane (ed.), *The Letters of Franklin K. Lane* (Boston, 1922), 254.
7. Hurley, *The Bridge to France*, 137.
8. *Ibid.*; *Shipping Board, Hearings*, testimony of Charles Piez, 4134-35, and testimony of Charles Schwab, 3968; *Iron Age*, April 23, 1925, 1191, reprinting statement by Bainbridge Colby.
9. Colby statement, *ibid.*
10. Whipple Notes, 187, 199.
11. *Shipping Board, Hearings*, text of statement, 4136.
12. Wilson Papers, April 16, 1918, file 484A.
13. Wilson Papers, file 484A.
14. *Emergency Fleet News*, I, April 29, 1918, 7.
15. *Literary Digest*, LVII (May 11, 1918), 11.
16. Josephus Daniels, *The Wilson Era* (Chapel Hill, N.C., 1946), 490.
17. Schwab to A. B. Farquhar, Aug. 5, 1918, Schwab Papers, National Archives, Emergency Fleet Corporation records, Record Group #32 (hereafter cited as NA/EFC). Schwab had minimal contact with Wilson; Hurley was their intermediary. Schwab to Ray Stannard Baker, Jan. 26, 1928, Baker Papers, Library of Congress, Series 1, Box 52.
18. Schwab Papers, NA/EFC., *passim*.
19. Schwab to Hurley, Aug. 1, 1918, *ibid.*
20. Geoffrey Creyke, Schwab's administrative assistant, to Wm. D. Galloway, May 17, 1918, *ibid.*
21. George W. Richardson to Schwab, June 12, 1918, *ibid.*
22. See Wilson Papers, file VI, folder 4555, expecially Grenville S. MacFarland to Wilson, Feb. 12, 1918.
23. Schwab Memo, May 15, 1918, Schwab Papers, NA/EFC.
24. Schwab to Samuel Mather, May 28, 1918, *ibid.*
25. Schwab to Warren G. Harding, June 26, 1918, *ibid.*, but also see Schwab to D. H. Cox, July 26, 1918, itemizing faults found by Inspectors, and Schwab to Harding, July 26, 1918, *ibid.*
26. Schwab to Harding, June 26, 1918, *ibid.*
27. Robert D. Heinl, head of the E.F.C.'s Publications Section, to Hurley, July 21, 1918, reporting on Schwab's West Coast tour, Schwab Papers, NA/EFC; also see telegrams between Hurley and Schwab, *ibid.*
28. Sherman Rogers, "When Charles M. Schwab Became Charlie," *The Outlook*, April 11, 1923, 654; Whipple Notes, 188.
29. *Emergency Fleet News*, July 18, 1918, 2.
30. Speech in San Francisco, July 4, 1918, reprinted in *Marine Engineering*, Aug. 1918, 440.
31. *Ibid.*; also see Charles M. Schwab, "The Shipbuilder's Job," *The Forum*, June 1918,

667–83, and Charles M. Schwab, "Our Industrial Victory," *National Geographic Magazine*, Sept. 1918, 212–23.
32. Schwab to F. F. Fletcher, May 29, 1918, and Schwab Memo, May 17, 1918, Schwab Papers, NA/EFC.
33. Schwab to Congressman N. J. Gould, May 22, 1918, *ibid.*
34. Mattox, *Building the Emergency Fleet*, 28–32, 205–31, and statistics of output, 106–7; also Hurley, *The Bridge to France*, 147–48, on ships built and delivered in 1918.
35. E. David Cronon (ed.), *The Cabinet Diaries of Josephus Daniels, 1913–1921* (Lincoln, Neb., 1963), 347.
36. Schwab to Carnegie, Sept. 5, 1918, ACLC, Vol. 237 #33687-89.
37. Carnegie to Schwab, Sept. 8, 1918, ACLC, Vol. 237 #44690.
38. Schwab to Wilson, Dec. 3, 1918, Wilson Papers, file 484A.
39. Wilson to Tumulty, Dec. 7, 1918, *ibid.*; also see Schwab's acknowledgement to Wilson, Dec. 9, 1918, *ibid.*
40. See press clippings in Schwab Papers, Bethlehem, Pa.

CHAPTER 13

1. New York *Times*, Dec. 22, 1916.
2. Schwab to Grace, Dec. 31, 1918, Grace Papers, Bethlehem, Pa.
3. *Ibid.*
4. Schwab to Wilson, Feb. 12, 1919, Wilson Papers, series 5A, Feb. 12–23, 1919.
5. *Iron Age*, Dec. 16, 1920, 1644.
6. Schwab to Grace, May 18, 1921, Grace Papers.
7. Interview, Edward H. Schwab, July 26, 1974.
8. Ruth Ayers, "My Boy Charlie," Pittsburgh *Press*, July 10, 1932.
9. "The Man Who Would Move a Mountain," Pittsburgh *Press*, May 2, 1933; undated photograph showing Schwab house on rollers, Schwab Papers, Bethlehem.
10. Altoona (Pa.) *Mirror*, Jan. 25, 1939; Johnstown (Pa.) *Tribune*, Sept. 19, 1949.
11. C. W. Leavitt, "The Gardens of Charles M. Schwab, Loretto, Pa.," *Country Life*, June 1920, 45–48; author's visit to Loretto.
12. Pittsburgh *Press*, Oct. 4, 1942; Johnstown *Tribune*, Sept. 19, 1949.
13. Interview with Gallitzin B. Moran, a life-long resident of Loretto and former employee on Schwab's Loretto estate, Jan. 24, 1967, Loretto, Pa.
14. John Poynton to Schwab, Aug. 13, 1919, Schwab Papers, Bethlehem; also see Schwab to Poynton, July 25, 1919, ACLC, Vol. 239 #44928.
15. Charles M. Schwab, "Address Delivered at the Memorial Service held in Honor of Andrew Carnegie," Pittsburgh, 1919, copy in Schwab Papers, Bethlehem; also see Whipple Notes, 84.
16. Dickson, *History of Carnegie Veteran Association, passim.*
17. Confidential source; obituary, New York *Times*, Feb. 18, 1922.
18. Obituary, New York *Times*, May 14, 1924; interviews with Edward H. Schwab.
19. New York *Times*, Nov. 18, 1922.
20. Hogan, *Iron and Steel Industry*, 556; New York *Times*, April 15, 1931.
21. Unless otherwise noted, the following section comparing Schwab and Grace is based on interviews with persons who knew them both: (1) Edward H. Schwab, July 26, 1974; (2) Mr. and Mrs. Arch B. Johnston, Jr., April 6, 1966 and July 16, 1974; (3) William H. Johnstone, Vice President and Chairman of the Finance Committee, Bethlehem Steel Corporation, April 6, 1966, Bethlehem, Pa.; (4) Clarence Randall, retired steel industry executive, May 16, 1966, New York, N.Y.; (5) John C. Long, re-

tired Manager of Public Relations, Bethlehem Steel Corp., Feb. 22, 1966, Princeton, N.J. Additional sources include the references to Grace made by Schwab in the Whipple Notes, *passim*. and Grace's comments on Schwab in his memorial lecture, *Charles M. Schwab, passim*.

22. B. C. Forbes, " 'Gene' Grace—Whose Story Reads like a Fairy Tale," *American Magazine*, XC, July 1920, 80, 84.

23. Allentown (Pa.) *Morning Call*, Aug. 23, 1957; Jack R. Ryan, "Grace 'Thinks Forward' at 80," New York *Times*, Aug. 26, 1956.

24. Forbes, " 'Gene' Grace," 16. Italics in original.

25. *Ibid.*, 84. Also see "A Man of Metals," *Metal Progress*, Feb. 1938, 149ff.; *Newsweek*, Aug. 8, 1960, 61; Jere Knight, "Men of Industry: Community Partners," Bethlehem *Globe-Times*, April 28, 1967.

26. Frank Parker Stockbridge article on Schwab, written for Doubleday-Page Syndicate, 1918, copy in Schwab Papers, Bethlehem.

27. Based on reading the files of Schwab's speeches in the Schwab Library, Bethlehem.

28. New York *World*, Jan. 25, 1918.

29. Baker to Dickson, Jan. 26, 1918, Papers of Wm. B. Dickson, Penn State.

30. Dickson to Schwab, Feb. 4, 1918, *ibid.*

31. For press reaction, see "The Workers to Rule the World," *Literary Digest*, LVI, Feb. 16, 1918, 12–13; "The Decadence of Mr. Schwab," *The Public*, 21, Feb. 1, 1918, 140.

32. Schwab's speech to the Golden Jubilee Anniversary Banquet of St. Andrew's Golf Club, Nov. 14, 1938, Schwab Papers, Bethlehem.

33. Whipple Notes, 44, 219.

34. Charles M. Schwab, "Straight Talk on Success in Life," *Engineering News-Record*, 84, April 1, 1920, 684, the text of Schwab's remarks at Princeton University on March 16, 1920.

35. This widely used opening first appears in Schwab's talk to the banquet of the Industrial Commission of Bethlehem, Pa., March 12, 1908, taken down in shorthand at the time of delivery, copy in Schwab Papers, Bethlehem.

36. *Ibid.*

37. Interview with Alfred D. McKelvey, Sept. 19, 1974.

CHAPTER 14

1. New York *Times*, Jan. 21, 1921.

2. *Ibid.*, Jan. 22, 1921.

3. U.S. Congress, House, *Select Committee on U.S. Shipping Board Operations, Hearings*, 66th Congress, 3rd session (1921), testimony of C. M. Schwab, 3996.

4. *Ibid.*, testimony of Eugene Grace, 4401–78, *passim.*; also New York *Times*, Jan. 25, 1921.

5. The letters were described and their number estimated by James H. Ward; see letter of Edward H. Schwab to C. M. Schwab, Feb. 1, 1921, Schwab Papers, Bethlehem. Arthur Brisbane to Schwab, Jan. 29, 1921, *ibid.*

6. *The Outlook*, Feb. 9, 1921.

7. Schwab to Mrs. Andrew Carnegie, March 18, 1921, Carnegie Papers, New York Public Library.

8. Eugene Grace letter to stockholders of Bethlehem Steel Corp., May 5, 1925, copy in Marvyn Scudder Financial Library, Columbia University.

9. New York *Times*, March 15, 1924.

10. Eugene Grace letter to stockholders, May 5, 1925, *loc. cit.*
11. *Report of Special Master and Referee*, Feb. 7, 1936, reprinted in U.S. Congress, Senate, *Munitions Industry*, Hearings before the Special Committee Investigating the Munitions Industry, 74th Congress, 2nd session (1936), Part 36, 12366–83.
12. Eugene Grace letter to stockholders, May 5, 1925, *loc. cit.; Report of Special Master*, 12367.
13. Eugene Grace letter to stockholders, May 5, 1925, *loc. cit.*
14. Charge quoted in *Report of Special Master*, 12368.
15. *Ibid.*, 12370.
16. *Ibid.*
17. *Iron Age*, April 23, 1925, 1245.
18. *Report of Special Master*, 12375; New York *Times*, Feb. 8, 1936.
19. 23 F. Supp. 676; 26 F. Supp. 259.
20. New York *Times*, Jan. 17, 1940; 113 F. (2d) 301.
21. *U.S. v. Bethlehem Steel Corp.*, U.S. Supreme Court, October term, 1941.
22. *Ibid.*
23. Hogan, *Iron and Steel Industry*, 549.
24. *Ibid.*, 900, quoting table in *Iron Trade Review*, Jan. 4, 1923, 10.
25. *Ibid.*, 902, quoting table in *Iron Age*, Dec. 8, 1921, 1492.
26. *Ibid.*, 899–910, 923–32.
27. Thomas R. Navin, "The 500 Largest American Industrials in 1917," *Business History Review* XLIV, Autumn 1970, 369.
28. Hogan, *Iron and Steel Industry*, 910–17, 932–41.
29. Whipple Notes, 121.
30. Hogan, *Iron and Steel Industry*, quoting an unpublished report of the Carnegie Steel Sales Dept., 1924, 887.
31. *Iron Trade Review*, Nov. 4, 1926, 1207.
32. Willis L. King to Schwab, Oct. 6, 1926, Whipple Notes, 122.
33. *Ibid.*, 122–23.
34. *Iron Trade Review*, 85, Oct. 12, 1926, 648.
35. Recollections of Eugene Grace, Whipple Notes, 257–58.
36. *Iron Trade Review*, Dec. 13, 1928, 1542–43; March 28, 1929, 887–88; June 13, 1929, 1604–5.
37. *Iron Age*, Nov. 21, 1929, 1302.
38. Whipple Notes, 213.
39. Hogan, *Iron and Steel Industry*, 987.
40. Marcus Gleisser, *Cyrus Eaton* (New York, 1964), 59–60.
41. New York *Times*, July 18, 1930.
42. *Ibid.*
43. *Ibid.*
44. Hogan, *Iron and Steel Industry*, 989–90, 1214–15.
45. *Ibid.*, 1215.
46. New York *Times*, Nov. 1, 1930.
47. Plaintiff's Brief, *Berendt et al. v. Bethlehem Steel Corp.*, New Jersey Court of Chancery, 1931, Dis. 47, Docket 82 #414, Superior Court Vault and Microfilm Library, State House Annex, Trenton, N.J., 9–23 and appendix, 15.
48. New York *Times*, Jan. 14, 1931.
49. Ivy Lee to Eugene Grace, Jan. 14, 1931, Ivy Lee Papers.
50. New York *Times*, Jan. 15, 1931; Pittsburgh *Press*, Jan. 15, 1931.
51. Ivy Lee to Grace, Jan. 19, 1931, Ivy Lee Papers.
52. New York *Times*, March 1, 1931. On the Protective Committee, see William H. Brown Collection, Yale University Library, New Haven, Conn.

53. Charles M. Schwab, *The Bethlehem Bonus System*, originally a letter to stockholders, March 2, 1931; New York *Times*, March 4, 1931.
54. Statement by Protective Committee, March 20, 1931, copy in Corporate Records Division, Baker Library, Harvard University.
55. Schwab, *Bethlehem Bonus System*, 4.
56. Statement by Protective Committee, April 29, 1931, copy in William H. Brown Collection, Yale University.
57. Schwab, *Bethlehem Bonus System*, 17. I am indebted to Professor George G. C. Parker of Stanford University's Graduate School of Business for discussing this point with me.
58. Schwab, *Bethlehem Bonus System*, *passim*.; New York *Times*, March 4, 1931.
59. Defendant's Brief, *Berendt et al. v. Bethlehem Steel Corp.*, *loc. cit.*; also New York *Times*, April 15, 1931, and July 2, 1931.
60. Quoted in statement of Protective Committee, April 29, 1931, *loc. cit.*
61. New York *Times*, April 15, 1931.
62. *Ibid.*
63. *Ibid.*
64. *Cases in Chancery* (1931), 108 New Jersey Equity 148–52, at 151–52; New York *Times*, March 27, 1931.
65. Statement of Protective Committee, July 3, 1931, Corporate Records Division, Baker Library, Harvard University; Schwab's letter to stockholders explaining the changes, July 3, 1931, William H. Brown Collection, Yale University Library; New York *Times*, July 2 & 3, 1931.
66. New York *Times*, July 3, 1931; dissolution order in case file of *Berendt et al. v. Bethlehem Steel Corp.*, *loc. cit.*
67. Quoted in Hiebert, *Courtier to the Crowd*, 161–62.

CHAPTER 15

1. *Yearbook of American Iron and Steel Institute* (hereafter cited as *Yearbook AISI*), 1927, 225–26.
2. *Ibid.*, 1928, 31–36.
3. *Ibid.*, 1928, 272–82.
4. *Ibid.*, 1929, 29–32.
5. Quoted in Pittsburgh *Press*, April 2, 1935.
6. Altoona (Pa.) *Mirror*, Jan. 1, 1930.
7. *Yearbook AISI*, 1930, 35; statements of 1930 quoted in Pittsburgh *Press*, April 2, 1935.
8. *Yearbook AISI*, 1930, 258–9, and 1931, 338.
9. Interview, John C. Long; Grace, *Charles M. Schwab*, 40; Hiebert, *Courtier to the Crowd*, 163.·
10. *Ibid.*, 1931, 301.
11. Interview with Clarence Randall, New York, May 16, 1966.
12. *Yearbook AISI*, 1931, 30.
13. New York *Times*, Feb. 19, 1933.
14. Herman Krooss, *Executive Opinion* (New York, 1970), 159, quoting New York *Times*, Oct. 19, 1932.
15. Schwab to Roosevelt, Sept. 14, 1933, Roosevelt Papers, Hyde Park, N.Y.
16. *Yearbook AISI*, 1933, 29–30.
17. *Ibid.*, 1933, 31–32, and 1934, 22–23; also see Ellis W. Hawley, *The New Deal and the Problem of Monopoly* (Princeton, 1966), 20–21, 26–29.

18. *Bethlehem Review*, March 1954, "Financial Summary, 1905–1953," 7.
19. *Yearbook AISI*, 1933, 28–29.
20. *Ibid.*, 1932, 32–33.
21. *Ibid.*, 1934, 27.
22. *Ibid.*, 28.
23. William E. Leuchtenburg, *Franklin D. Roosevelt and the New Deal* (New York, 1963), 150–52, 203–5, 243; Whipple Notes, *passim*.
24. Interviews with Edward H. Schwab and Rana Ward; Schwab to Grace, Sept. 20, 1922, ". . . what I am really looking for is a good income during my lifetime. As you know I have no heirs, so that it does not make a great deal of difference to me whether I leave anything or not." Grace Papers, Bethlehem.
25. Interviews with Edward H. Schwab, Rana Ward, John C. Long, and Mr. and Mrs. Arch B. Johnston, Jr.
26. Whipple Notes, 40.
27. Interview with Alfred D. McKelvey.
28. John K. Winkler, "Profile: Charles M. Schwab," *New Yorker*, May 2, 1931, 28. Winkler's essay is riddled with inaccuracies.
29. Interviews with Edward Schwab and Rana Ward.
30. Whipple Notes, *passim*.
31. *Ibid.*
32. Interviews with Edward Schwab and Rana Ward; interview with Mother Superior, Monastery of St. Theresa Lisieux, Jan. 25, 1967, Loretto, Pa.; Whipple Notes, 6; recollections of Father Paul, T.O.R., June 1966, file on local history, St. Francis College Library; New York *Times*, Dec. 26, 1928, 19, and Sept. 23, 1930, 3; obituary, Altoona (Pa.) *Mirror*, Dec. 27, 1954.
33. Interviews with Edward Schwab; Edward Schwab, "My Business Life," privately printed booklet, 1973.
34. *Ibid.*
35. *Ibid.*
36. *Ibid.*
37. *Yearbook AISI*, 1930, 228.
38. Schwab to Roosevelt, Sept. 14, 1933, Roosevelt Papers, Hyde Park, N.Y.; also see Schwab to Roosevelt, Nov. 17, 1937, urging retroactive lowering of excess profits taxes.
39. Interview with Edward Schwab.
40. *Ibid.*
41. Unless otherwise noted, the following section is based on interviews with Edward H. Schwab, Rana Ward, and Alfred D. McKelvey.
42. Interview with Mrs. Arch B. Johnston, Jr., July 16, 1974.
43. Mrs. Schwab's niece, Rana Ward, was her most frequent companion on these excursions to Tiffany's.
44. New York *Times*, Dec. 22, 1934.
45. John Edward Wiltz, *In Search of Peace: The Senate Munitions Inquiry, 1934–36* (Baton Rouge, La., 1963).
46. *Ibid.*, 23.
47. See Walter Sulzbach, *Capitalistic Warmongers: A Modern Superstition* (Chicago, 1943), and Ludwig von Mises, *The Anti-Capitalistic Mentality* (Princeton, 1956).
48. U.S. Congress, Senate, *Munitions Industry, Hearings, loc. cit.*, testimony of Eugene Grace, Feb. 25, 1935, part 21, 5776.
49. New York *Times*, March 29, 1935.
50. *Ibid.*, Feb. 19, 1936.

51. Quoted in *Literary Digest*, LVII, May 11, 1918, from a 1916 story in the Seattle *Post-Intelligencer*.
52. O'Connor, *Steel—Dictator*, 77–85.
53. New York *Times*, May 29, 1936; Robert Forsythe, "Charley," *Fight*, Aug. 1936, 26.
54. New York *Times*, May 29, 1936.
55. *Ibid.*, May 7, 1937.
56. *Ibid.*, April 11, 1934.
57. *Ibid.*
58. New York *Sun*, April 9, 1935.
59. New York *Herald Tribune*, April 10, 1935.
60. Altoona (Pa.) *Mirror*, April 10, 1935.
61. New York *Times* and Altoona *Mirror*, April 14, 1937.
62. New York *Times*, April 13, 1938.
63. Lewis D. Gilbert, *Dividends and Democracy* (New Rochelle, N.Y., 1954), 39.
64. Dickson, *History of Carnegie Veteran Association*, 159–60.
65. *Ibid.*, 160–61.
66. B. C. Forbes, syndicated column, New York *American*, July 2, 1936.
67. Interview with Gallitzin B. Moran, Jan. 24, 1967, Loretto, Pa.
68. New York *Times*, Dec. 19, 1936.
69. Interview with Alfred D. McKelvey.
70. Winkler, "Profile: Charles M. Schwab," *New Yorker*, May 2, 1931, 28.
71. Schwab to Rev. Father John P. J. Sullivan, Dec. 1, 1936, St. Francis College Library, Loretto, Pa.
72. Interview with Edward Schwab.
73. Winkler, "Profile," *New Yorker*, May 2, 1931, 28; Arthur Strawn, "A Man of Heart," *American Mercury*, XII, Oct. 1927, 143; New York *Times*, Oct. 5, 1924, Sec. 9.
74. Interview with Edward Schwab.
75. Interview with Rana Ward.
76. Interview with Alfred D. McKelvey.
77. Interview with Edward Schwab.
78. Hampden E. Tener to Wm. B. Dickson, Sept. 5, 1939, quoting New York *Evening Sun*, Aug. 31, 1939, and James H. Ward to Wm. B. Dickson, Sept. 13, 1939, Dickson Papers, Penn State.
79. Interview with Edward Schwab.
80. New York *Herald Tribune* and Allentown *Call*, April 17, 1943.
81. Information from Real Estate Dept., Prudential Insurance Co.
82. Philadelphia *Record*, Feb. 22, 1942; interview with Rana Ward.
83. Author's visit to Loretto.

APPENDIX A

1. Interview with John C. Long, Princeton, N.J., Feb. 22, 1966; Mrs. John C. Long to the author, July 18, 1974, recording her husband's answers to written questions; interview with Alfred D. McKelvey, Atherton, Calif., Sept. 12, 1974; S. B. Whipple to J. C. Long, July 16, 1936, transferring all of his notes and 135 pages of his draft biography (the latter subsequently disappeared), letter in Schwab Papers, Bethlehem. The Metropolitan Weekend Magazine Section of the N.Y. *World-Telegram*, Dec. 12, 1936, carried an article by John Lowell, dealing with the December 1900 dinner that resulted in the formation of U.S. Steel.

APPENDIX B

1. Gustavus Myers, *History of the Great American Fortunes* (New York, 1907; reprinted 1937), 596–97, 600.
2. John K. Winkler, *Incredible Carnegie* (New York, 1931), 226–34.
3. Matthew Josephson, *The Robber Barons* (New York, 1934; reprinted 1962), 391–92.
4. U.S. Congress, House, Special Subcommittee of the Committee on Naval Affairs, *Investigation of Armor-Plate Contracts*, 1894. Testimony of C. M. Schwab, 674–75.
5. George Seldes, *Iron, Blood and Profits: An Exposure of the World-Wide Munitions Racket* (New York, 1934), 236–37.
6. *Ibid.*, 360–61.
7. H. C. Engelbrecht and F. C. Hanighen, *Merchants of Death: A Study of the International Armament Industry* (New York, 1934), 53–55.
8. Burton J. Hendrick, *The Life of Andrew Carnegie* (Garden City, N.Y., 1932), II, 401–6: also see Wall, *Andrew Carnegie*, 650–52.
9. Harvey O'Connor, *Steel: Dictator* (New York, 1935), 69–70, 341.
10. Philip Noel-Baker, *The Private Manufacture of Armaments* (London, 1936), 316.
11. Ben B. Seligman, *The Potentates: Business and Businessmen in American History* (New York, 1971), 182.
12. R. Gordon Wasson, *The Hall Carbine Affair: A Study in Contemporary Folklore* (New York, 1941); third edition, *The Hall Carbine Affair: An Essay in Historiography* (Danbury, Conn., 1971). The erroneous accounts cited by Wasson include Myers, *op. cit.*, 170–76; John K. Winkler, *Magnificent Morgan* (New York, 1930), 56–57; Josephson, *op. cit.*, 59–62; Engelbrecht and Hanighen, *op. cit.*, 59–61; Seldes, *op. cit.*, 227–28.
13. Seligman, *The Potentates*, 99.
14. John Wiltz, *In Search of Peace* (Baton Rouge, La., 1963), 3–24, *passim*.

Index

343